Environmental Management of Marine Ecosystems

Applied Ecology
and Environmental Management

A SERIES

Series Editor
Sven E. Jørgensen
Copenhagen University, Denmark

Environmental Management of Marine Ecosystems, *Md. Nazrul Islam and Sven E. Jørgensen*

Ecotoxicology and Chemistry Applications in Environmental Management, *Sven E. Jørgensen*

Ecological Forest Management Handbook, *Guy R. Larocque*

Handbook of Environmental Engineering, *Frank R. Spellman*

Integrated Environmental Management: A Transdisciplinary Approach, *Sven E. Jørgensen, João Carlos Marques, and Søren Nors Nielsen*

Ecological Processes Handbook, *Luca Palmeri, Alberto Barausse, and Sven E. Jørgensen*

Handbook of Inland Aquatic Ecosystem Management, *Sven E. Jørgensen, Jose Galizia Tundisi, and Takako Matsumura Tundisi*

Eco-Cities: A Planning Guide, *Zhifeng Yang*

Sustainable Energy Landscapes: Designing, Planning, and Development, *Sven Stremke and Andy Van Den Dobbelsteen*

Introduction to Systems Ecology, *Sven E. Jørgensen*

Handbook of Ecological Models Used in Ecosystem and Environmental Management, *Sven E. Jørgensen*

Surface Modeling: High Accuracy and High Speed Methods, *Tian-Xiang Yue*

ADDITIONAL VOLUMES IN PREPARATION

Environmental Management of Marine Ecosystems

Edited by
Md. Nazrul Islam
Sven Erik Jørgensen

CRC Press
Taylor & Francis Group
Boca Raton London New York

CRC Press is an imprint of the
Taylor & Francis Group, an **Informa** business

CRC Press
Taylor & Francis Group
6000 Broken Sound Parkway NW, Suite 300
Boca Raton, FL 33487-2742

First issued in paperback 2020

ISBN-13: 978-0-367-57194-8 (pbk)
ISBN-13: 978-1-4987-6772-9 (hbk)

Library of Congress Cataloging-in-Publication Data

Names: Islam, Md. Nazrul, 1975- | Jørgensen, Sven Erik, 1934-
Title: Environmental management of marine ecosystems / [edited by] Md. Nazrul
    Islam and Sven Erik Jørgensen.
Description: Boca Raton : CRC Press, [2018] | Includes bibliographical
    references and index.
Identifiers: LCCN 2017020721 | ISBN 9781498767729 (hardcopy : alk. paper)
Subjects: LCSH: Marine ecosystem management. | Marine ecosystem health. |
    Marine ecology.
Classification: LCC QH541.5.S3 E5845 2018 | DDC 333.95/6--dc23
LC record available at https://lccn.loc.gov/2017020721

**Visit the Taylor & Francis Web site at
http://www.taylorandfrancis.com**

**and the CRC Press Web site at
http://www.crcpress.com**

To

My Wife

SAHANAJ TAMANNA

and

My Daughter

SABABA MOBASHIRA ISLAM

Contents

Preface...ix

Acknowledgments xv

Editors...xvii

Contributors... xix

1. Integrated Coastal Zone Monitoring in Support of Ecosystem-Based
   Management of Marine Ecosystem Services 1
   Thomas C. Malone and Anthony H. Knap

2. Inclusive Impact Index 'Triple I' 31
   Koji Otsuka, Fengjun Duan, Toru Sato, Shigeru Tabeta and M. Murai

3. Biophysical Modelling of Marine Organisms: Fundamentals and
   Applications to Management of Coastal Waters 65
   Dmitry Aleynik, Thomas Adams, Keith Davidson, Andrew Dale, Marie Porter,
   Kenny Black and Michael Burrows

4. Oceanic and Fisheries Response in the Northwest Pacific Marginal Seas with
   Climate Variability...99
   Chung Il Lee, S. M. Mustafizur Rahman, Hae Kun Jung, Chang-Keun Kang
   and Hyun Je Park

5. Environmental Effects and Management of Oil Spills on Marine Ecosystems.... 119
   Piers Chapman, Terry L. Wade and Anthony H. Knap

6. Modeling for Management of Marine Ecosystem Around an Artificial
   Structure in a Semiclosed Bay 147
   Daisuke Kitazawa, Shigeru Tabeta and Masataka Fujino

7. The Economics of Ecosystem-Based Fisheries Management 157
   Hans Frost, Lars Ravensbeck, Ayoe Hoff and Peder Andersen

8. Climate Change Impacts on Marine Ecosystems in Vietnam.............................209
   Nguyen Quang Hung, Hoang Dinh Chieu, Dong Thi Dung, Le Tuan Son,
   Vu Trieu Duc and Do Anh Duy

9. Management Strategies of St. Martin's Coral Island at Bay of Bengal in
   Bangladesh..237
   Md. Nazrul Islam and Mamunur Roshid

10. Habitat Complexity of Tropical Coastal Ecosystems: An Ecosystem
    Management Perspective 263
    Chandrashekher U. Rivonker, Vinay P. Padate, Mahabaleshwar R. Hegde
    and Dinesh T. Velip

11. Biomonitoring Ecosystem Health: Current State of Malaysian
    Coastal Waters..287
    *W. O. Wan Maznah, Khairun Yahya, Anita Talib, M. S. M. Faradina Merican
    and S. Shuhaida*

12. Conservation and Management of Saltwater Crocodile (*Crocodylus porosus*)
    in Bhitarkanika Wildlife Sanctuary, Odisha, India....................................307
    Lakshman Nayak, Satyabrata Das Sharma and Mitali Priyadarsini Pati

13. Management Challenges of Sinking of Oil Tanker at
    Shela Coastal River in Sundarbans Mangroves in Bangladesh 323
    Md. Nazrul Islam, Md. Al Amin and Md. Noman

Index 343

Preface

The marine ecosystem is the largest aquatic system on the planet. Life on our planet is dependent upon the oceans, which are the sources of wealth, opportunity and abundance. About 71% of the surface of this planet is covered by salt water. The oceans provide us with food, energy and water and sustain the livelihoods of hundreds of millions of people. They cover most of our planet, are the source of most life on Earth, regulate our weather and climate, provide most of our oxygen, and feed much of the human population. Marine ecosystems, which include coral reefs, mangroves and deep-sea beds, are threatened by a number of factors, including unsustainable development and fishing practices. One of the most holistic approaches to protecting marine and coastal ecosystems is *ecosystem-based management (EBM)*, which recognizes the need to protect entire marine ecosystems instead of individual species. After decades of pollution, habitat degradation and overfishing, now climate change and ocean acidification threaten the health of the oceans in unprecedented ways. Organisms living in the oceans develop special adaptations to the chemical and physical characteristics of seawater. Basic producers of food and their distribution in the oceans rely on the availability of sunlight and nutrients. It is important that scientists find out as much as possible about ecosystems all over Earth and how they work in order to protect them. This is because humans can have large negative impacts on marine ecosystems. Readers will learn more about anthropogenic impacts on marine habitats from this book. All chapters in the book have some relevance to the physical and chemical composition of marine ecosystems, modeling, policy and management and were written by academics from around the world working in a variety of disciplines.

The aim of this book is to provide information to scientists, practitioners, academicians and government and non-government policymakers to help them better understand the particularities of marine ecosystem management. By 'environmental management of marine ecosystems' we mean management using strategies that aim to mitigate and adapt our activities in marine ecosystems. How can we manage marine ecosystems in such a way that we avoid polluting them? This book will also illustrate the current status, trends and effects of climate, natural disturbances and anthropogenic stressors on marine ecosystems. It shows how to integrate different management tools and models in an up-to-date, multidisciplinary approach to environmental management. Through several case studies, it provides a powerful framework for identifying management tools and their applications in coral reefs, fisheries, migratory species, marine islands and associated ecosystems such as mangroves and seagrass beds, including their ecosystem services, threats to their sustainability, and the actions needed to protect them.

Marine ecosystems are composed of biological communities and their ambient physicochemical parameters within a specific habitat. Tropical coastal ecosystems are diverse, each characterized by unique features, biotic community and ecological processes. The coexistence of these ecosystems within a given locale promotes heterogeneity, and the diverse interactions among their respective communities enhance species diversity. Rapid changes in land-use patterns to meet the ever-increasing demands of burgeoning coastal and tourist populations are exerting immense pressure on these fragile ecosystems. Furthermore, these ecosystems are being used largely as dumping sites for different sources of waste material from continuously changing land-use patterns. In view of these considerations, it is imperative that coastal ecosystems be managed in a sustainable manner.

Several events occurred in the northwest Pacific marginal seas (NPMS) that caused wholesale changes in marine environments and fishery assemblage patterns, namely climate regime shift (CRS) after the 1970s. Each sea in the NPMS responded differently with each CRS, and the major driving factors that caused changes in the marine environments and fisheries were different. The common squid regime, the sardine regime and the pollock regime from the 1950s to the 2000s were easily identified in the seas. Demersal fisheries were enhanced in the colder regime after the 1976/1977 CRS, whereas large predatory and small pelagic fisheries generally expanded during warm regimes following the 1988/1989 CRS. The mean trophic level (MTL) and biomass diversity index (DI) of fishery resources decreased gradually from 1950 to 2010. A poleward latitudinal shift of commercial species fishing grounds, such as hairtail and small yellow croaker, occurred with the CRS.

One of our goals with this book was for it to have broad application to EBM approaches. The EBM approaches will change current methodologies used for resource assessment and future regulation of marine resources. Conventional fishery economic theory focuses on single species, hence neglecting externalities from the fisheries on marine ecosystems, for example, the food web and bottom structures. In addition, it does not include recreational values and benefits from non-commercial species. EBM requires that these additional elements be taken into account. However, the models need not always assess many species and many fishing fleets. According to the scope of the assessment, relatively simple models may be sufficient.

Take, for example, our ability to assess the fact that the increasing world population cannot be supported without new infrastructures based on utilization of the oceans, which store almost all water, carbon and energy resources on Earth. It is important, however, to assess the associated technologies and systems from the viewpoint of sustainability since the goal is to develop sustainable water, food and energy resources from the ocean rather than just another means of conventional land-based productions. The Inclusive Impact Index (Triple I) is a metric that was developed to assess environmental sustainability and economic feasibility of ocean utilization technologies to predict their public acceptance. Ecological footprint and environmental risk assessment, which are the basis of the Triple I, are introduced in the first part of this chapter. The concept and the structure of the Triple I are then explained in the second part of the chapter. Then the Triple I calculation method and results for several ocean utilization technologies are described, and examples of an application of the Triple I metric are presented.

We did not wish to present a description of marine ecosystems management, in the form of a handbook on the overall environmental management design of marine ecosystems. Understanding the dynamics of populations of marine organisms is a challenge. The ocean is vast, and the tiny larvae of many organisms become dispersed in a patchy manner by both their behaviour and the hydrodynamic features that they encounter. Recent years have seen the development and application of what are known as coupled bio-physical models for marine species' dispersal. These models use a representation of oceanographic conditions in conjunction with encoded biological behaviour to predict the spread and abundance of dispersing organisms over space and time and have been applied to a diverse range of systems from coral reefs to estuarine waters.

We also describe the general foundations of such a modelling approach. Making reference to a study system in the complex coastal waters of the west coast of Scotland, we examine the relative importance of various physical and biological components and assumptions for model behaviour. We then describe the application of the model framework to several contemporary management issues in our study region: (1) modelling of harmful algal bloom dynamics; (2) understanding the dispersal of sea lice (a parasite of salmon), with implications for management procedures; and (3) possible impacts on intertidal communities for the provision of novel offshore habitats. Each example benefits greatly from insights that would be difficult or impossible to attain without specific application of the model framework.

When an artificial structure is installed in a semi-closed bay, its environmental impact should be assessed before construction. Modelling and numerical simulation of the physical, chemical and biological processes around the artificial structure are powerful tools for the management of marine ecosystem around the structure. The results of numerical simulation can be utilized not only for environmental impact assessment but for promoting the mitigation of impacts. The present chapter presents example results of numerical simulations of a marine ecosystem around a floating platform in Tokyo Bay, Japan, using a hydrodynamic and ecosystem coupled model. The reason for the changes in water quality around the floating platform is elucidated by numerical simulations under several computational conditions, which stem from the existence of the platform.

Coastal marine ecosystems worldwide continue to be under unrelenting pressure from sustained urban development along shorelines; industrial pollution; cumulative exposure to trace hazardous substances; extensive fishing and overfishing; habitat destruction, especially of wetlands (saltmarshes, mangroves, seagrasses) and coral reefs; the continued introduction of exotic or non-indigenous species and other threats to oceanic biodiversity; and increasing numbers of outbreaks of toxic algal blooms. Governmental agencies and the general public have become increasingly concerned about maintaining the quality of these aquatic resources. Physico-chemical measurements provide quantitative data on the presence and levels of aquatic pollution and degradation, but these parameters do not reflect the extent of environmental stress affecting living organisms and their subsequent effects. In this chapter, case studies will be highlighted involving the utilization of bioindicators such as plankton and benthic organisms in monitoring the condition of coastal ecosystems in Malaysia at present and the immediate future. Variation in the occurrence and abundance of these bioindicators could be linked to climate change, eutrophication, overfishing, invasions and marine construction. In our search for indicators of coastal ecosystem health, data-based modelling that is knowledge-driven is needed for the translation of knowledge from ecological research to ecosystem management

Different organisms in marine ecosystems respond very differently to pollution by oil, which means that recovery of oiled environments also varies considerably, depending on the amount of oil spilled, the substrate on which the organisms are living, and the duration of exposure. These different responses mean that those charged with managing a marine oil spill need to consider the physical environment as much as the organisms – effective sandy beach clean-up is very different on pebble beaches than on marshes, for example – and this can determine the best methods to use during the cleanup process. During a major spill, however, responders are frequently under political pressure to be seen to be 'doing something', regardless of whether it is the best option ecologically. We discuss several different spills where different cleanup techniques have been used, and provide some basic recommendations that are accepted by many national response groups.

Following the recommendation made by Contracting Parties to the SPAW Protocol during the last Conference of the Parties (2014) to 'support more research on origins, impacts and ways to control the brown algae, as well as to improve the development of models in response to the increasing incidence of such events which affect the marine environment, biodiversity and health in the region', work and research is now on-going to address this matter through the SPAW-RAC. This is being facilitated through a number of regional initiatives in order to enhance regional cooperation and encourage collaboration and group discussions. Efforts are also being made to support more research on the origins and impacts of Sargassum and methods to control it, as well as to improve the development of models in response to the increasing incidence of events that affect the marine environment, biodiversity and health in the region.

One of the important case studies discussed in this book is that of the Sundarbans Mangrove Forest, which is a comprehensive ecosystem comprising one of the world's three largest single tracts of mangrove forests. But we are losing the value of the natural beauty of Sundarbans through our unwise actions. The Padma oil tanker *Southern Star*, which sank at the Shela River in Mangrove Sundarbans in Bangladesh, is a recent example (9 December 2014). It was almost a disaster for the surrounding forest and created a serious threat for the habitat and ecosystem. Many people are involved in river- and forest-related work such as fishing, rowing, honey and comb collection, wood and crab selling, and so forth. When the problem began, 2500 people were made jobless. The case study on this event reported in this book will help to decrease this type of accident by creating awareness and helping to develop the policies and systems to protect mangrove forest and waterways. But it is also necessary to rehabilitate the people. The case study will serve as an example to guide policymakers, regulators, environmental experts, and geographers and, above all, will offer a national and international negotiations between 'environmental researchers' to assess integrated phenomena like oil spills in a meaningful way.

St. Martin's Island is the only place in Bangladesh where coral colonies are found and is a natural treasure of Bangladesh that attracts thousands of tourists. The island also holds important ecological value as one of the few remaining nesting places in the region for several species of globally threatened marine turtles, as well as being a flyway and wintering site for migratory birds of the East Asian and Australasian region. St. Martin's Island is not only significant for its biodiversity value but also for Bangladesh in defining its Exclusive Economic Zone and delineating its sea boundary in accordance with the United Nations Convention on the Law of the Sea. Unfortunately, unregulated tourism has become detrimental for the health of this unique ecosystem. Unless tourists visiting St. Martin's Island quickly adopt ecologically responsible behaviour, the unique flora and fauna of the island, which have experienced tragic changes over the last two decades, will continue to be degraded. Against this backdrop, this report provides key information including on biodiversity, problems and conservation challenges. It also discusses the necessity of enabling policies and programming actions. It is our sincere hope that this chapter will also raise awareness about this unique island and the formidable challenges it faces in generating the necessary policy debates and actions in support of sustainable solutions.

Each chapter of this book brings fresh ideas to this new, emerging scientific frontier of environmental management of marine ecosystems. The book presents viewpoints of the authors on the challenges involved in the design and implementation of sound environmental management of marine ecosystems and might be valuable to both academics and practitioners wishing to deepen their knowledge in the field of marine ecosystems and approaches to their management. We offer a formal and heartfelt thank you to all the authors for providing their collaborative insights and for putting up with us during the editorial phase in producing this book.

We greatly appreciate the superb editorial work and patience of Irma Shagla Britton and Claudia Kisielewicz and her staff. We extend a heartfelt thank you to (in alphabetical order) Drs. Chandrashekher U. Rivonker, Chung Il Lee, Daisuke Kitazawa, Dmitry Aleynik, Hans Frost, Koji Otsuka, Lakshman Nayak, Md. Nazrul Islam, Nguyen Quang Hung, Piers Chapman, Shigeru TABETA, Thomas C. Malone and Wan Maznah Wan Omar for writing such high-quality manuscripts and for maintaining the proper focus in their chapters. This book would not have been written without decades of collegial interactions and community engagement with our peers, students and mentors and our forward-thinking community, stakeholders and researchers who have advanced the concepts of environmental management of marine ecosystems for marine science, the environment and society.

Dr. Md. Nazrul Islam

Dr. Sven Eric Jørgensen

Acknowledgments

The editors would like to acknowledge the help of all the people involved in this book project and, more specifically, the authors and reviewers who took part in the review process. Without their support, this book would not have become a reality. The editors wish to express their gratitude to the many people who provided support, offered comments, allowed themselves to be quoted, furnished data and information and assisted in the book's editing, proofreading and design.

First, the editors thank all of the authors for their contributions. Our sincere gratitude to all the authors who contributed their time and expertise to this book. Second, the editors wish to acknowledge the valuable contributions of the reviewers, which greatly improved the quality, coherence and content presentation of the chapters. Most of the authors also served as referees; we greatly appreciate their effort in fulfilling both their roles.

The late Prof. Sven Eric Jørgensen, co-editor of this book, was my higher-ranked mentor and deserves much credit. His creative and supportive influence was strongly felt and was extremely useful, especially during the preparation of the contents of this book and in the reviewing/editing of the chapters. Unfortunately, we are saddened to report the death of our friend and colleague, Sven Erik Jørgensen (29 August 1934–5 March 2016) during the editorial process of this book. Sven was a great scientist, and those of us fortunate to say we knew him well will also remember what a warm, generous person he was. He always wore a smile on his face and had an undying curiosity for learning something new, though he had strong ideas about issues. He always found time for young scientists and displayed a youthful enthusiasm for new ideas. We will sorely miss him. I extend my condolences to Prof. Jørgensen's family, friends and colleagues.

I thank Irma Shagla Britton, Senior Editor, Environmental and Engineering, CRC Press and Taylor & Francis Group, for enabling me to publish this book. I thank Claudia Kisielewicz, Editorial Assistant, CRC Press for helping me in the process of selection, editing and production.

Above all, I thank my beloved wife, Sahanaj Tamanna, my loving daughter, Sababa Mobashira Islam, and the rest of my family, who supported and encouraged me despite all the time it took me away from them. It was a long and difficult journey for them.

Last, but not least, I beg the forgiveness of all those who have been with me over the years but whose names I have failed to mention.

I hope the end result meets your approval.

Md. Nazrul Islam
Jahangirnagar University, Savar, Dhaka, Bangladesh

Editors

Dr. Md. Nazrul Islam is an Associate Professor of the Department of Geography and Environment in Jahangirnagar University, Savar, Dhaka-1342, Bangladesh. His fields of interest are environmental and ecological modelling of climate change impact on aquatic ecosystems, phytoplankton transition, harmful algae and environmental management of marine ecosystems regarding hydrodynamic ecosystems-coupled model on coastal seas, bays, and estuaries. Dr. Nazrul completed his PhD (environmental systems innovation) at the Graduate School of Engineering at the University of Tokyo, Japan entitled 'Numerical Modeling for Predictive Assessment and Mitigation of Cyanobacteria Toxins in a Eutrophic Lake'. He has also completed two years as a standard JSPS postdoctoral research fellow on marine ecosystems and engineering at the University of Tokyo, Japan with novel and interesting research on 'Numerical Assessment of the impacts of toxic materials on marine ecosystem by MEC and ECOPATH coupled model in Kamaishi Bay, Japan'. He has also studied with the Environmental Systems Analysis group at Wageninegen University, the Netherlands. Dr. Nazrul also visited as a faculty member and as an invited speaker in several foreign universities in Japan, USA, Australia, UK, Canada, China, South Korea, Germany, France, the Netherlands, Taiwan, Malaysia, Singapur, and Vietnam. He has been awarded the 'Best Young Researcher Award' by the International Society of Ecological Modeling (ISEM) for his outstanding contribution to the ecological modelling field and he has also been awarded 'Best Paper Presenter Award' for his paper entitled 'Cyanobacteria Bloom and Toxicity of Lake Kasumigaura in Japan' by SautaiN in Kyoto, Japan. He has made more than 45 scholarly presentations in many countries around the world, authored more than 65 peer-reviewed articles and authored 15 books and research volumes. Dr. Islam is presently honoured as an executive editor-in-chief of the journal *Modeling Earth Systems and Environment*, Springer International Publications in Germany and USA (journal no. 40808).

Sven Erik Jørgensen was a professor of environmental chemistry at Copenhagen University. He earned a doctorate in engineering in environmental technology and a doctorate of science in ecological modelling. He is an Honourable Doctor of Science at Coimbra University, in Coimbra, Portugal, and at Dar es Salaam University, in Dar es Salaam, Tanzania. He served as editor-in-chief of *Ecological Modelling* from the journal's inception in 1975 until 2009 and as editor-in-chief of the *Encyclopedia of Ecology*. He was president of the International Society for Ecological Modelling and chairman of the International Lake Environment Committee from 1994 to 2006. In 2004, Dr. Jørgensen received the prestigious Stockholm Water Prize and the Prigogine Prize. In 2005, he was awarded the Einstein Professorship by the Chinese Academy of Science.

In 2007, he received the Pascal medal and was elected member of the European Academy of Sciences. He authored more than 360 papers, most of which were published in international peer-reviewed journals, and edited or wrote 70 books. Dr. Jørgensen gave lectures and courses worldwide in ecological modelling, ecosystem theory and ecological engineering.

Contributors

Thomas Adams
Scottish Association for Marine Science
Scottish Marine Institute
Oban, United Kingdom

Md. Al Amin
Department of Geography and Environment
Jahangirnagar University
Savar, Dhaka, Bangladesh

Dmitry Aleynik
Scottish Association for Marine Science
Scottish Marine Institute
Oban, United Kingdom

Peder Andersen
Department of Food and Resource
  Economics
University of Copenhagen
Frederiksberg, Denmark

Kenny Black
Scottish Association for Marine Science
Scottish Marine Institute
Oban, United Kingdom

Michael Burrows
Scottish Association for Marine Science
Scottish Marine Institute
Oban, United Kingdom

Piers Chapman
Department of Oceanography
and
Geochemical and Environmental Research
  Group
Texas A&M University
College Station, Texas

Hoang Dinh Chieu
Research Institute for Marine Fisheries
  (RIMF)
Ministry of Agriculture and Rural
  Development (MARD)
Hai Phong, Vietnam

Andrew Dale
Scottish Association for Marine Science
Scottish Marine Institute
Oban, United Kingdom

Keith Davidson
Scottish Association for Marine Science
Scottish Marine Institute
Oban, United Kingdom

Fengjun Duan
The Canon Institute for Global Studies
Japan

Vu Trieu Duc
Research Institute for Marine Fisheries
  (RIMF)
Ministry of Agriculture and Rural
  Development (MARD)
Hai Phong, Vietnam

Dong Thi Dung
Research Institute for Marine Fisheries
  (RIMF)
Ministry of Agriculture and Rural
  Development (MARD)
Hai Phong, Vietnam

Do Anh Duy
Research Institute for Marine Fisheries
  (RIMF)
Ministry of Agriculture and Rural
  Development (MARD)
Hai Phong, Vietnam

M. S. M. Faradina Merican
School of Biological Sciences
Universiti Sains Malaysia (USM)
Penang, Malaysia

Hans Frost
Department of Food and Resource
  Economics
University of Copenhagen
Frederiksberg, Denmark

Masataka Fujino
University of Tokyo
Japan

Mahabaleshwar R. Hegde
Department of Marine Sciences
Goa University, Taleigao Plateau
Goa, India

Ayoe Hoff
Department of Food and Resource
    Economics
University of Copenhagen
Frederiksberg, Denmark

Nguyen Quang Hung
Research Institute for Marine Fisheries
    (RIMF)
Ministry of Agriculture and Rural
    Development (MARD)
Hai Phong, Vietnam

Md. Nazrul Islam
Department of Geography and
    Environment
Jahangirnagar University
Savar, Dhaka, Bangladesh

Hyun Je Park
Department of Marine Bioscience
Gangneung-Wonju National University
Gangwon, South Korea

Chang-Keun Kang
Gwangju Institute of Science and
    Technology
Gwangju, South Korea

Daisuke Kitazawa
Institute of Industrial Science
University of Tokyo
Japan

Anthony H. Knap
Geochemical and Environmental Research
    Group
Texas A&M University
College Station, Texas

Har Kun Jung
Department of Marine Bioscience
Gangneung-Wonju National University
Gangwon, South Korea

Chung Il Lee
Department of Marine Bioscience
Gangneung-Wonju National University
Gangwon, South Korea

Thomas C. Malone
Horn Point Laboratory
University of Maryland Center for
    Environmental Science
Cambridge, Maryland

S. M. Mustafizur Rahman
Research Institute for Dok-do and
    Ulleung-do
Kyungpook National University
Bukgu, Daegu, South Korea

M. Murai
Graduate school of information and
    sciences
Yokohama National University
Japan

Lakshman Nayak
P.G. Department of Marine Sciences
Berhampur University, Bhanja Bihar
Berhampur, Odisha, India

Md. Noman
Department of Geography and
    Environment
Jahangirnagar University
Savar, Dhaka, Bangladesh

Koji Otsuka
Department of Sustainable System Sciences
Osaka Prefecture University
Sakai, Japan

Vinay P. Padate
Department of Marine Sciences
Goa University, Taleigao Plateau
Goa, India

Marie Porter
Scottish Association for Marine Science
Scottish Marine Institute
Oban, United Kingdom

Mitali Priyadarsini Pati
P.G. Department of Marine Sciences
Berhampur University, Bhanja Bihar
Berhampur, Odisha, India

Lars Ravensbeck
Department of Food and Resource
    Economics
University of Copenhagen
Frederiksberg, Denmark

Chandrashekher U. Rivonker
Department of Marine Sciences
Goa University, Taleigao Plateau
Goa, India

Mamunur Roshid
Department of Geography and Environment
Jahangirnagar University
Savar, Dhaka, Bangladesh

Toru Sato
Department of Ocean Technology, Policy,
    and Environment
University of Tokyo
Japan

Satyabrata Das Sharma
CSIR-Institute of Minerals and Materials
    Technology
Berhampur, Odisha, India

S. Shuhaida
School of Biological Sciences
Universiti Sains Malaysia (USM)
Penang, Malaysia

Le Tuan Son
Research Institute for Marine Fisheries
    (RIMF)
Ministry of Agriculture and Rural
    Development (MARD)
Hai Phong, Vietnam

Shigeru Tabeta
Department of Environment
    Systems
University of Tokyo
Japan

Anita Talib
School of Distance Education
Universiti Sains Malaysia (USM)
Penang, Malaysia

Dinesh T. Velip
Department of Marine Sciences
Goa University, Taleigao Plateau
Goa, India

Terry L. Wade
Department of Oceanography
and
Geochemical and Environmental
    Research Group
Texas A&M University
College Station, Texas

W. O. Wan Maznah
School of Biological Sciences
and
Centre for Marine and Coastal Studies
    (CEMACS)
Universiti Sains Malaysia (USM)
Penang, Malaysia

Khairun Yahya
School of Biological Sciences
and
Centre for Marine and Coastal Studies
    (CEMACS)
Universiti Sains Malaysia (USM)
Penang, Malaysia

1

Integrated Coastal Zone Monitoring in Support of Ecosystem-Based Management of Marine Ecosystem Services

Thomas C. Malone and Anthony H. Knap

CONTENTS

1.1 Introduction 2
1.2 Framework for Designing and Assessing the Performance of an SoS 3
    1.2.1 Coastal Ecosystem Services 4
    1.2.2 Coastal Ecosystem States and Indicators 4
    1.2.3 Pressures on Coastal Ecosystems 6
1.3 Informing Integrated Ecosystem Assessments 7
1.4 An Integrated System of Systems 9
    1.4.1 Boundary Conditions and Time Scales 11
    1.4.2 Sampling Ecosystems and the Problem of Undersampling 11
    1.4.3 Linking Observations and Models 12
    1.4.4 Performance Assessments 13
    1.4.5 Integrated Coastal Governance 14
1.5 Sustained and Integrated Coastal Zone Observing System of Systems 15
    1.5.1 Chesapeake Bay Program (USA) 16
        1.5.1.1 Program Overview 16
        1.5.1.2 Ecosystem Recovery Targets 17
        1.5.1.3 Data Management and Communications 18
        1.5.1.4 Governance for Integration (Figure 1.7) 18
    1.5.2 Ecosystem Health Monitoring Program (Australia) 19
        1.5.2.1 Program Overview 19
        1.5.2.2 Ecosystem Recovery Targets 21
        1.5.2.3 Integrating Programs 21
        1.5.2.4 Governance for Integration 22
1.6 Conclusions and Lessons Learned 23
    1.6.1 Programmatic Strengths 23
        1.6.1.1 Chesapeake Bay Program 23
        1.6.1.2 Ecosystem Health–Monitoring Program 24
    1.6.2 Problems and Challenges 24
        1.6.2.1 Chesapeake Bay Program 24
        1.6.2.2 Ecosystem Health Monitoring Program 26
References 26

1.1 Introduction

Sustainable development and human well-being depend on healthy ecosystems that provide services valued by society (Costanza et al. 1997, 2014; Hassan et al. 2005; Millennium Ecosystem Assessment 2005; Adams 2006; Scott-Cato 2009; Sidlea et al. 2013; Malone et al. 2014a,b). Globally, people and ecosystem services are concentrated in the coastal zone (Small and Cohen 2004; Martinez et al. 2007; Barbier et al. 2011) where the health of marine ecosystems is most at risk due to convergent anthropogenic pressures associated with human expansion and climate change (Jackson et al. 2001; Lotze et al. 2006; Nicholls et al. 2007; Montoya and Raffaelli 2010). The scientific community has responded by calling for ecosystem-based management (EBM) of ecosystem services for sustainable development (Browman et al. 2004; Garcia and Cochrane 2005; Leslie and McLeod 2007; Murawski 2007; Kaplan and Levin 2009; Palumbi et al. 2009; Foley et al. 2010; de Suarez et al. 2014; Sherman 2014), an integrated approach to restoring and sustaining ecosystem services that has been embraced in both national and international environmental policies (Malone et al. 2010). Marine spatial planning is a form of EBM (Foley et al. 2010).

To be effective, EBM must be guided by frequently updated integrated ecosystem assessments (IEAs) that take into account temporal and spatial patterns of pressures on marine ecosystems, changes in marine ecosystem states and the impacts of these changes on marine ecosystem services (Browman et al. 2004; Leslie and McLeod 2007; Murawski 2007; UNEP 2007; Kaplan and Levin 2009; Levin et al. 2009; Montoya and Raffaelli 2010). Developing this capability depends on the design, implementation and sustained operation of an observing and prediction* system of systems (SoS)† for concurrent monitoring of key states, pressures and impacts. The operational goals of such an SoS are to (1) decrease the lag time between changes in states and integrated assessments of the impacts of such changes on services and human well-being and (2) increase the ability to anticipate changes in states with sufficient lead-time to engage in adaptive EBM for sustainable development (Figure 1.1).

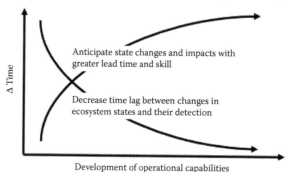

FIGURE 1.1
The operational goals of the SoS are to provide data at rates and in forms required to decrease the time lag between changes in ecosystem states and their detection and to anticipate state changes and their impacts with greater lead time and skill.

---

* Predictions include estimates of past, current and future ecosystem states and services.
† An SoS is a set of integrated systems that exhibits emergent behavior while enabling the delivery of needed data and information that individual systems cannot (Maier 1998; DOD 2008; Jamshidi 2009). It consists of complex, multidisciplinary, multiscale, interoperable and distributed systems operating in concert. For our purposes, the SoS includes observations of marine and estuarine ecosystems, data telemetry, data management and communications, and ecosystem models that inform the preparation of IEAs.

1.2 Framework for Designing and Assessing the Performance of an SoS

The driver-pressure-state-impact-response (DPSIR) model (Figure 1.2) provides a framework for designing an SoS that informs IEAs based on observations and models of pressures on ecosystems, indicators of ecosystem states and impacts on ecosystem services (Malone et al. 2014a). For our purposes, drivers, pressures, states, impacts and responses are defined as follows:

- *Drivers* are the primary sources of pressures on marine and estuarine ecosystems.
- *Pressures* are human interventions and external forces of nature that cause changes in coastal marine ecosystem states.
- *Ecosystem states* are measures of ecosystem health that are sensitive to pressures and underpin the capacity of ecosystems to support services.
- Changes in ecosystem states *impact* the well-being of human populations through changes in the provision of services.
- Such changes lead to human *responses* or social and political actions including the formulation of ocean policies, integrated ecosystem assessments, ecosystem-based approaches (to managing human pressures and adapting to or mitigating impacts changes in states), and performance assessments.

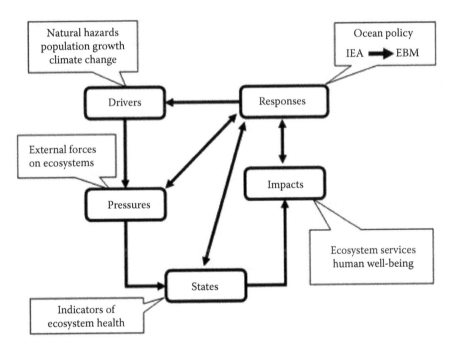

FIGURE 1.2
A DPSIR conceptual model. IEAs inform EBM by assessing and reporting on (1) ecosystem states and changes in states, (2) external pressures (natural and anthropogenic) on ecosystems that perturb ecosystem dynamics and lead to changes in states, and (3) impacts of ecosystem state changes on ecosystem services and human well-being. (Adapted from Malone, T.C. et al. (2014a) *Marine Policy*, 43: 262–272.)

Rapid detection of changes in ecosystem states and timely anticipation of their impacts on services begin with the operational* provision of multidisciplinary data (from observations and from models) on pressures, key state indicators and the status of ecosystem services (UNESCO 2012).

1.2.1 Coastal Ecosystem Services

Ecosystems provide a range of services that are fundamentally important to human well-being (Millennium Ecosystem Assessment 2005). Costanza et al. (2014) estimated the mean value of these services to be approximately US$125 trillion globally (2007 USD). By comparison, services from the Earth's coastal marine ecosystems (estuaries, seagrass beds, kelp forests, coral reefs, tidal marshes, mangrove forests and continental shelves) are valued at approximately $52 trillion year. Thus, while coastal ecosystems occupy less than 1% of the Earth's surface area, they account for more than 40% of ecosystems services. Those services that are most threatened by changes in coastal ecosystem states include the following (Malone et al. 2014b):

- Food supplies (landings of fin fish and shellfish)
- Climate regulation (carbon sequestration into biomass, air/sea fluxes of heat and carbon dioxide)
- Maintenance of water quality (nutrient cycling, filter feeding, sediment trapping)
- Protection against coastal erosion and flooding (coastline stabilization, dampening wave action and storm surge, buffers to flooding)
- Refugia for biodiversity (biologically structured habitats[†])
- Aesthetic value, recreation and tourism (coral reefs, salt marshes, mangrove forests and iconic species)

Lost protection against coastal erosion and flooding is arguably the most costly in terms of human well-being. For example, it is likely that sea level rise and the loss of biologically structured habitats will combine to more than triple the number of people vulnerable to coastal flooding by 2070 (OECD 2007). Flooding events will become more frequent and severe and subsequent runoff events will degrade water quality and increase risks of public exposure to waterborne pathogens and toxic chemicals.

1.2.2 Coastal Ecosystem States and Indicators

Key ecosystem states changes which have significant impacts on the provision of services include the following: (1) spatial extent and condition of biologically structured habitats (habitat), (2) species diversity (diversity), (3) abundance of living marine resources (LMRs), (4) net primary production (net PP), (5) sea level (SL), (6) vertical stratification of water column, (7) heat content of water column (heat), (8) acidity of water column (acid), (9) concentration of anthropogenic chemical toxins (toxins), (10) concentration of waterborne pathogens (path), (11) hypoxia, (12) abundance of harmful (toxic) algal species (HABs), (13) abundance of invasive species (invasive), and (14) number and magnitude of mass mortalities of marine animals (Table 1.1).

---

* Routine and sustained provision of quality-controlled data and information.
† Biologically structured habitats include mangrove forests, salt marshes, seagrass beds, kelp forests and coral reefs.

TABLE 1.1

Coastal Marine Ecosystem States and Services that are Directly Impacted by Changes in State

Key States		Refugia	Food	Climate	WQ	Flood	Tour
		Key Services at Risk					
Beneficial	Habitat	+	+	+	+	+	+
	Diversity	+	+	+	+	+	+
	LMRs		+		+		+
	Net PP		±	+	±		
Jeopardize	SL	−	−		−	−	−
	Stratification		±	−	−		
	Heat	−	−		−		
	Acid	−	−		−		
	Toxins	−	−		−		−
	Path		−		−		−
	Hypoxia	−	−	−	−		
	HABs		−		−		−
	Invasive	−	−		−		
	Mortality		−		−		−

Source: Modified from UNESCO. (2012). *Requirements for Global Implementation of the Strategic Plan for Coastal GOOS. GOOS Report 193*. Paris: Intergovernmental Oceanographic Commission (www.ioc-goos.org/index.php?option=com_oe&task=viewDocumentRecord&docID=7702&lang=en); Malone, T.C. et al. (2014a). *Marine Policy*, 43: 262–272.

Note: Services include refugia for biodiversity, food supply, climate regulation, maintenance of water quality (WQ), protection from coastal erosion and flooding (flood), and tourism/recreation/aesthetic value (tour). Ecosystem states may be divided into two categories, those for which increases are likely to be beneficial to services (+) (habitat – spatial extent and condition of biologically structured habitats, diversity – biodiversity, LMRs – abundance of living marine resources), and those that are likely to jeopardize services (−) (SL – mean sea level, heat content, acidity, concentration of chemical toxins, path – concentration of waterborne pathogens, hypoxia – temporal and spatial extent of oxygen-depleted bottom water, HABs – abundance of toxic algal species, abundance of invasive species and mass mortalities of marine animals. Note that increases in two states (net PP – net annual primary production, Stratification – vertical stratification of the water column) may benefit or jeopardize (±) services depending on the environmental conditions of specific ecosystems. For example, an increase in net phytoplankton PP may enhance food supplies if the increase is passed up the food chain to commercial fish species but may jeopardize food supplies if the increase fuels bacterial metabolism and oxygen depletion. (e.g., hypoxia). Likewise, increases in net seagrass PP may increase habitat space for fish and shellfish populations.

Given that ecosystems are characterized by diverse interactions among many populations of species and environmental parameters, only some of which can be monitored, it is important to select a small number of indicators of ecosystem states. These should include (1) the spatial extent and condition of biologically structured habitats; (2) species richness;* (3) abundance of harvestable LMRs; (4) net annual PP by phytoplankton, seagrass beds, mangrove forests and salt marshes; (5) annual mean local SL; (6) distribution of pycnocline strength; (7) temperature fields; (8) aragonite saturation fields; (9) chemical toxin fields; (10) dissolved oxygen fields; (11) abundance of toxic phytoplankton; (12) abundance of invasive species; and (13) frequency and magnitude of mass mortalities of marine birds, mammals,

---

* Species richness is an unweighted list of species present in an ecosystem that is especially important to monitor because it is the simplest indicator of species diversity, and it does not discount rare species, which are often the primary concern.

TABLE 1.2

Essential State Variables to be Monitored for Routine and Repeat IEAs

Essential State Variables			
Geophysical	**Chemical**	**Biological**	**Biophysical**
Temperature	Dissolved inorganic nutrient concentrations (N, P, Si)	Spatial extent of coral reefs, seagrass beds, kelp beds, salt marshes and mangrove forests	Water leaving radiance
Salinity	Dissolved oxygen concentration	Species richness	Downwelling irradiance
Currents	Aragonite saturation state	Biomass of harvestable fish stocks	
Surface waves	Chemical contaminants in sediments and sea food (e.g., mercury, lead, PCBs, DDT and dieldrin)	Biomass of primary producers (phytoplankton, seagrass beds, kelp beds, marsh grasses and mangrove trees)	
Local mean sea level		Coral skeletal density	
Shoreline position		Abundance and size of apex predators (fish, sharks, mammals, and birds)	
Bathymetry		Abundance of toxic phytoplankton species	
		Abundance of waterborne pathogens	
		Species composition and abundance of invasive species	
		Magnitude of mass mortalities of marine mammals, fish and birds	

Source: Adapted from UNESCO. (2012). *Requirements for Global Implementation of the Strategic Plan for Coastal GOOS. GOOS Report 193.* Paris: Intergovernmental Oceanographic Commission (www.ioc-goos.org/index.php?option=com_oe&task=viewDocumentRecord&docID=7702&lang=en); Malone, T.C. et al. (2014b). *Natural Resources Forum,* 38: 168–181.

fin fish, and shellfish. This core set of states and indicators may be modified or enhanced based on priority ecosystem services to be sustained and the characteristics of targeted ecosystems. However, we emphasize that declines in species diversity (as indicated by declines in species richness) and habitat loss (as indicated by decreases in the spatial extent of salt marshes, mangrove forests, seagrass beds, kelp forests and coral reefs) impact all services and are therefore of fundamental importance to the capacity of ecosystems to provide services. Essential state variables that should be monitored to estimate indicators of states have been tabulated (Table 1.2).

1.2.3 Pressures on Coastal Ecosystems

Changes in ecosystem states reflect the interplay between *internal* interactions among their biotic and abiotic components (ecosystem dynamics) and the effects of multiple pressures that impinge upon ecosystems. Pressures can be parsed into two broad categories (Malone et al. 2014a): (1) *fast* pressures associated with the near term (days – months) effects of human expansion and (2) *slow* pressures associated with anthropogenic climate change (years – decades) (Table 1.3). In addition to their direct impact on services and human well-being, slow pressures make coastal ecosystem services and human populations increasingly vulnerable to the impacts of fast pressures.

TABLE 1.3

Pressures on Coastal Marine Ecosystems Associated with Human Expansion, Climate Change and Natural Hazards

Major Pressures on Coastal Ecosystems	
Fast	Coastal development (urbanization, agriculture, shoreline hardening)
	Erosion
	Spatial and temporal extent of flooding
	Land-based inputs of nutrients (nitrogen and phosphorus), sediments and toxic chemicals (e.g., herbicides, pesticides, carcinogens) and pathogens
	Extraction of living marine resources (fin fish, crustaceans, mollusks, cephalopods, macroalgae)
	Invasions of non-native species
	Maritime operations (e.g., commerce; oil and gas exploration, extraction and transport; wind farms)
	Mariculture (fin fish, crustaceans, mollusks, macroalgae)
Slow	Net annual net flux across air/sea interface
	Net annual CO_2 flux across air/sea interface
	Strengthening hydrological cycle (increase in wet precipitation and river discharge)
	Melting glaciers, sea ice and polar ice caps
	Increase in tropical storm intensity
	Natural ocean-atmosphere climate modes (e.g., El Niño-Southern Oscillation, Pacific Decadal Oscillation, Atlantic Multidecadal Oscillation)

1.3 Informing Integrated Ecosystem Assessments

To be useful to decision makers and the public, IEAs must be based on multidisciplinary observations and be repeated frequently at rates tuned to the time scales on which decisions need to be made to achieve desired goals (Waltner-Toews et al. 2003; Chapin et al. 2009; Malone et al. 2014a). Few, if any, monitoring programs serve multidisciplinary data on appropriate time and space scales for the routine provision of IEAs – and many are not sustained (Duarte et al. 1992; Christian et al. 2006; Carstensen 2014).* Time scales range from near real-time (e.g., nowcasts and forecasts of the time-space extent of coastal flooding, oil spill trajectories; alerts of potential exposure to algal toxins and waterborne pathogens) and annual assessments (e.g., set quotas for fish landings, manage land-based nutrient loads from point and diffuse sources, assess compliance for point source dischargers, evaluate progress toward meeting management targets for ambient water quality parameters) to decadal assessments of trends in the provision of ecosystem services (cf. Costanza et al. 2014).

Procedures for developing and sustaining the SoS must enable the establishment of an integrated ocean observing system that serves the multidisciplinary data streams required for adaptive[†] EBM. The process begins with the establishment of a multi-sector[‡] stakeholder forum to (1) reach consensus on priority objectives for sustaining and restoring

---

* http://www.seagrasses.org/handbook/european_seagrasses_low.pdf

† A form of structured decision making that emphasizes the need to reduce uncertainty over time in order to improve the efficacy of management decisions. Uncertainty occurs because the variability of pressures on ecosystems is often unpredictable; management actions often have unintended consequences; observations and models of ecosystem states and services are often insufficient or inadequate to detect trends with known certainty; and there is a lack of understanding of ecosystem dynamics.

‡ Sectors include, for example, management of land-uses; management of natural resources and water quality; extraction of nonliving marine resources; environmental regulation; marine operations and commerce; marine conservation; coastal and marine spatial planning, education and research and technical groups responsible for designing, implementing and operating the SoS.

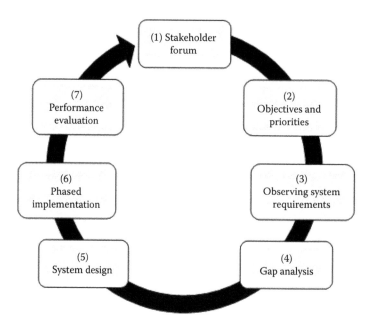

FIGURE 1.3
A stakeholder-driven, sustained and iterative life cycle for designing, implementing, evaluating and improving the SoS. In effect, the stakeholder forum is a *community of practice*.* (Adapted from UNESCO. (2012). *Requirements for Global Implementation of the Strategic Plan for Coastal GOOS. GOOS Report 193.* Paris: Intergovernmental Oceanographic Commission [www.ioc-goos.org/index.php?option=com_oe&task=viewDocumentRecord&docID=7702&lang=en.])

healthy ecosystems and the services they provide; (2) identify priority pressures and ecosystem state indicators; (3) identify systems to be integrated into an SoS; these include observations, data telemetry, data management and communications (DMAC) and models; (4) perform gap analyses of capabilities against objectives; (5) facilitate partnerships among data providers (e.g., technical groups that monitor and model coastal terrestrial and marine ecosystems, manage associated data streams, prepare IEAs) and users (e.g., government agencies, land-use planners, marine spatial planners, conservation groups, scientists and educators); and (6) oversee an iterative process for reviewing objectives and designing, implementing, evaluating and improving the emerging SoS to achieve objectives more effectively (Figure 1.3).

The major operational objectives of the forum should include the following:

1. Establish an integrated governance structure for the SoS that enables (i) oversight and coordination of all systems including close linkages between observations, data management and modeling; (ii) interoperability among systems and programs; (iii) timely feedback from user groups to tune data products (IEAs, nowcasts, forecasts, predictions) to their needs; (iv) effective use of science and technology advisory panels to guide adaptive management through an iterative processes of design, implementation and improvement of the SoS; (v) routine, timely and frequent communications and reporting to targeted audiences; (vi) public support; and (vii) transparency, accountability and performance assessments of the SoS.

---

* http://net.educause.edu/ir/library/pdf/nli0531.pdf

2. Serve, process and report data and information on changes in ecosystem states and impacts in as close to real-time* as possible.

3. Conduct observations and modeling in the larger context of regional- and global-scale observations of pressures on the targeted ecosystems and impacts of changes in ecosystem states.

4. Integrate data streams from both *in situ* and remote sensing in near real time to assess time-dependent changes in ecosystem states in two (benthic communities) and three dimensions (pelagic communities).

5. Minimize errors associated with undersampling and improve the skill of predictions (including estimates of current and future ecosystem states), establish synergy between observations and modeling by building a DMAC system that provides rapid access to multidisciplinary data from multiple databases and sources (*in situ* measurements, remote sensing and models).

6. Specify requirements and standards for observations, data telemetry, data assimilation, models and data management; and conduct gap analyses (assess current technical expertise, infrastructure assets, management capabilities and licenses against observing system requirements and objectives) to inform the stakeholder forum on priorities for improving the SoS to achieve objectives more effectively.

7. Prepare and update IEAs at rates that inform evidence-based decisions in a timely fashion. Given the time-space scales of the fast and slow pressures and the complex and dynamic nature of coastal ecosystems (Steele 1985; Barber and Chavez 1986; Powell 1989; Costanza et al. 1993), observations and modeling must be sustained in perpetuity at sufficient resolution in time and space to capture the broad spectrum of variability that characterizes coastal ecosystems.

8. Build and sustain strong public and political support by engaging stakeholders and incorporating volunteer monitoring programs into the SoS (Duarte et al. 1992; Christian et al. 2006). This requires public and political awareness of ecosystem services and their vulnerability to changes in ecosystem states, an awareness that can only come through timely, frequent and continuous provision of information to the public about pressures, states and impacts (and, therefore, on human well-being). A diversity of stakeholders must be actively engaged in the sustained implementation and evolution of the SoS.

These challenging objectives can only be achieved through a stepwise, iterative process that promotes, builds on and complements existing programs, partnerships and initiatives over time.

1.4 An Integrated System of Systems

Developing and sustaining an SoS goes beyond monitoring *per se* by enabling strong feedbacks between user needs and an end-to-end system of observations, data management,

---

* The provision of data in near real-time is important for quality control, rapid detection of instrument failures, including calibration issues and improving the skill of model-based predictions.

and modeling (Christian et al. 2006). Thus, design plans for the SoS include (1) specification of data and information requirements for IEAs (e.g., Table 1.2), (2) identification of assets* to be integrated into an end-to-end SoS (Figure 1.4), (3) specification of boundary conditions and time scales of resolution for observations and models, and (4) formal linkages between observations and models.

The link between observations and modeling and the primary mechanism for integration (in concert with modeling), DMAC is of central importance to the development of an integrated, interoperable SoS. To this end, DMAC must (1) process and archive data on the essential variables according to well-documented standards and formats; (2) archive all relevant data types (*in situ* measurements, remote sensing and model outputs) in near real time and delayed mode as required; (3) enable close coupling between observations and modeling (for data assimilation, model calibration and validation); and (4) enable rapid and easy access to these data and derived products (e.g., IEAs, predictions and early warnings) by users. In short, the DMAC infrastructure must evolve to reduce the time required for users to discover, acquire, process and analyze multidisciplinary data of known quality from multiple sources. DMAC development as part of the U.S. Integrated Ocean Observing System[†] and the Australian Integrated Marine Observing System[‡] (Proctor et al. 2010) provide good illustrations of the way forward.

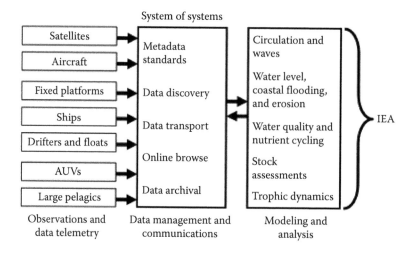

FIGURE 1.4

An end-to-end system that enables exploitation of multidisciplinary data from many sources. Observations from both remote and *in situ* sensing (including autonomous underwater vehicles (AUVs) and large pelagic animals as platforms) are integrated via a data management and communications system. (For details see UNESCO. (2012). *Requirements for Global Implementation of the Strategic Plan for Coastal GOOS. GOOS Report 193.* Paris: Intergovernmental Oceanographic Commission [www.ioc-goos.org/index.php?option=com_oe&task= viewDocumentRecord&docID=7702&lang=en.])

---

* Note that *assets* may include existing programs (that are integrated without affecting their operations or products) as well as new programs needed to fill gaps.
† https://ioos.noaa.gov/
‡ http://www.adv-geosci.net/28/11/2010/adgeo-28-11-2010.pdf

1.4.1 Boundary Conditions and Time Scales

The marine environment from estuaries to the open ocean consists of nested, interact-ing ecosystems in which boundaries may overlap; state changes in one may be pres-sures in another, and state changes may impact services that span multiple ecosystems. Ecosystems are spatially bound (in either a Lagrangian or Eulerian frame of reference) dynamic networks of organisms (including people) interacting with each other and their abiotic environment. Their scale is determined by the spatial extent of the dynamic net-works of interest as well as by management jurisdictions. Thus, although the focus here is state changes occurring in coastal ecosystems, a coastal ocean SoS must encompass a broad range of spatial scales from ocean basins to coastal watersheds (catchments) and coastal ecosystems in between.

From a Eulerian perspective, spatial scales of marine ecosystems range from the ocean basins (1–15×10^7 km^2) (e.g., de Young et al. 2004) to coastal ecosystems that may be less than 10 km^2 (e.g., tidal marshes, coral reefs) or as large as 1–5×10^5 km^2 (e.g., Large Marine Ecosystems and a few large marine protected areas) (Sherman et al. 2007; Wood et al. 2008). Temporal scales that must be resolved to detect secular trends in pressures, states and impacts also exhibit a broad spectrum of variability, i.e., hourly–decadal (Steele 1985; Levin 1992; Costanza et al. 1993; Petersen et al. 2009). To complicate matters, rapid and unan-ticipated changes ('ecological surprises' or regime shifts) in the abundance of one or more species are common, especially in pelagic communities (Steele, 2004; de Young et al. 2008; Petersen et al. 2008; Kraberg et al. 2011). These considerations underscore the importance of establishing spatial boundary conditions for ecosystems and of sustained observations with sufficient temporal resolution to resolve secular trends from the natural variability that characterize ecosystem states and the pressures on them.

1.4.2 Sampling Ecosystems and the Problem of Undersampling

Ideally, data on pressures, services and essential state variables should be collected syn-optically in time and space over a wide range of scales (Figure 1.5). However, given the dynamics and ecological complexity of marine ecosystems, undersampling will always be a problem that can only be minimized through the strategic use of a mix of platforms that enable both remote and *in situ* sensing as well as feedback between observations and mod-eling optimized to minimize errors and increase the accuracy of model-based predictions. In short, the emergence of a cost-effective SoS depends on leveraging synergies between models, sensors, platforms and sampling regimes.

Remote sensing (space-based and airborne) is most useful for providing time-series observations of spatially synoptic surface fields for some essential state variables (shoreline position, vector winds, surface currents and waves, turbidity, light attenuation, tempera-ture, salinity, chlorophyll a, spatial extent of seagrass beds, salt marshes and mangrove forests). The main challenges here are continuity, validation and increasing temporal, spa-tial and spectral resolution.

Historically, *in situ* observations have depended on ships, moorings, small boats, piers and divers. Moorings and piers equipped with *in situ* sensors can provide multidisciplinary, long-term (months to years), high-resolution time series. However, spatial resolution is generally poor unless large numbers of moorings are deployed simultaneously (which is usually too expensive). Research vessels provide controlled laboratory environments for precise and accurate measurements of all essential variables and serve as platforms for underway measurements, vertical profiling and benthic surveys. But ship surveys are

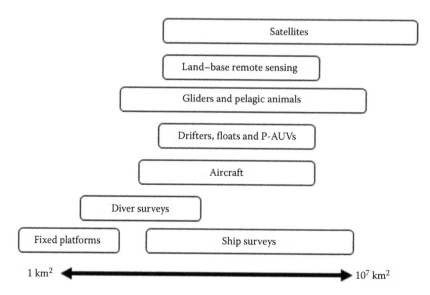

FIGURE 1.5
Changes in ecosystem states reflect biotic and abiotic interactions that occur over a spectrum of spatial scales that span more than seven orders of magnitude that can only be captured using a mix of platforms (P-AUV, powered autonomous underwater vehicles). (For details see UNESCO. (2012). *Requirements for Global Implementation of the Strategic Plan for Coastal GOOS. GOOS Report 193.* Paris: Intergovernmental Oceanographic Commission [www.ioc-goos.org/index.php?option=com_oe&task=viewDocumentRecord&docID=7702&lang=en.])

slow and expensive. Ferries and other ships of opportunity are especially important for deploying sensors with relatively high power requirements and accuracy and for providing data on boundary conditions for numerical modeling. Remote underwater autonomous vehicles, gliders, autonomous surface vehicles and *in-situ* floats with diverse sensor packages are providing new observation capabilities. Diver-controlled underwater vehicles equipped with high-definition cameras are especially useful for mapping submerged habitats such as coral reefs.*

In situ measurements of the essential variables (Table 1.2) should be made at a representative network of sentinel and reference sites (in the form of fixed stations, transects, and grids).† These, in concert with remote sensing, are intended to populate a sampling regime that enables rapid detection of state changes and modeling relationships between pressures, state changes and services that are needed to anticipate potential impacts with sufficient lead time to respond appropriately.

1.4.3 Linking Observations and Models

Observing marine ecosystems and modeling them are mutually dependent processes. Indicators of ecosystem states cannot be observed directly; they are computed from models that provide predictions (hindcasts, nowcasts and scenarios of future states that are

---

* For example, see http://catlinseaviewsurvey.com/science/technology
† A sentinel site is a location at which in-depth observations can be used to predict larger-scale patterns in time or space. Reference sites are selected because the environment is relatively undisturbed by anthropogenic pressures. Observations from such sites can be used as baselines from which deviations (anomalies) caused by anthropogenic pressures can be detected.

not observed directly) of states that are grounded in observations. Models include simple statistical relationships (e.g., multiple and multivariate regression models), more sophisticated statistical constructs (e.g., geospatial information systems, neural networks, network analysis), dynamical models based on first principles (e.g., numerical circulation, storm surge and ecosystem models in Lagrangian or Eulerian forms) and, coupled models of the biotic and abiotic components of the marine ecosystem (e.g., coupled atmosphere-ocean circulation-trophic dynamical models).

Of central importance for the provision of IEAs is the computation of gridded reconstructions of past and current ecosystem states using data assimilation techniques that also provide estimates of the errors associated with interpolation and extrapolation (Wang et al. 2000; Bell et al. 2009). Data assimilation techniques are most advanced for numerical predictions of weather and physical oceanographic states (sea surface temperature, currents and waves) and least advanced for predictions of chemical and biological ecosystem states. An analysis is the computation of estimates of ecosystem states based on a set of observations that typically under sample the ecosystem. The analysis is informed by 'baseline' information such as a long-term average of ecosystem state. Knowledge of mean states of marine ecosystems over specific periods is fundamental to resolving short-term variability from longer-term trends and to predicting changes in ecosystem states and their impacts.

An important application of hindcasting that is especially relevant to the development of an SoS is the computation of means and first moments of variability for the essential variables in targeted marine ecosystems. Unfortunately, historical data on most essential variables (especially nonphysical variables) are grossly inadequate for most marine and estuarine ecosystems, and it is essential that observations required to compute climatologies for key indicators be a high initial priority (Smith and Koblinsky 2001).

For the purposes of SoS development, analyses can be used to provide comprehensive and internally consistent diagnostics of ecosystem states as input data to another operation (e.g., as the initial state for a numerical forecast of a current field), as a reference against which observations can be compared for quality control, and as data retrieval for an observing system simulation experiment (OSSEs). The latter is particularly important for optimizing sampling schemes for observing marine ecosystems on local to global scales (Schiller and Brassington 2011). The importance of selected sentinel and reference sites for the accuracy of a prediction can be assessed by observing system experiments (OSEs) in which existing observations (e.g., variable-specific data from selected locations, data on a particular variable) are removed from a standard database. The impact of future instruments can be assessed using hypothetical data in OSSEs. OSSEs use data-assimilating models to specify the optimal mix of observations (locations, spatial extent and variables measured) for model-based predictions and can accelerate the transition of observations from newly developed instruments to operational use. In fisheries, and increasingly in coastal management, model-based scenarios may be used to assess the likely performance of EBM strategies, combining observing strategies and management decision rules.

1.4.4 Performance Assessments

The sustained evolution of an SoS requires a systematic and rigorous process for periodic evaluations of performance against objectives. Performance metrics fall into two broad categories: (1) system performance and (2) user satisfaction. System performance includes measures of data quality, continuity of data streams, integration of data from both *in situ* and remote sensing, and the accuracy of model predictions. User satisfaction is measured in terms of user demand and the timely provision of data and information that inform IEAs. User satisfaction may also be assessed in terms of the efficacy of management actions in achieving specific outcomes.

As recommended by the Intergovernmental Panel on Climate Change, examples of metrics for sea level observations (system performance) include the following:

1. Complete the installation of real-time, remote reporting of tidal gauges and colocated permanent Global Positioning System (GPS) receivers at 62 stations for documenting long-term trends, and 30 stations for altimeter drift calibration, as part of the international Global Sea Level Observing System.

2. Establish the permanent infrastructure needed to process and analyze satellite altimetry, tidal gauge and GPS data for the routine provision of annual sea-level-change reports with (i) estimates of monthly mean sea level for the past 100 years with 95% confidence, (ii) variations in the relative annual mean sea level for the entire record for each instrument, and (iii) estimates of absolute global sea-level-change accurate to 1 mm per year.

1.4.5 Integrated Coastal Governance

Governance, not the science and technology behind the development of an SoS, is the weak link in the implementation chain for IEAs and EBM (Browman et al. 2004). Two problems must be addressed: (1) sector-specific management and (2) mismatches between the scale of ecosystems and the scale of management.

Historically, governments have responded to changes in ecosystem states and their impacts in an *ad hoc* fashion focusing on sector-specific management (e.g., water quality, fishery land use, water, transportation, oil and gas extraction, wind energy, tourism, and recreation) rather than on an integrated, holistic strategy for managing human uses of ecosystem services (Carpenter et al. 2009; Palumbi et al. 2009; Foley et al. 2010; Tallis et al. 2012). Limitations of sector-specific approaches to governance are compounded by inherent complexities in managing a growing number of interacting activities across different levels of government, across agencies (ministries) within governments, and (for ecosystems that span multiple national jurisdictions) among governments. As a result, competition for funding among government agencies often inhibits needed collaborations and can result in policy choices that are detrimental to ecosystems and their services and dependent coastal communities.

The boundaries of marine ecosystems and the spatial scales of pressures and services typically do not conform to the jurisdictional boundaries of sector-specific management (Perry and Ommer 2003; Weeks 2014). For example, large pelagic predators such as sharks and tuna often migrate on the scale of the ocean basins, marine larvae can travel hundreds of kilometers before settling, land-based inputs of nutrients occur on the scale of coastal watersheds, and climate-driven ocean warming is occurring globally. Mismatches between the scales of marine ecosystems and the spatial scales at which marine monitoring, management, and conservation are conducted can lead to ineffective management actions and consequent degradation of the capacity of marine ecosystems to provide services. Implementation of EBM will help to tune monitoring and management to the scales of marine ecosystems.

What is needed is the development of institutional frameworks that promote a shift from sector-by-sector management and regulation to a more holistic approach that considers

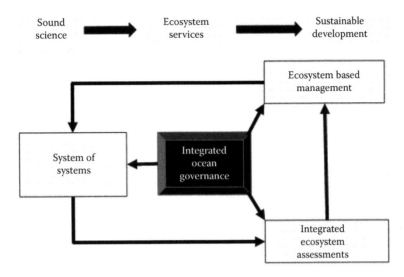

FIGURE 1.6
Protecting and sustaining ecosystem services depends on the sustained development of an SoS that continuously provides data and information required for IEAs used to guide EBM. (Malone, T.C. et al. (2014a). *Marine Policy*, 43: 262–272.)

the entire DPSIR framework concurrently, that is, the development of integrated ocean governance that oversees and enables efficient linkages between (1) establishing an SoS, (2) routine and repeated IEAs based on these observations, and (3) the use of IEAs to help guide the sustainable implementation of EBM (Figure 1.6). The flow of data and information among these activities must enable an iterative process of evaluating performance against objectives that leads to an SoS that informs routine preparation of effective IEAs and frequent updates.

Finally, it is important to recognize that the governance structure should be directly linked to the strategies and processes an organization employs (Imperial and Koontz 2007). Creating a watershed management structure for an SoS is a laborious task, and the risk of organizational failure is at its highest in the early stages. If not well planned, program governance structures can become complicated, confusing, fragmented and, thus, ineffective.

1.5 Sustained and Integrated Coastal Zone Observing System of Systems

Two programs are discussed here to illustrate the importance of developing an SoS and the challenges of doing so effectively. The U.S. Chesapeake Bay Program (CBP) is an example of a mature sustained and integrated observing system relative to the Australian Ecosystem Health Monitoring Program (EHMP), which is in its early stages of development. Both focus on an economically valuable estuarine system (Chesapeake Bay and Moreton Bay, respectively) with due consideration of land-based inputs of pollutants.

1.5.1 Chesapeake Bay Program (USA)*

1.5.1.1 Program Overview

Initiated in 1983, the CBP is among the longest running estuarine restoration efforts in the United States. It is a multistate program with a holistic approach to monitoring the bay, its ten major tributaries, and its watershed. Program development is guided by Chesapeake Bay Agreements, the first of which established an Executive Council (EC) to coordinate the work of restoring and protecting Chesapeake Bay.[†] Program development is guided by commitments (Chesapeake Bay Agreements) made in concert by the governors of Maryland, Virginia and Pennsylvania, the mayor of the District of Columbia, the chair of the Chesapeake Bay Commission[‡], and the administrator of the U.S. Environmental Protection Agency (representing the federal government), who constitute the EC.

The council oversees a program that aims to (1) control land-based inputs of nutrients, sediments, and toxic chemicals to the bay and its tributaries; (2) restore essential habitats and living marine resources; and (3) ensure ecologically sound land-use in the watershed. To achieve these goals, the council established the Chesapeake Bay Monitoring Program (CBMP) and outreach and education programs to ensure public awareness and stakeholder involvement in the restoration effort. The budget for the CBP as a whole is around US$2.5 billion a year (2001–2007, 80% federal funds, 20% state funds).

There have been five Chesapeake Bay Agreements (1983, 1987, 1992, 2000 and 2014). The first committed the parties 'to improve and protect the water quality and living resources of the Bay system,[§] to accommodate population growth and development in an ecologically sound manner, to ensure a continuing process of public participation, and to facilitate regional cooperation in the management of the Bay.' Importantly, the agreement also led to the implementation of the CBMP in 1984.

The CBMP's activities are performed by a consortium of governmental agencies (state and federal), universities, and private companies. The CBMP[¶] encompasses the bay's main stem, tidal and nontidal portions of major tributaries, and the bay's watershed. Surface run-off and associated inputs of nutrients (N, P) and sediments are monitored by a non-tidal water quality network of gauged stations. Water-quality parameters** are monitored throughout the year. Benthic monitoring (abundance and biomass of macrofauna, percentage silt-clay content, carbon and nitrogen concentrations) is used to estimate the proportion of benthic communities that fail to meet the CBP's restoration goals. Stock assessments of anadromous fish (spawn in freshwater), marine-spawning fish, blue crabs, and oysters are conducted annually as are surveys of the spatial extent of seagrass beds. Monitoring data are archived and fed into an extensive modeling program that has expanded the spatial and temporal scale of IEAs which are based on both monitoring data and models. In addition, monitoring data are used to calibrate and validate model-based predictions.

Near real-time *in situ* observations are provided by the Chesapeake Bay Observing System (CBOS), a network of moored instruments that continuously transmit data

---

* www.chesapeakebay.net/; https://www.google.com/#q=Chesapeake+watershed+agreement+2014; http://www.gao.gov/assets/250/248291.pdf

[†] http://msa.maryland.gov/msa/mdmanual/38inters/html/05chesef.html

[‡] http://msa.maryland.gov/msa/mdmanual/38inters/html/04chesb.html

[§] The bay, its tributaries and its watershed; watershed area = 166,600 km², area of the bay and its major tributaries = 11,600 km²

[¶] http://dnr.maryland.gov/Waters/bay/Pages/default.aspx

** Temperature, salinity, nutrients, pH, dissolved oxygen, total suspended solids, chlorophyll a, phytoplankton species and abundance, and zooplankton species and abundance. Anthropogenic toxins in sediments and fish are also monitored.

(vector winds, air temperature, water temperature, salinity, currents, dissolved oxygen, dissolved nitrate, and chlorophyll a).* In 2007, the National Oceanic and Atmospheric Administration (NOAA) initiated the Chesapeake Bay Interpretive Buoy System (CBIBS)[†] as a contribution to CBOS. CBOS is a contribution to a regional observing system (operated by the Middle Atlantic Coastal Ocean Observations Regional Association) of the U.S. Integrated Ocean Observing System and is operated by a consortium of universities and federal agencies.[‡]

The Chesapeake Bay Remote Sensing Program[§] (CBRSP) produced a 25-year time series (1989–2013) of ocean color measurements from light aircraft. When integrated with *in situ* measurements, this valuable, long-term data set has been used to assess the effects of climate (variations in the hydrological cycle and associated variations in nutrient loading) on phytoplankton biomass in the bay – critical information for assessing the efficacy of management actions to control land-based inputs (point and diffuse) of nutrients (Adolf et al. 2006; Miller et al. 2006; Harding et al. 2014).

1.5.1.2 Ecosystem Recovery Targets

Priority areas for the CBP are water quality, biologically structured benthic habitats (seagrass beds, tidal and nontidal wetlands), and LMRs (fish and shellfish). The Bay Health (water-quality) Index is an average of seven indicators (chlorophyll a, dissolved oxygen, water clarity, total nitrogen, total phosphorus, aquatic grasses, and benthic index of biological integrity[¶]) that provides an IEA of the health of the bay and its tributaries.[**] In this context, the 2000 Agreement articulates six goals with quantitative targets for some:

- Restore and protect the finfish, shellfish and other living resources, their habitats and ecological relationships to sustain all fisheries and provide for a balanced ecosystem. At a minimum, achieve a 10-fold increase in native oysters based on a 1994 baseline by 2010; identify and rank nonnative, invasive aquatic, and terrestrial species that are causing or have the potential to cause significant negative impacts on the bay's aquatic ecosystem by 2001; and prepare and implement management plans for those species deemed problematic to the restoration and integrity of the bay's ecosystem by 2003.
- Restore and protect habitats that are vital to the survival and diversity of living resources. Recommit to protecting and restoring 46,000 ha of submerged aquatic vegetation and 10,000 ha of tidal and nontidal wetlands by 2010.
- Continue efforts to achieve and maintain the 40% nutrient reduction goal agreed to in 1987; by 2010, correct the nutrient- and sediment-related problems sufficiently to remove the bay and its tributaries from the list of impaired waters under the Clean Water Act.[††]
- Commit to fulfilling the 1994 goal of a Chesapeake Bay free of toxins by reducing or eliminating inputs from all controllable sources.

---

* http://www.cbos.org/
[†] http://buoybay.noaa.gov/
[‡] http://www.ioos.noaa.gov/
[§] http://www.cbrsp.org/
[¶] http://sci.odu.edu/chesapeakebay/data/benthic/BIBIcalc.pdf
[**] http://baystat.maryland.gov/current-health/
[††] http://www.epw.senate.gov/water.pdf

- By 2012, reduce the rate at which forests and agricultural land is developed by 30% using the 5-year mean for 1992–1997 as the baseline.
- Foster stewardship and community engagement, for example, by 2001, develop and maintain a clearinghouse for information on local watershed restoration efforts, including financial and technical assistance, and, beginning in 2005, provide a meaningful bay or stream outdoor experience for every school student in the watershed.

The 2014 Agreement* modified the 2000 Agreement in two significant ways. On the positive side, it reaffirmed the commitment of the six watershed states and the District of Columbia to implement the Total Maximum Daily Loads (TMDLs) setting enforceable limits on inputs of nitrogen, phosphorus, and sediments to the bay and its tributaries.[†] However, it dropped the commitment to a toxin-free bay.

1.5.1.3 Data Management and Communications

The CBP's data hub offers access to monitoring and modeling data, past and present (water quality and living resources), data management programs, guidance for data management and data management tools.[‡] The hub is the CBP's primary tool for searching and downloading environmental data for the Chesapeake Bay watershed. Databases can be queried based upon user-defined inputs such as geographic region and date range. Data on living resources are parsed into databases for plankton (phytoplankton and zooplankton taxonomic composition, abundance and biomass estimates, phytoplankton primary production rates, and plankton indicators), chlorophyll a fluorescence (surface transects and vertical profiles), benthic animals (taxonomic composition and abundance, biomass estimates, sediment images and analysis, bottom layer water quality), and submerged aquatic vegetation (areal coverage, taxonomic composition, and relative densities). The benthic and plankton databases also provide indicators derived from the monitoring data (e.g., Llansó 2002; Lacouture et al. 2006). Additional bay-related databases from other programs and research projects are also maintained at the CBP Data Center.

1.5.1.4 Governance for Integration (Figure 1.7)[§]

An EC establishes policy for the restoration and protection of the bay and its living resources; exerts leadership to engage public support for the bay effort; signs directives, agreements and amendments that set goals for bay restoration; and is accountable to the public for progress made under the bay agreements. The EC meets annually. Its Principals' Staff Committee meets as needed to facilitate communication among the Implementation Committee, the advisory committees (Citizens Advisory Committee,[¶] Local Government

* https://www.google.com/#q=2014+chesapeake+bay+agreement
† http://www.epa.gov/reg3wapd/pdf/pdf_chesbay/BayTMDLFactSheet8_26_13.pdf
‡ https://www.chesapeakebay.net/what/data; http://datahub.chesapeakebay.net/
§ http://msa.maryland.gov/msa/mdmanual/38inters/html/05chesef.html
¶ The Citizens Advisory Committee is composed of representatives from agriculture, business, conservation, industry, and civic groups. Since 1984, this group has provided a non-governmental perspective on the bay cleanup effort and on how bay program policies affect citizens who live and work in the Chesapeake Bay watershed.

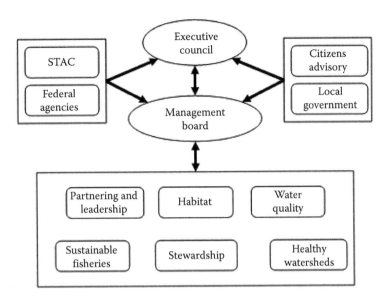

FIGURE 1.7
Governance structure for CBP. As the governing body of the CBP, the EC is responsible for oversight and the Chesapeake Bay Agreements. The management board advises the EC on implementing the CBP and is responsible for preparing the EC's annual work plan and budget, for technical and computer support, and for public outreach. Four advisory committees report to the management board (quarterly) and to the EC (annually). Six subcommittees report to the board and are informed by a variety of work groups (e.g., for monitoring, modeling, data management and communications, storm water, wastewater treatment, agriculture).

Advisory Committee,* and the Scientific and Technical Advisory Committee†) and the EC. Each state has the freedom to decide how it will achieve the goals and meet regulatory requirements.

1.5.2 Ecosystem Health Monitoring Program (Australia)‡

1.5.2.1 Program Overview

Initiated in 1999 by the South East Queensland (SEQ) Healthy Waterways Partnership, the Ecosystem Health Monitoring Program (EHMP) is one of the most comprehensive fresh-water, estuarine and marine monitoring programs in Australia. National and international collaboration is maintained through the involvement of staff in projects and networks outside of SEQ, in particular through the work of the International Water Centre (a joint venture between the Queensland State Government and four Australian universities).§

---

* The Chesapeake Bay Local Government Advisory Committee (LGAC) is a body of officials appointed by the principals to improve the role local governments play in bay restoration efforts and develop strategies to broaden local government participation in the CBP.
† Through the Science and Technical Advisory Committee (STAC), the science community identifies the issues that are most critical to the functioning of a healthy estuary. The STAC establishes peer review systems for all CBP-funded competitive research, reviews and comments on all proposed budget items, holds symposia and carries out technical reviews of key scientific issues.
‡ http://www.healthywaterways.org/ehmphome.aspx, http://www.healthywaterways.org/EcosystemHealth MonitoringProgram/AboutEHMP.aspx
§ https://publications.qld.gov.au/storage/f/2014-05-08T23%3A08%3A29.955Z/integrated-waterways-monitor-ing-framework.pdf; http://www.watercentre.org/about

The EHMP has five themes: integrated water management; water, sanitation and hygiene; healthy river basins; sustainable urban communities; and leadership. The Monitoring Program provides a regional assessment of the ecosystem health for each of South East Queensland's major catchments, river estuaries and Moreton Bay zones. Healthy waterways synthesizes the data and information to produce an annual report card, which provides a clear understanding of the health of their waterways, and highlights any issues that require intervention.

Ecological and biological indicators are monitored and used for IEAs of the health of the region's waterways* based on indicator trends. This information is then used to identify issues in need of management intervention. EHMP (2007) provides detailed information on the methods used in both the estuarine/marine and freshwater programs. Monitoring data are processed by two – catchment and receiving – water-quality models for calibration, validation and prediction: the Environmental Management Support System (EMSS) and the Receiving Water Quality Model (RWQM).[†] The EMSS is used to estimate daily runoff, sediment, and nutrient loads. The RWQM is used to relate sources and sinks in coastal receiving waters (concentrations in the water column and sediments), for example, the eutrophication submodel predicts fields of nutrients, chlorophyll, and dissolved oxygen.

The SEQ Healthy Waterways Strategy 2007–2012 (the Strategy) gives expected outcomes of management actions to restore and sustain the health of SEQ's catchments and waterways and a set of 12 action plans (Figure 1.8) for achieving these outcomes.[‡] The Strategy, a living document, was developed through extensive consultation to align goals and expected outcomes with plans and priorities of the partners. Monitoring and evaluation of progress is to be carried out by the South East Queensland Healthy Waterways Partnership Office.

After over 10 years of operation, the EHMP was expanded in 2011, creating the Monitoring and Evaluation (M&E) Program.[§] The M&E Program includes nested local

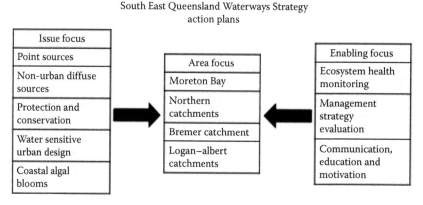

FIGURE 1.8
The 12 action plans of the integrated SEQ Healthy Waterways Strategy. The issue-based and area-based action plans are supported by three enabling action plans.

* https://wetlandinfo.ehp.qld.gov.au/wetlands/resources/tools/assessment-search-tool/12/index.html; http://hlw.org.au/initatives/monitoring
[†] http://ian.umces.edu/blog/2013/08/27/healthy-waterways-healthy-catchments-making-the-connection-in-south-east-queensland-australia/
[‡] http://nrmonline.nrm.gov.au/catalog/mql:325 (strategy)
[§] http://www.coraltriangleinitiative.org/mewg

to regional monitoring that targets regional conditions (continuation of the EHMP), local conditions, human health risks, pressures, runoff events (e.g., stormwater), and the efficacy of management actions.

1.5.2.2 Ecosystem Recovery Targets

The Healthy Waterways Strategy identifies priority outcomes to be achieved by 2026. These include:

- Point source pollution – 100% of nutrient inputs from point sources are prevented from entering receiving waterways
- Nonurban diffuse source – Inputs to Moreton Bay* are 50% of currently predicted loads
- Protection and conservation – Waterways, wetlands and vegetated areas that make significant contributions to waterway health are protected and conserved
- Water-sensitive urban design – All urban development meets consistent regional performance standards for water reclamation†
- Coastal algal blooms – The intensity, frequency and time-space extent of algal blooms are reduced in all estuarine and marine waterways with their impacts minimized (relative to current pattern)

The goal is to ensure the long-term sustainability of the region's waterways and the services they provide through enabling action plans for EHMP, management strategy evaluation, and communication action plans (Figure 1.8).

1.5.2.3 Integrating Programs

The Management Strategy Evaluation Action Plan‡ calls for establishing a DMAC system that provides rapid access to multidisciplinary data from many sources, the heart of which is the Health-e-Waterways information management system.§ Its objectives are to (1) enable information-based decision-making by providing scientists, urban planners and policymakers with fast, web-based access to data and models describing all water-related data; (2) develop frameworks and services that provide streamlined access to real-time, near-real-time and static data sets with collaborative tools that will establish an online community of practice; and (3) collate Queensland's water information making it universally accessible and useful (Figure 1.9).

Health-e-Waterways supports the preparation of annual ecosystem health report cards (IEAs) via an online map interface connected to the underlying databases. Users can quickly access both current and historical report card grades for the waterways of SEQ. In addition, Health-e-Waterways supports a diversity of communication products, including annual technical reports, electronic newsletters, and online information.¶ Once fully operational, Health-e-Waterways will be a distributed system with interoperable data-bases

---

* Moreton Bay is a high-ecological-value area and an international Ramsar wetlands site.
† A holistic approach to urban development that aims to minimize negative impacts on the natural water cycle and protect the health of aquatic ecosystems. It promotes the integration of storm water, water supply and sewage management within a development precinct.
‡ http://nrmonline.nrm.gov.au/catalog/mql:325
§ https://www.uq.edu.au/news/article/2009/10/qld-invention-keeps-healthy-eye-waterways
¶ http://www.healthywaterways.org/EcosystemHealthMonitoringProgram/ProductsandPublications.aspx

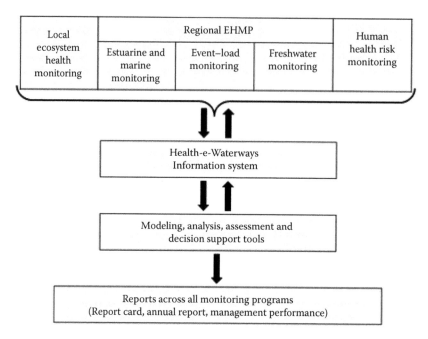

FIGURE 1.9
The Healthy Waterways Monitoring and Evaluation framework for monitoring SEQ's catchments and water-ways and for assessing the effectiveness of management actions in protecting and restoring ecosystem health. The Health-e-Waterways data management and communications system is key to the development of an integrated SoS. Based on the scientific data server and data cubes approach developed for the Berkeley Water Center, this is a geographically distributed network of hydrologic data sources and functions, integrated using web services, functioning as a seamless, integrated whole.

hosted by a variety of government agencies and research organizations. The goal is an information management system that will enable scientists, policymakers, managers, and the general public to access diverse data and information from multiple sources rapidly through a single online portal using search and query engines.

1.5.2.4 Governance for Integration

Healthy Waterways Ltd. (HWL), a nonprofit nongovernmental organization (NGO) established in 2008, works with the Healthy Waterways Network partners* on behalf of the Queensland Government to restore and protect the health of SEQ's catchments and water-ways.[†] A Board of Directors provides strategic guidance and oversees the management and operations of Healthy Waterways Ltd., which acts as the hub in a network of member organizations and individuals, the Healthy Waterways Network. The network provides a forum for representatives from state agencies, local government, water utilities, industry, and SEQ catchment organizations[‡] to work together to ensure an integrated, scientifically sound, regional approach to restoring and sustaining healthy waterways. To these ends, the work of Healthy Waterways is informed by Scientific Expert Panels (SEPs) and is

---

* A consortium of government, public, industrial, and academic research organizations and institutions.
† https://www.ehp.qld.gov.au/water/caring-for-water/seq-healthy-waterways.html
‡ http://www.seqcatchments.com.au/

validated by independent academic peer review. SEPs provide advice on (1) the design and conduct of research projects, (2) emerging scientific issues, and (3) advances in scientific understanding and technology. There are seven technical SEPs:

- EHMP (peer review for the annual report card and proposed modifications to the EHMP program, including indicators and methods used as well as a spatial and temporal sampling strategy)
- Catchment Source (issues of nonurban catchment diffuse pollution management)
- Coastal Ecosystems (monitoring the effects of pollutants on ecosystem resilience, spatial planning for sustainable fisheries, ecosystem services and biodiversity conservation, restoration of biologically structured habitats, and carbon storage in coastal ecosystems)
- Socioeconomics (measurement of social and economic aspects of waterways, e.g., how individual and organizational values affect attitudes, awareness, and behavior toward waterways and the value of ecosystem services)
- Decision Support (decision support tools that are currently being developed by HWL on behalf of its members and advice on methods for application of these tools to meet members' needs)
- Integrated Urban Water (assessments of erosion and sediment control, water-sensitive urban design, water treatment and storm water offsets, all within the context of total water cycle management in urban spaces)
- Human Health (promoting health and well-being benefits of recreational use of waterways as well as advice on managing the associated human health risks)

An Executive SEP populated by the chairs of the seven SEPs coordinates SEP activities, integrates their reports, and provides advice and guidance to the board of directors. Under the guidance of the board, the EHMP is implemented by a team of experts from the Queensland Government, Australian universities, and the Commonwealth Scientific and Industrial Research Organization.

1.6 Conclusions and Lessons Learned

1.6.1 Programmatic Strengths

1.6.1.1 Chesapeake Bay Program

The CBMP has successfully collected high-quality, scientifically credible, multidisciplinary data for 30 years. The EC achieved a consensus on ecosystem recovery targets, and science-based assessments have informed policy decisions. Each member state has set caps on certain pollutants based on the monitoring data. A benthic index of biotic integrity (B-IBI) is now required in the listing and de-listing of impaired water bodies.

The partnership successfully integrated academia into its programs by offering universities long-term contracts, research funding and input in decision-making. Academic participation has contributed to the success of the partnership. Initially, the CBMP used analytical methods designed for wastewater. Collaboration with universities resulted in partner laboratories adopting common standards and lower detection levels needed for the waters of the bay system.

Public support has been strong from the beginning. The monitoring program has contributed to this success by offering readily accessible data and annual IEAs. The *Chesapeake Bay Journal** has been an important reason for sustained public support. The journal is published monthly by Chesapeake Media Service to inform the public about issues and events that affect the bay and its watershed. Subscriptions are free, and, with a print circulation of around 50,000, the journal has a broad impact.

1.6.1.2 Ecosystem Health–Monitoring Program

The SEQ Healthy Waterways Strategy includes clear objectives and indicators dealing with ecology and hydrology with a focus on the social acceptance of outcomes. It is a holistic strategy that addresses key interactions among ecological, social and economic systems. In this context, objectives, targets, and action plans align well with environmental policies.

Standards have been established for collecting and disseminating data. The process used to generate annual ecosystem health report cards ensures that the data are available in useable formats. Monitoring is linked to management objectives and quality appears to be assured through regular reviews of its science, the methods, site selection, and stakeholder feedback.

Management responsibilities and accountabilities are clearly defined. The program is very strong in terms of its approach to adaptive management and the involvement of stakeholders in setting objectives and implementing the Strategy. The diversity of the program's partners is exceptional (government, research, private, and community organizations). There is a strong community focus and stakeholder buy-in is high. In terms of cost-effectiveness and public perception, the risk of failure is relatively low. The partnership has been successful in raising awareness of environmental conditions in Moreton Bay and elsewhere. The Annual Report Card appears to be an effective vehicle for informing stakeholders on trends in ecosystem health, the effects of land-use practices, and the efficacy of management actions. Briefings are conducted for politicians and senior officials before and after the report card is released. The combined effect of the briefings and the report card provides important guidance for managers responsible for protecting and restoring SEQ's waterways.

In a related effort, the Communication, Education and Motivation Program has established a water educators and communications network (SUCCESS) that fosters collaborative and cooperative partnerships among stakeholders to improve the delivery of water and catchment education. Meetings (approximately every 2 months) of the SUCCESS network rotate around the region to share ideas, resources, and best practices, collaborate on projects, and promote activities. The membership of this growing network includes teachers, teacher trainers, staff from environmental education centers, water educators (from government agencies, industry, councils, and water utilities), and community catchment group representatives. This is, in effect, a 'community of practice.'

1.6.2 Problems and Challenges

1.6.2.1 Chesapeake Bay Program

Following the 2000 Chesapeake Bay Agreement, the scientific community and NGOs such as the Chesapeake Bay Foundation raised concerns that the CBP was overstating

* http://www.bayjournal.com/about

progress made in restoring the bay's health, especially the water-quality-restoration goal.* Subsequently, the federal government initiated evaluations by the Government Accountability Office and prepared a strategic plan for restoring the Chesapeake Bay watershed.[†]

The 2005 GAO report concluded that the CBP did not have a comprehensive, coordinated implementation strategy to enable it to achieve the goals of the 2000 Chesapeake Bay Agreement.[‡] What is needed is a realistic, comprehensive, coordinated implementation strategy for providing the data and information required to prepare robust annual IEAs that are scientifically sound and clearly articulate the progress that has been made (or lack thereof) toward achieving the goals set forth in the 2000 Chesapeake Bay Agreement.

Since the bay and its tributaries remained impaired[§], an executive order was issued in 2009 establishing a Federal Leadership Committee[¶] charged with preparing a strategy for protecting and restoring the bay. The *Strategy for Protecting and Restoring the Chesapeake Bay Watershed* includes 4 broad goals, 12 specific quantitative environmental outcomes with deadlines, and 116 actions to restore the bay.** Federal agencies (with EPA in the lead) were given the authority to work with the states to ensure that they take appropriate actions to achieve these goals and outcomes by 2025. Among the most controversial actions to achieve the desired outcomes is the federal requirement to implement the Chesapeake TMDL, which sets limits on point source and diffuse inputs of nutrients and sediments to the bay and its tributaries. As part of the Clean Water Act,[††] the Chesapeake Bay TMDL became law in 2010 when the EPA completed its assessment.[‡‡] The TMDL limits nitrogen, phosphorus, and sediment loads from the watershed to the bay and its tributaries to 84.3 million kg N, 5.7 million kg phosphorus, and 2.9 billion kg sediment per year. Achieving these limits requires a 25% reduction in nitrogen, 24% reduction in phosphorus and 20% reduction in sediment inputs. The pollution limits are further parsed among the five watershed states and major river basin based on state-of-the-art models, monitoring data, peer-reviewed science and collaboration among partners in each jurisdiction.[§§]

Subsequently, the GAO was directed to conduct performance assessments of progress made on bay restoration. The first assessment released in 2011 examined (1) the extent to which the strategy included measurable goals for restoring the bay that are shared by stakeholders and actions to attain these goals; (2) the key factors, if any, federal and state officials identified that might reduce the likelihood of achieving strategic goals and actions; and (3) agency plans for assessing progress made in implementing the strategy and restoring bay health.[¶¶] It was concluded that critical factors that reduce the likelihood of achieving the goals and outcomes of the Strategy are (1) poor collaboration among the watershed states, (2) funding constraints, (3) climate change, and (4) the need for adaptive management.

---

* http://ian.umces.edu/ecocheck/report-cards/chesapeake-bay/2011/indicators/water_quality_index/
† https://federalleadership.chesapeakebay.net/file.axd?file=2009%2f11%2fChesapeake+Bay+Executive+Order+Draft+Strategy.pdf
‡ www.gao.gov/new.items/d0696.pdf
§ Under the federal Clean Water Act, impaired waters are estuaries, rivers, lakes, or streams that do not meet one or more water-quality standards and are considered too polluted for their intended uses.
¶ EPA (chair) with the departments of Agriculture, Commerce, Defense, Homeland Security, Interior, and Transportation
** http://executiveorder.chesapeakebay.net/file.axd?file=2010%2F5%2FChesapeake+EO+Strategy%20.pdf
†† Clean Water Act sec. 303(d), 33 U.S.C. § 1313(d)
‡‡ https://www.govtrack.us/congress/bills/111/s1816
§§ http://www.epa.gov/reg3wapd/pdf/pdf_chesbay/BayTMDLFactSheet8_26_13.pdf
¶¶ http://www.gao.gov/assets/330/323256.pdf

Beginning in 2011 the CBP began implementing an adaptive management strategy intended to enable restoration plans and actions to be modified and updated based on advances in scientific understanding and new data from monitoring and models. In addition, several initiatives have been launched that are intended to address issues of transparency, coordination for better integration, and timely access to data and information. These include Eco-check,* the Bay Barometer,† Eyes on the Bay,‡ and ChesapeakeStat.§

1.6.2.2 Ecosystem Health Monitoring Program

The EHMP is focused on managing nutrient and sediment loads to improve water quality independently of the management of living resources. While conceptual models relating changes in water quality to the management of natural resource are given, quantitative dynamical models are needed to inform integrated EBM of water quality and living resources. What is needed is an integrated approach to EBM for water quality and living resources.

Although the annual Report Card provides a year-to-year comparison of ecosystem health in terms of water quality, it does not normalize for interannual seasonal differences. Consequently, it is difficult to assess the efficacy of management actions and detect trends due to interannual variations in rainfall and, therefore, land-based sources of nutrients and sediment.

The EHMP is focused on localized sampling (135 freshwater sites and 254 estuarine and marine sites) with insufficient consideration of larger-scale pressures such as those associated with climate change, for example, a strategy for adapting to the effects of ocean warming and acidification, sea-level rise, basin-scale oscillations, and a strengthening hydrological cycle on ecosystem states and services. In this context, the EHMP does not appear to integrate space-based Earth observations¶ with *in situ* observations to monitor larger-scale pressures and improve the accuracy of estimates of temperature, chlorophyll a, surface current, and surface wave fields.

Finally, mechanisms do not appear to be in place to trigger policy or regulatory responses to adverse IEAs. Partnering with the Integrated Marine Observing System (IMOS)** would provide critical information on larger-scale, climate-related changes of local and regional importance and increase the value of IMOS observations and modeling.

References

Adams, W.M. (2006). *The Future of Sustainability: Re-thinking Environment and Development in the Twenty-first Century*. Report of the IUCN Renowned Thinkers Meeting (http://cmsdata.iucn.org/downloads/iucn_future_of_sustanability.pdf).

* http://ian.umces.edu/ecocheck/
† https://www.google.com/webhp?sourceid=chrome-instant&rlz=1C1GGGE_enUS468US582&ion=1&espv=2&ie=UTF-8#q=Chesapeake%20Bay%20Barometer
‡ http://mddnr.chesapeakebay.net/eyesonthebay/
§ http://stat.chesapeakebay.net/
¶ http://www.eohandbook.com/
** http://imos.org.au/

Adolf, J.E., C.L. Yeager, W.D. Miller, M.E. Mallonee and L.W. Harding, Jr. (2006). Environmental forcing of phytoplankton floral composition, biomass, and primary productivity in Chesapeake Bay, USA. *Estuarine, Coastal and Shelf Science*, 67: 108–122.

Barber, R.T. and F.P. Chavez. (1986). Ocean variability in relation to living resources during the 1982–83 El Niño. *Nature*, 319: 279–285.

Barbier, E.B., S.D. Hacker, C. Kennedy, E.W. Koch, A.C. Stier and B.R. Silliman. (2011). The value of estuarine and coastal ecosystem services. *Ecological Monographs*, 81: 169–193.

Bell, M.J., M. Lefèbvre, N. Smith and K. Wilmer-Becker. (2009). GODAE: The global ocean data assimilation experiment. *Oceanography*, 22(3): 14–21.

Browman, H.I., K.I. Stergiou, P.M. Cury, R. Hilborn, S. Jennings, H.K. Lotze, P.M. Mace et al. (2004). Perspectives on ecosystem-based approaches to the management of marine resources. *Marine Ecology Progress Series*, 274: 269–303.

Carpenter, S.R., H.A. Mooney, J. Agard, D. Capistrano, R.S. DeFriese, S. Díaz, T. Dietzg et al. (2009). Science for managing ecosystem services: Beyond the Millennium Ecosystem Assessment. *Proceedings of the National Academy of Sciences USA*, 106: 1305–1312.

Carstensen, J. (2014). Need for monitoring and maintaining sustainable marine ecosystem services. *Frontiers in Marine Science*, 1: 1–4 (doi: 10.3389/fmars.2014.00033).

Chapin, F.S., G.P. Kofinas and C. Folke. (2009). *Principles of Ecosystem Stewardship*. Springer: New York, USA.

Christian, R.R., P.M. DiGiacomo, T.C. Malone and L. Taulaue-McManus. (2006). Opportunities and challenges of establishing coastal observing systems. *Estuaries and Coasts*, 29(5): 871–875.

Costanza, R., W.M. Kemp and W.R. Boynton. (1993). Predictability, scale and biodiversity in coastal and estuarine ecosystems: Implications for management. *Ambio*, 22: 88–96.

Costanza, R., R. d'Arge, R. de Groot, C. Farberk, M. Grasso, B. Hannon, K. Limburg et al. (1997). The value of the world's ecosystem services and natural capital. *Nature*, 387: 253–260.

Costanza, R., R. de Groot, P. Sutton, S. van der Ploeg, S.J. Anderson, I. Kubiszewski, S. Farber and R.K. Turner. (2014). Changes in the global value of ecosystem services. *Global Environmental Change*, 26: 152–158.

de Suarez, J.M., B. Cicin-Sain, K. Wowk, R. Payet and O. Hoegh-Guldberg. (2014). Ensuring survival: Oceans, climate and security. *Oceans and Coastal Management*, 90: 27–37.

de Young, B., M. Barange, G. Beaugrand, R. Harris, R.I. Perry, M. Scheffer and F. Werner. (2008). Regime shifts in marine ecosystems: Detection, prediction and management. *Trends in Ecology and Evolution*, 23: 402–409.

de Young, B., M. Heath, F. Werner, F. Chai, B. Megrey and P. Monfray. (2004). Challenges of modeling ocean basin ecosystems. *Science*, 304: 1463–1466.

DOD. (2008). Systems Engineering Guide for Systems of Systems, Version 1.0. Office of the Deputy Under Secretary of Defense for Acquisition and Technology, Systems and Software Engineering, Washington, DC (http://www.acq.osd.mil/se/docs/SE-Guide-for-SoS.pdf).

Duarte, C.M., M.J. Cebrián and N. Marbà. (1992). Uncertainty of detecting sea change. *Nature*, 356: 190.

EHMP. (2007). Ecosystem Health Monitoring Program Annual Technical Report 2005–06. South East Queensland Healthy Waterways Partnership (www.healthywaterways.org/EcosystemHealthMonitoringProgram/ProductsandPublications/AnnualTechnicalReports.aspx).

Foley, M.M., B.S. Halpern, F. Micheli, M.H. Armsby, M.R. Caldwell, C.M. Crain, E. Prahler et al. (2010). Guiding ecological principles for marine spatial planning. *Marine Policy* 34: 955–966.

Garcia, S.M. and K.L. Cochrane. (2005). Ecosystem approach to fisheries: A review of implementation guidelines. *ICES Journal of Marine Science* 62: 311–318.

Harding, L.W., Jr., R.A. Batiuk, T.R. Fisher, C.L. Gallegos, T.C. Malone, W.D. Miller, M.R. Mulholland, H. Paerl, E.S. Perry and P. Tango. (2014). Scientific bases for numerical chlorophyll criteria in Chesapeake Bay. *Estuaries and Coasts*, 37: 134–148.

Hassan, R., R. Scholes, and N. Ash (Eds.). (2005). *Ecosystems and Human Well-being: Current State and Trends*. Island Press: Washington, DC.

Imperial, M.T. and T. Koontz. (2007). Evolution of Collaborative Organizations for Watershed Governance: Structural Properties, Life-cycles, and Factors Contributing to the Longevity of Watershed Partnerships. (http://people.uncw.edu/imperialm/Instructor/Papers/APPAM_07_ Imperial_Koontz_Final_11_5_07.pdf)

Jackson, J.B.C., M.X. Kirby, W.H. Berger, K.A. Bjorndal, L.W. Botsford, B.J. Bourque, R.H. Bradbury et al. (2001). Historical overfishing and the recent collapse of coastal ecosystems. *Science*, 293: 629–643.

Jamshidi, M. (Ed) (2009). *Systems of Systems Engineering: Principles and Applications*. CRC Press: Boca Raton, FL, USA, 480 pp.

Kaplan, I.C. and P.S. Levin. (2009). Ecosystem based management of what? An emerging approach for balancing conflicting objectives in marine resource management. In: Beamish, R.J. and B.J. Rothschild (Eds.). *The Future of Fisheries in North America*. Springer: New York, pp. 77–95.

Kraberg, A.C., N. Wasmund, J. Vanaverbeke, D. Schiedek, K.H. Wiltshire, and N. Mieszkowsk. (2011). Regime shifts in the marine environment: The scientific basis and political context. *Marine Pollution Bulletin*, 62: 7–20.

Lacouture, R.V., J.M. Johnson, C. Buchanan and H.G. Marshall. (2006). Phytoplankton index of biotic integrity for Chesapeake Bay and its tidal tributaries. *Estuaries*, 29(4): 598–616.

Leslie, H.M. and K.L. McLeod. (2007). Confronting the challenges of implementing marine ecosystem-based management. *Frontiers in Ecology and the Environment*, 5: 540–548.

Levin, S.A. (1992). The problem of pattern and scale in ecology. *Ecology*, 73: 1943–1967.

Levin, P.S., M.J. Fogarty, S.A. Murawski and D. Fluharty. (2009). Integrated ecosystem assessments: Developing the scientific basis for ecosystem-based management of the ocean. *PLoS Biology* 7: e1000014 (doi:10.1371/journal.pbio.1000014).

Llansó, R.J. (2002). Methods for Calculating the Chesapeake Bay Benthic Index of Biotic Integrity, Versar (http://www.baybenthos.versar.com).

Lotze, H.K., F. Micheli, S.R. Palumbi, E. Sala, K.A. Selkoe, J.J. Stachowicz and R. Watson. (2006). Impacts of biodiversity loss on ocean ecosystem services. *Science*, 314 (5800): 787–790.

Maier, M. (1998). Architecting principles for systems-of-systems. *Systems Engineering*, 1: 267–284.

Malone, T.C., M. Davidson, P. DiGiacomo, E. Gonçalves, T. Knap, J. Muelbert, J. Parslow, N. Sweijd, T. Yanagai and H. Yap. (2010). Climate change, sustainable development and coastal ocean information needs. *Procedia Environmental Sciences*, 1: 324–341.

Malone, T.C., P.M. DiGiacomo, E. Gonçalves, A.H. Knap, L. Talaue-McManus and S. de Mora. (2014a). A global ocean observing system framework for sustainable development. *Marine Policy*, 43: 262–272.

Malone, T.C., P.M. DiGiacomo, E. Gonçalves, A.H. Knap, L. Talaue-McManus, S. de Mora and J. Muelbert. (2014b). Enhancing the Global Ocean Observing System to meet evidence based needs for the ecosystem-based management of coastal ecosystem services. *Natural Resources Forum*, 38: 168–181.

Martinez, M.L., A. Intralawan, G. Vázquez, O. Pérez-Maqueo, P. Sutton and R. Landgrave, (2007). The coasts of our world: Ecological, economic and social importance. *Ecological Economics*, 63, 254–272.

Millennium Ecosystem Assessment. (2005). *Ecosystems and Human Well-being: Synthesis*. Island Press: Washington, DC, 137 pp.

Miller, W.D., L.W. Harding, Jr. and J.E. Adolf. (2006). Hurricane Isabel generated an unusual fall bloom in Chesapeake Bay. *Geophysical Research Letters*, 33: L06612, doi:10.1029/2005GL025658.

Montoya, J.M. and D. Raffaelli. (2010). Climate change, biotic interactions and ecosystem services. *Philosophical Transactions of the Royal Society B: Biological Sciences*, 365: 2013–2018.

Murawski, S.A. (2007). Ten myths concerning ecosystem approaches to marine resource management. *Marine Policy*, 31: 681–690.

Nicholls, R.J., P.P. Wong, V.R. Burkett, J.O. Codignotto, J.E. Hay, R.G. McLean, S. Ragoonaden and
    C.D. Woodroffe. (2007). Coastal systems and low-lying areas, p. 315–356. In: Parry, M.L., O.F.
    Canziani, J.P. Palutikof, P.J., van der Linden and C.E. Hanson (Eds.). *Climate Change 2007:
    Impacts, Adaptation and Vulnerability. Contribution of Working Group II to the Fourth Assessment
    Report of the Intergovernmental Panel on Climate Change.* 2007. Cambridge University Press:
    Cambridge.
OECD. (2007). Ranking of Port Cities with High Exposure and Vulnerability to Climate Extremes:
    Interim Analysis of Exposure Estimates. Organization for Economic Cooperation and
    Development Working Papers (www.oecd.org/env/workingpapers).
Palumbi, S.R., P.A. Sandifer, J.D. Allan, M.W. Beck, D.G. Fautin, M.J. Fogarty, B.S. Halpern et al. (2009).
    Managing for ocean biodiversity to sustain marine ecosystem services. *Frontiers in Ecology and
    the Environment,* 7: 204–211.
Perry, R.I. and R.E. Ommer. (2003). Scales in marine ecosystems and human interactions. *Fisheries
    Oceanography,* 12: 513–522.
Petersen, J.E., V.S. Kennedy, W.C. Dennison and W.M. Kemp (Eds.). (2009). *Enclosed Experimental
    Ecosystems and Scale: Tools for Understanding and Managing Coastal Ecosystems.* Springer: New
    York, 222 pp.
Petersen, J.K., J.W. Hansen, M.B. Jaursen, P. Clausen, J. Carstensen and D.J. Conley (2008). Regime
    shift in a coastal marine ecosystem. *Ecological Applications,* 18: 497–510.
Powell, T. (1989). Physical and biological scales of variability in lakes, estuaries, and the coastal
    ocean. In: Roughgarden, J., R.M. May and S.A. Levin (Eds.). *Perspectives in Theoretical Ecology.*
    Princeton University Press: Princeton, NJ, pp. 157–180.
Proctor, R., K. Roberts and B.J. Ward. (2010). A data delivery system for IMOS, the Australian
    Integrated Marine Observing System. *Advances in Geosciences,* 28: 11–16.
Schiller, A. and G.B. Brassington (Eds.). (2011). *Operational Oceanography in the 21st Century.* Springer:
    Berlin, 762 pp.
Scott-Cato, M. (2009). *Green Economics: An Introduction to Theory, Policy and Practice.* Earthscan: London
    (www.gci.org.uk/Documents/128075741-Green-Economics-an-Introduction-to-Theory-Policy-
    and-Practice.pdf).
Sherman, K. (2014). Adaptive management institutions at the regional level: The case of large marine
    ecosystems. *Ocean & Coastal Management,* 90: 38–49.
Sherman, K., M.C. Aquarone and S. Adams. (2007). Global applications of the Large Marine
    Ecosystem concept 2007–2010. NOAA Technical Memorandum, NMFS-NE-208: 1–71.
Sidlea, R.C., W.H. Bensonb, J.F. Carrigerb and T. Kamaic. (2013). Broader perspective on ecosys-
    tem sustainability: Consequences for decision making. *Proceedings of the National Academy of
    Sciences, USA,* 110(23): 9201–9208.
Small, C. and J.E. Cohen. (2004). Continental physiography, climate, and the global distribution of
    human population. *Current Anthropology,* 45(2): 269–279.
Smith, N.R. and C.J. Koblinsky. (2001). The ocean observing system for the 21st Century: A consen-
    sus statement. In: Koblinsky, C.J. and N.R. Smith (Eds.). *Observing the Oceans in the 21st Century.*
    GODAE Project Office and Bureau of Meteorology: Melbourne, Australia, p. 1–25.
Steele, J. (2004). Regime shifts in the ocean: Reconciling observations and theory. *Progress in
    Oceanography,* 60: 135–141.
Steele, J.H. (1985). A comparison of terrestrial and marine ecological systems. *Nature,* 313: 355–358.
Tallis, H., H. Mooney, S. Andelman, P. Balvanera, W. Cramer, D. Karp, S. Polasky et al. (2012). A
    global system for monitoring ecosystem service change. *BioScience,* 62: 977–986.
UNEP. (2007). *Global Marine Assessments: A Survey of Global and Regional Assessments and Related
    Activities of the Marine Environment.* UNEP/UNESCO-IOC/UNEP-WCMC: Paris. (http://www.
    unga-regular-process.org/images/Documents/unep%202007%20gma.pdf).
UNESCO. (2012). *Requirements for Global Implementation of the Strategic Plan for Coastal GOOS. GOOS
    Report 193.* Paris: Intergovernmental Oceanographic Commission (www.ioc-goos.org/index.
    php?option=com_oe&task=viewDocumentRecord&docID=7702&lang=en).

Waltner-Toews, D., J.J. Kay, C. Neudeorffer and T. Gitau. (2003). Perspective changes everything: managing ecosystems from the inside out. *Frontiers in Ecology and the Environment*, 1: 23–30.

Wang, B., X. Zou and J. Zhu. (2000). Data assimilation and its applications. *Proceedings National Academy of Sciences USA*, 97(21): 11143–11144.

Weeks, R. (2014). How MPA networks can address scale mismatches. *Marine Ecosystems and Management*, 7 (5): 1–3.

Wood, L.J., L. Fish, J. Laughren and D. Pauly D. (2008). Assessing progress towards global marine protection targets: shortfalls in information and action. *Oryx*, 42: 1–12.

2

Inclusive Impact Index 'Triple I'

Koji Otsuka, Fengjun Duan, Toru Sato, Shigeru Tabeta and M. Murai

CONTENTS

2.1 Concepts of the Inclusive Impact Index...32
    2.1.1 Introduction...32
    2.1.2 Ecological Footprint..33
    2.1.3 Environmental Risk...34
    2.1.4 Inclusive Impact Index..34
        2.1.4.1 Triple I...34
        2.1.4.2 Triple I Light 35
        2.1.4.3 Triple I Star...35
    2.1.5 Conclusions...36
2.2 Simulation-based Evaluation of Ecological Footprint and Biocapacity in the
    Inclusive Impact Index 36
    2.2.1 Introduction...36
    2.2.2 Action Plans and Impact Simulation 36
        2.2.2.1 Water Purification Technology Application Plan....................36
        2.2.2.2 Environmental Impact Simulation 36
    2.2.3 Water Purification Effects and Triple I...38
        2.2.3.1 Water Purification Effects 38
        2.2.3.2 Triple I...38
    2.2.4 Conclusion 40
2.3 Evaluation of Ecological and Human Risks in Inclusive Impact Index.............41
    2.3.1 Introduction...41
    2.3.2 Scenario Definition 42
    2.3.3 Ecological Footprint..42
        2.3.3.1 LCA of Transport and Injection of CO_2 42
        2.3.3.2 Calculation of EF 43
    2.3.4 Human Risk...43
    2.3.5 Cost and Benefit 44
    2.3.6 Ecological Risk 44
        2.3.6.1 Cause-Effect Relationships 44
        2.3.6.2 Occurrence Probability Using Cross-Impact Method 45
        2.3.6.3 Quantifying Endpoint using Environmentally Changed Area........48
        2.3.6.4 Calculation of ER..49
    2.3.7 Triple I..49
    2.3.8 Conclusions...50
2.4 Scaling of Ecological and Economic Values in the Inclusive Impact Index 50
    2.4.1 Introduction...50
    2.4.2 Scaling of Ecological and Economic Values...51

2.4.3   Application to Artificial Upwelling Technology...53
2.5   Spatial Analysis of Inclusive Impact Index...56
     2.5.1   Introduction..56
     2.5.2   Object...56
           2.5.2.1   The EF in a Life of SPAR-Type Wind Turbines 57
           2.5.2.2   Comparison of III_{light} by LCA Among Power Generating Systems ... 59
     2.5.3   Spatial Analysis of III of Floating Offshore Wind Turbine..............................60
References.. 61

2.1 Concepts of the Inclusive Impact Index

2.1.1 Introduction

The world population has increased rapidly since the Industrial Revolution, reaching seven billion in 2012. Several forecasts estimate that this number will rise to about eight billion in 2025 (e.g., FAO STAT), and it is not unreasonable to expect global water, food, and energy shortages in the future. To date, growth in the production of these necessities has been accomplished through tremendous use of freshwater, chemical fertilizers, and fossil fuels. This, in turn, has resulted in the depletion of freshwater resources, expanding areas of infertile land, climate change, and other problems.

It is well known that 70% of Earth's surface area is covered by ocean. Moreover, while the average elevation of dry land is just 840 m above sea level, the average depth of the ocean is 3800 m. The ocean is abundant in water and carbon resources, which are fundamental materials of life, containing 97% of all water and 85% of carbon on Earth (Otsuka, 2007). Therefore, it will likely be necessary for land-based conventional production systems to evolve into ocean utilization systems to support human life as the world population increases. It is important, however, to assess such ocean utilization technologies and systems from the viewpoint of sustainability since the goal is to develop sustainable water, food, and energy resources rather than to just expand conventional land-based production methods that exploit and degrade the environment.

The Inclusive Marine Pressure Assessment and Classification Technology (IMPACT) Research Group was established in 2002 by the Research Committee on Marine Environment of the Japan Society of Naval Architects and Ocean Engineers. During its first four years, the group reviewed several ocean utilization technologies (e.g., deep ocean water utilization, CO_2 ocean sequestration, ocean fertilization) and their environmental impact assessment methods and studied global or long-term assessment and management methodologies (Otsuka and Sato 2005).

In the second stage, the IMPACT participants investigated fundamental knowledge related to global water, food, and energy resource capacities and sustainability indicators for predicting public acceptance of ocean utilization technologies. In September 2006, they met to review and select methodologies to assess both environmental sustainability and economic feasibility and proposed a new assessment metric, the Inclusive Impact Index, or Triple I (Otsuka 2009), which is based on ecological footprint (Wackernagel and Rees, 1996) and environmental risk assessments (Nakanishi, 1995). Then the group conducted case studies for several ocean utilization technologies, such as deep ocean water utilization, CO_2 ocean sequestration, ocean energy, and deep-sea mineral resource development. They also tried to popularize the Triple I and the associated assessment method by way of publications, symposium presentations, and dissemination of the results of the case studies on a website, for example. In this chapter, the ecological footprint and the environmental risk approaches are briefly reviewed, and the concept and structure of the Triple I are introduced.

2.1.2 Ecological Footprint

Ecological footprint (Wackernagel and Rees, 1996) is a parameter that accounts for the impact of human activity on the global environment. It was developed by the research group of Prof. William Rees at the University of British Columbia in the 1990s. Dr. Marthis Wackernagel, one of the group's members, has been promoting this parameter as a representative of the Global Footprint Network. In ecological footprint accounting (Ewing et al., 2010), all human activities, such as utilization of natural resources and generation of wastes, are converted into areas of several types of productive land and ocean, such as cropland, forest, grazing land, marine and inland water, and built-up land (Table 2.1). These areas are summed, considering their production efficiency using an 'equivalence factor' to obtain the global averaged productive area in units of global hectares (gha). In national or regional ecological footprint accounting, a so-called yield factor, which is country-specific and varies by land-use type and year, is also considered. This factor may account for countries' different levels of productivity.

Two calculation methods exist for ecological footprint accounting. One is the compound method, which is applied to calculate an ecological footprint of each country. In this method, the ecological footprint of biological resource consumption is obtained from data on domestic production and import and export statistics. The ecological footprint of energy consumption, which is estimated by summing up domestic energy production and imported energy contained in various goods, is also considered. The component method, on the other hand, is used to determine the ecological footprint of limited (smaller-scale) activities, which can be obtained by an inventory analysis. A life-cycle assessment (LCA) technique is applicable in this method. In both methods, carbon emissions due to energy consumption are converted into the amount of forest area required to absorb the emitted CO_2.

Ecological footprint (EF) can be used to measure the sustainability of human activities by comparison with the regenerative natural capacity, the so-called biocapacity (BC). The biocapacity accounting of each country is usually associated with the compound method. The World Wide Fund for Nature (WWF), the world's largest nongovernmental organization (NGO) for wildlife and environmental conservation, biennially publishes its Living Planet Report (WWF, 2012), which includes information on humanity's ecological footprint. The

TABLE 2.1

Equivalence Factors for Various Types of Productive Land and Ocean

Area Type	Equivalence Factor (gha ha^{-1})
Cropland	2.51
Forest	1.26
Grazing land	0.46
Marine and inland water	0.37
Built-up land	2.51

Source: Ewing, B et al. (2010). Calculation Methodology for National Footprint Accounts, 2010 Edition, Global Footprint Network, Oakland.

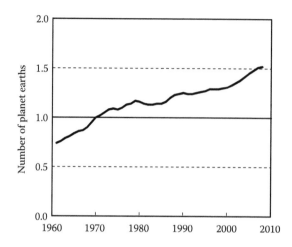

FIGURE 2.1
Ratio of humanity's ecological footprint to global biocapacity. (World Wide Fund For Nature (WWF) (2010). Living Planet Report 2012, WWF International, Switzerland.)

latest data shown in Figure 2.1 illustrate that humanity's ecological footprint has exceeded global biocapacity since the early 1970s; it is now over 50% greater than global biocapacity.

2.1.3 Environmental Risk

The environmental risk assessment method (Nakanishi, 1995) was proposed by Prof. Junko Nakanishi of the University of Tokyo (presently Director, Research Center for Chemical Risk Management, National Institute of Advanced Industrial Science and Technology) in the 1990s. It is considered an effective method for assessing long-term environmental impact issues covering wide areas, such as the effects of global warming and chemical substances on the human body. According to this method, the impact on economic losses and on the health of an individual and the impact on the ecosystem at present and in the future are treated as environmental impacts, present and future costs are treated as resource consumption items, present and future profits are treated as benefits, and each of these is an item considered for economic impact assessment. These can be summarized as cost (C), health risk to humans (HR), and ecological risk (ER) and assessed by the final benefit (B) and risk ratio.

2.1.4 Inclusive Impact Index

2.1.4.1 Triple I

Ecological footprint (EF) is used in many countries since it is a simple and easy-to-understand index. However, EF is estimated based on currently available data and, hence, cannot be employed to predict future scenarios. On the other hand, the environmental risk method can be applied in long-term environmental assessments that consider future risk. Therefore, if its results can be converted into EF, then this would yield a robust index with which to evaluate the feasibility and sustainability of new ocean technologies.

   In Triple I accounting, it is assumed that the value of the ecosystem in the future including risk can be converted into EF by considering biological productivity. This means ecological risk (ER) is integrated into EF. It is also assumed that human risk (HR) is integrated with cost (C), as shown in the environmental risk assessment theory. It should be noted that HR is treated as the health risk to humans in environmental risk assessment theory, but in Triple I accounting, it is treated as human risk, which includes social risks such as economic loss due to disaster, for example. Accordingly, the Triple I (*III*) can be expressed by the following equation:

$$III = [(EF - BC) + \alpha ER] + \gamma[(C - B) + \beta HR] \tag{2.1}$$

where α, β and γ are the conversion factors from ER to EF, from HR to C, and from economic value to environmental value, respectively. In some case studies we have performed, the ratio of EF to the gross domestic product (GDP) of the country or region where an ocean utilization technology is applied, EF_{region}/GDP_{region}, was introduced as the conversion factor γ.

In *III* accounting, the terms EF, C, ER, and HR, which express environmental and economic resource consumptions including risks, are assigned positive values, while the terms BC and B, which show environmental and economic revenues, are assigned negative values. This means that sustainable technologies or systems will have negative values using this indicator.

2.1.4.2 Triple I Light

The equation for Triple I includes ER and HR, which are calculated in terms of probability. The accuracies of these terms are sometimes lower than those of other terms, such as EF, BC, C, and B. Identifying appropriate methods to estimate ER is the focus of a number of ongoing investigations. In consideration of this, a simpler inclusive impact index, so-called Triple I light (III_{light}), is proposed:

$$III_{light} = (EF - BC) + \gamma(C - B) \tag{2.2}$$

2.1.4.3 Triple I Star

Triple I and Triple I light can express both sustainability (positive or negative value) and the level of environmental loading or biocapacity increase. For policymakers and the public, however, the ratio of the environmental load to biocapacity may be a more intuitive metric; hence, the normalized indicators Triple I star (III^*) and Triple I light star (III_{light}^*) are defined as follows:

$$III^* = \frac{EF + \alpha ER + \gamma(C + \beta HR)}{BC + \gamma B} \tag{2.3}$$

$$III_{light}^* = \frac{EF + \gamma C}{BC + \gamma B} \tag{2.4}$$

Using these indicators, sustainable technologies or systems have values of less than one.

2.1.5 Conclusions

Inclusive Impact Index Triple I and its simplified and normalized versions Triple I light and Triple I star are proposed as metrics to assess the environmental sustainability and economic feasibility of ocean utilization technologies. These indicators are based on ecological footprint and environmental risk assessment concepts. Sustainable technologies or systems have negative Triple I (III) and Triple I light (III_{light}) values and Triple I star (III^*) and Triple I light star (III_{light}^*) values of less than one.

2.2 Simulation-based Evaluation of Ecological Footprint and Biocapacity in the Inclusive Impact Index

Application to Water-Quality Improvement in an Estuary

2.2.1 Introduction

Tokyo Bay, an enclosed estuary area located in the southern Kantō region of Japan, has been facing serious environmental problems since the mid-1900s. To prevent further pollution, several laws were enacted starting in the 1970s to reduce the external environmental load to the ocean area. However, cleaner inflow did not significantly improve the environment (Ministry of Environment, 2005). Therefore, recently researchers (Horie, 1987; Kozuki et al., 2001; Kimura et al., 2002) have begun to pay more attention to internal load reduction and self-cleaning capacity improvement using engineering and bioengineering technologies.

All of these technologies were demonstrated to be useful within the local sea area. However, deciding how to introduce them in practice is difficult from the viewpoint of sustainability because their environmental and social functions are conventionally assessed individually. Therefore, a case study was carried out to meet the challenge using a numerical model to predict the environmental impacts caused by assumed technology application plans, thereby estimating the Triple I of each plan to evaluate its sustainability performance.

2.2.2 Action Plans and Impact Simulation

2.2.2.1 Water Purification Technology Application Plan

Based on careful investigations of the natural environment and their historical status, artificial tidal flat creation, eelgrass field restoration, and sewage treatment enhancement were selected as the potential technologies to be used in the study area. At the same time, a geographic information system (GIS) database that includes the oceanography of the area, the geography and geology of the seabed, land use in the coastal area, drainage information on incoming rivers, and average meteorological conditions, was compiled to explore suitable sites for the application of these technologies. The technology application plans (Table 2.2, Figure 2.2) were generated based on the database and the applicable conditions according to previous researches (Kozuki et al., 2001; Kanto Regional Development Bureau, MLIT, 2005; Fukuoka City, 2006).

2.2.2.2 Environmental Impact Simulation

The environmental impacts of the assumed application plans were estimated through numerical projections using an ocean-ecosystem coupled model (Kitazawa, 2001; Kitazawa

TABLE 2.2

Assumed Technology Application Plans

Plan	Description
T1–T6	To create artificial tidal flat of 20 ha at each site
E1–E4	To restore eelgrass field of 1 ha at each site
S1	To achieve the sewage treatment target of government regulations
S2	To achieve 10% more than government target
S3	To achieve 20% more than government target

et al., 2003). Several modifications to the model were performed for the purpose of simulating the mass circulations in the Tokyo Bay (Duan et al., 2011).

The water purification functions of the artificial tidal flat and eelgrass field were integrated into the original computational scheme. The water purification mechanism of the artificial tidal flat was expressed in a dynamic scheme. According to previous researches (Okura, 1993; Artificial Tideland Investigation Committee, 1998; Kimura et al., 2002), 80% of the suspended organics and nutrients in the water mass that flows over the area would be cleaned up. On the other hand, it was established that the function of the eelgrass field was to absorb the nutrients and produce oxygen through their growth. Following Okura (1993), Horie (1987), and Duarte (1990), the eelgrass within one square meter can, at maximum, fix 619 g carbon, 93.7 g nitrogen, and 6.26 g phosphorus and produce 2080 g oxygen annually. The function of the sewage treatment was expressed by adjusting the water quality of the runoff from the rivers.

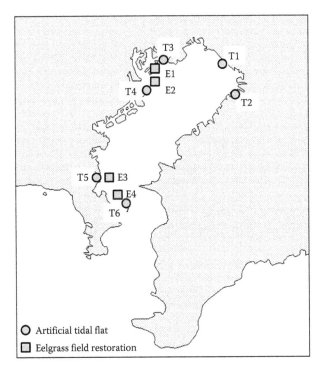

FIGURE 2.2

Assumed technology application plans.

Ten years of numerical experiments were conducted for all of the assumed action plans based on the current environmental status of the Tokyo Bay and under the mean meteorological conditions of the past 50 years. The simulation results of the tenth year were used to assess impacts.

2.2.3 Water Purification Effects and Triple I

Fourteen numerical experiments were carried out, including 6 artificial tidal flat creations, 4 eelgrass field restorations, 2 sewage treatment enhancing cases, and a reference case with no action taken. The environmental impacts in the following discussion are defined as the differences between action plans and reference case.

2.2.3.1 Water Purification Effects

The simulated annual Chemical Oxygen Demand (COD) reductions of the assumed action plans in the tenth year are summarized in Table 2.3. The results indicate that all of the assumed action plans generate a water purification effect. The differences among the action groups lie in the different scales of application; they cannot be compared in a straightforward manner. However, differences also occur among action plans using the same technology.

The differences among the three sewage treatment enhancing plans indicate that further reduction of the external load could reduce the environmental pressure almost linearly. This result suggests an overload situation in Tokyo Bay, in which the internal mass circulation is not capable of dealing with the input. The effects of the four eelgrass field restoration plans show a relatively average water purification capacity despite the fact that a slightly larger one occurred in the northwest area (Figure 2.2). This indicates that the nutrient conditions in all of the assumed sites can accommodate the growth of the eelgrass. The large differences among the six artificial tidal flat creation plans tell us that this water purification function is strongly location dependent. Actions in the inner part of the bay are more efficient, especially in the northwest corner (Figure 2.2). These differences are due to the nutrient distribution in Tokyo Bay and the local ocean current system.

2.2.3.2 Triple I

EF accounting in the case study is based on the total CO_2 emissions of the assumed action plans. The manufacture of material, transportation to the site, construction, and annual maintenance are taken into account for the artificial tidal flat. The processes of seeding sheet manufacture and installation and subsequent monitoring are included for eelgrass field restoration. The CO_2 emission intensities are based on industrial statistics (CEIS, 2005), and the total emissions are calculated according to the assumed application scale. For sewage treatment, the CO_2 emission intensity comes from life-cycle research (Nakajima and Nakajima, 2003), and the necessary treatments are calculated based on the current status of the flow of sewage (Matsumura and Ishimaru, 2004) and assumed action plans.

TABLE 2.3

Water Purification Effect of Assumed Action Plans (1000 t)

T1	T2	T3	T4	T5	T6	E1	E2	E3	E4	S1	S2	S3
0.3	0.9	3.5	0.8	0.5	0.2	3.0	2.0	1.8	1.8	7.4	10.4	13.6

The change in biocapacity *(BC)* caused by the assumed action plans are estimated by comparing the simulated primary productivity with the current primary productivity. For the artificial tidal flat and eelgrass field, additional local primary productivity is also taken into account based on previous studies (Horie, 1987; Duarte, 1990; Okura, 1993; Kuwae et al., 2000). On the other hand, the decrease in productivity due to reclamation for creating the artificial tidal flat is also reflected in the calculations.

Ecological risk (ER) in this case study is defined as the risk of extinction of species in the local area. The probability of this is analyzed using a statistical relationship between primary productivity and species richness (Dodson et al., 2006) and then converting this into changes in the productive ocean area based on the species-area relation (Arrhenius, 1921). The conversion factor α can be calculated as 1 because the estimated ER is expressed by changes in the productive ocean area.

Human risk (HR) is set as the number of human deaths in a probable marine accident. This accounting is conducted only for the artificial tidal flat because the flat provides additional opportunities for recreational use of the sea, which increases the probability of an accident. The estimated number of deaths is converted into an economic value according to insurance data. The conversion factor β can also be determined as 1 because the estimated HR has a unit of economic value.

The cost (C) of an individual action plan is estimated based on previous practices in Japan (Kozuki et al., 2001; Tokyo Bureau of Sewerage, 2009), and benefit (B) accounting is only conducted for the artificial tidal flat by estimating the additional recreational value using a Travel Cost Method (TCM) (Tamaki, 2003).

The conversion factor γ is set to the ratio of EF and GDP of Japan in 2005 because most of the economic data used in this case study are from the first half of the 2000s. According to the WWF (2008), the total EF of Japan was 628 Mgha in 2005. In reference to the real GDP in the same year, the value of γ is $1.0.3 \times 10^{-6}$ gha/Yen. The estimated Triple I and its terms are shown in Table 2.4.

Despite the fact the differences among the absolute values of different technologies are not comparable owing to the different application scales, several common features can

TABLE 2.4

Estimated Triple I and Its Terms for Assumed Action Plans

	EF	BC	ER	HR	C	B	Triple I
Plan		(1000 gha)			(Billion Yen)		(1000 gha)
T1	3.25	39.14	−0.53	0.04	5.00	0.62	−31.86
T2		17.35	−1.13				−10.68
T3		−166.75	−6.55				168.00
T4		24.35	−0.94				−17.48
T5		−8.43	−1.86				14.38
T6		31.10	−0.75				−24.05
E1	0.15	21.45	0.58	−	0.69	−	−20.01
E2		91.15	2.41				−87.88
E3		101.15	2.67				−97.62
E4		102.15	2.69				−98.60
S1	57.90	−52.10	−11.97	−	492	−	604.79
S2	115.00	−58.00	−16.60		980		1165.80
S3	172.00	−67.00	−22.35		1468		1728.69

TABLE 2.5

Assessment Results for Assumed Action Plans (gha/t-COD)

T1	T2	T4	T6	E1	E2	E3	E4
−94.3	−12.5	−22.9	−133.0	−6.7	−44.0	−53.9	−54.7

still be grasped. The sewage treatment enhancing technology is not sustainable when it is assessed by its effects in the ocean area. On the other hand, the artificial tidal flat and eelgrass field are reasonable candidates for water purification in Tokyo Bay, except for a few assumed plans. In general, the environmental term is much larger than the economic one for the artificial tidal flat and eelgrass field, so it dominates the final assessment. In contrast, the economic term of the Triple I for the sewage treatment is much larger than the environmental one, and the difference increases with application scale. Therefore, economic feasibility is essential in the final assessment.

The differences among the three sewage treatment enhancing plans indicate that large-scale application costs more, and the increasing costs suggest there would be difficulties associated with further action. The assessments for the eelgrass field restoration plans show a relatively stable impact and indicate the homogeneous effectiveness of this technology.

However, the assessment result is quite different from site to site for the artificial tidal flat creation. The applications along the northeast coast (Figure 2.2, T1) and near the channel in the south (Figure 2.2, T6) show relatively high performance. The applications along the west coast are also sustainable (Figure 2.2, T2 and T4), but the applications in the northwest (Figure 2.2, T3) and southwest (Figure 2.2, T5) corners are not good plans. In particular, the T3 plan, which is assumed to be applied in the large estuary, shows very poor sustainability performance.

A comparison of the different technologies is conducted after normalizing the Triple I by COD reduction to remove the scale effects. The results of all the sustainable plans are shown in Table 2.5.

Unlike the Triple I itself, the normalized values suggest that the artificial tidal flat created in a proper site can generate an optimized sustainability and water purification effect. This result indicates that a balanced uptake of organics and nutrients is the most profitable. However, the effects of the eelgrass field might be below estimates because only the plant itself was taken into account, and the eelgrass habitat was neglected due to a lack of data.

2.2.4 Conclusion

A case study on the inclusive assessment of water purification technologies was carried out in Tokyo Bay. The self-cleaning technologies of artificial tidal flat creation and eelgrass field restoration were assessed together with the external load reduction technology of the sewage treatment enhancement.

The assessment results using Triple I suggest that efforts in the sea are more effective than those on land, and artificial tidal flat creation can have larger effects than eelgrass field restoration. However, careful environmental analysis is necessary to determine the application site.

At the same time, the case study illustrated Triple I's ability to perform inclusive assessments in the ocean. However, further research is necessary to improve the indicator, for example, by developing approaches to risk evaluation, analyzing the sensitivity of conversion factors, and conducting more applications to examine the indicator's performance.

2.3 Evaluation of Ecological and Human Risks in Inclusive Impact Index

Application to CO_2 Ocean Sequestration

2.3.1 Introduction

Climate change caused by an increase in greenhouse gases (GHGs) in the atmosphere has become an international problem. In particular, CO_2, whose atmospheric concentration rises rapidly as a result of human activities, causes not only global warming but also ocean surface acidification (OSA) and exerts a large influence on ecosystems both on land and in the seas (Orr et al., 2005). CO_2 ocean sequestration (COS) in middle-depth layers (1500–2500 m) at sea sites with water depths of around 5000–6000 m has been proposed as one of its countermeasures (Intergovernmental Panel on Climate Change (IPCC), 2005).

The benefits of COS are an artificial acceleration of the carbon fixation process in the ocean and an ease of OSA, though there is concern over its risks: impacts on ecosystems by CO_2 exposure (e.g., Sato, 2004; Sato et al., 2005; Kamishiro and Sato, 2007). Therefore, an evaluation of the benefits and risks of COS is required, considering the serious nature of global warming and the capacity of renewable energies. To perform such an evaluation, some sort of indices would be useful for judging whether a large-scale technology that uses oceanic or coastal space like COS is effective and acceptable from both economic and environmental points of view.

Recently, the Triple I was developed for this purpose (Otsuka, 2006). Triple I is defined by

$$III = \Delta EF + \Delta ER + \frac{\Sigma EF}{\Sigma GDP}[\Delta HR + \Delta(C - B)], \qquad (2.5)$$

where *III* is the Triple I, EF is the ecological footprint in global hectares (gha), ER is the ecological risk described by EF in gha, $\Sigma EF / \Sigma GDP$ is the ratio of domestic EF to GDP (which actually means the coefficient between ecological and economical activities), HR is the human risk including the risks to human health and social resources in a currency like the Japanese yen (JPY), C is the cost, and B is the financial benefit. Δ indicates the difference between those with and without the implementation of target technology, the latter of which means the so-called business-as-usual (BaU), for instance. Readers interested in the details of EF may consult the literature (e.g., Wackernagel and Rees, 1996; Chambers et al., 2000).

Triple I is a novel concept because ER, which is difficult to quantify in currency, is treated as a footprint (area) and added to EF in global hectares, and those in yen are combined using $\Sigma EF / \Sigma GDP$.

This study aims to propose a methodology to calculate Triple I by picking up COS as an example of a large-scale technology in the ocean. In particular, the quantification of ER is a focus in this study. Because ER is not easy to obtain in many cases, sometimes Triple I without ER and HR, also known as Triple I light, is used for simplicity. In general, ER is defined by the production of the quantified damage of an endpoint and its occurrence probability (Nakanishi, 1995). The endpoint of ER is usually assumed to be the extinction of a particular species. In this study, as an endpoint of ER of both OSA and COS, species extinction of marine organisms in three oceanic spaces, namely, coastal seas, surface pelagic ocean, and the deep ocean, is considered.

2.3.2 Scenario Definition

Once OSA takes place, there must be environmental impacts on the ecosystems in the deep ocean, where the food web starts with the flux from the surface layer. This must also lead to risks in the middle-depth layer.

In this study, Triple I is obtained for the difference between OSA and COS. It was assumed that if we keep using fossil fuels, we have two options: to emit CO_2 into the atmosphere or to sequester CO_2 in the deep ocean. Based on this, we came up with the following two scenarios:

1. Keep emitting CO_2 into the atmosphere as BaU, where its atmospheric concentration becomes 1000 ppm by 2100, following the A1F1 scenario of the IPCC (2001).
2. Inject CO_2 into the deep ocean all over the world and bring atmospheric CO_2 concentrations below 550 ppm and additional CO_2 concentrations (ΔPCO_2) in the deep ocean below 500 ppm by 2100 (Sato, 2004; Sato et al., 2005).

The second scenario may sound unrealistic because the method to reduce atmospheric CO_2 from 1000 ppm to 550 ppm is only supposed to be the COS. However, this assumption is an extreme case.

According to the Central Environment Council, Japan (2005), it is necessary to reduce CO_2 emissions by 71 Gt-C from that in the BaU case for the 100 years starting with 2000 to stabilize its atmospheric concentrations at 550 ppm or less. If this is simply averaged over 100 years, the annual reduction rate is 2.7×10^{10} t-CO_2/year. As described later in this chapter, because the energy penalty, which is the ratio of emitted CO_2 in the Carbon dioxide Capture and Storage (CCS) process to sequestered CO_2, is 18%, the net reduction should be 3.3×10^{10} t-CO_2/year globally.

2.3.3 Ecological Footprint

2.3.3.1 LCA of Transport and Injection of CO_2

It is assumed that ships used to transport and inject liquefied CO_2 are almost the same as a current LNG carrier and the LCA of the transport and injection of CO_2 was conducted using the data on the environmental impact of an LNG carrier (Yokoyama and Kudo, 1995).

Since the deadweight tonnage of an LNG127 is 67,554 t and its mean load factor is 87%, we can obtain the environmental impact per set of transport ship and injection ship by assuming that the average navigation distance between a capture facility and a storage site is 1500 km, with an injection rate of 360 t/h, and the service speed of injection ships is 5.83 knot (3.0 m/s). To inject 58,800 t of CO_2, an injection ship needs to travel 1760 km for 163 h. To supply CO_2 to this injection ship, CO_2 transporters need to travel 3000 km. Because the injection rate in the entire world is 3.3×10^{10} t-CO_2/year, there must be 561,000 sets composed of a transport ship and an injection ship.

Resource consumption and disposed material with their environmental impacts are stated in Tables 2.6 and 2.7, respectively, where the resource consumption per unit of LNG transport (t · km) was obtained from Hiraoka and Kameyama (2005). The ratio of loaded cargo to maximum loading capacities of CO_2 transport ships was set at 50% because the ships were empty on their return trips, and that for injection ships was 50% because the load linearly reduces to zero during injection.

Emissions of CH_4 and N_2O, whose greenhouse coefficients are 21 and 310, respectively, are estimated as 4.3×10^6 t-CO_2/year and 7.3×10^6 t-CO_2/year. On the other hand, according to the Ecological Footprint Network (EFN) (2006), CO_2 emission from the capture process

TABLE 2.6

Resource Consumption of LNG127 (kg)

	Bauxite	Coal	Oil	Natural Gas
Per unit transport (t · km)	1.1×10^{-6}	1.8×10^{-5}	3.0×10^{-3}	2.5×10^{-3}
Per set of CO_2 Transport + injection (kg)	3.2×10^{2}	5.1×10^{3}	8.3×10^{5}	7.1×10^{5}
Per global COS (kg/year)	1.8×10^{8}	2.9×10^{9}	4.7×10^{11}	4.0×10^{11}

TABLE 2.7

Disposed Material of LNG127 (kg)

	CO_2	CH_4	N_2O	NO_2	SO_2	VOC
Per unit transport (t · km)	1.6×10^{-2}	1.3×10^{-5}	1.5×10^{-7}	7.6×10^{-5}	1.7×10^{-4}	3.5×10^{-13}
Per 1 set of CO_2 Transport + injection (kg)	4.5×10^{6}	3.6×10^{2}	4.2×10^{1}	2.1×10^{4}	4.8×10^{4}	9.8×10^{-5}
Per global COS (kg/year)	2.5×10^{12}	2.0×10^{8}	2.4×10^{5}	1.2×10^{10}	2.7×10^{9}	5.5×10^{1}

for 3.3×10^{10} t-CO_2/year is 3.6×10^{9} t-CO_2/year, when the CO_2 emission of unit electricity generation is 0.555 kg-CO_2/kWh and the power reduction ratio is 10%. Therefore, CO_2 emissions due to capture and transport are 6.1×10^{9} t-CO_2/year, which means the energy penalty of COS is 18%.

2.3.3.2 Calculation of EF

In the EF, emitted CO_2 is compensated by forest area that absorbs the CO_2 at an absorption rate of 5.2 t-CO_2/ha/year and an equivalence factor of 1.34 (Ecological Footprint Network (EFN), 2006). Hence, the EF for the capture, transport, and injection of CO_2 in COS is estimated as 1.6×10^{9} gha because the total CO_2-converted GHG emission is 6.1×10^{9} t-CO_2/year. Since the real land area necessary for capture is 4.2 ha for a facility that has a capacity of 245 t-CO_2/h, the area necessary around the world is estimated to be 6.0×10^{4} ha. The equivalence factor of the productive area obstruction, which is the land occupied by buildings, is 2.21 (EFN, 2006), and consequently, the EF of the capture facility becomes 1.3×10^{5} gha.

As a result, annual sequestered CO_2 of 3.3×10^{10} t-CO_2/year is estimated to be -8.5×10^{9} gha and the total EF of the COS becomes -6.9×10^{9} gha. It seems that, in general, the CO_2 footprint or, in other words, the energy footprint, occupies a large portion of the EF, and components other than CO_2 can almost be ignored. This may be an EF-intrinsic problem that should be discussed. However, technologies for directly reducing CO_2 are regarded as negative EF, which means that such technologies supply surplus or extra land area to the Earth by the current EF and the COS is, of course, one of those technologies.

2.3.4 Human Risk

To calculate HR, the results of the life-cycle impact assessment method (LIME) (Itsubo and Inaba, 2005) were adopted. The LIME estimates economical values per unit damage against human health, social resources, biodiversity (BD), primary production, and so forth. Using the damage costs per unit emission shown in Table 2.8 and the emission in Table 2.7, the HR caused by global warning, in which emitted CO_2 is subtracted from sequestered CO_2, becomes -4.5×10^{13} JPY.

TABLE 2.8

Damage Costs (JPY/kg) by LIME

Damage	Material	Human Health	Society Resource	Total
Global warming	CO_2	1.19×10^0	5.48×10^{-1}	1.66×10^0
Acidification	SO_2		6.08×10^1	6.08×10^1
	NO_2		4.34×10^1	4.34×10^1
Resource consumption	Bauxite		4.67×10^{-3}	4.67×10^{-3}
(interest rate 3%)	Coal		1.61×10^{-1}	1.61×10^{-1}
	Oil		3.55×10^0	3.55×10^0
	Natural gas		2.11×10^0	2.11×10^0
Photochemical oxidant	VOC	6.78×10^1	2.72×10^1	9.50×10^1

Source: Itsubo, T. and Inaba, A. (2005). Life Cycle Environmental Assessment Method – LIME. Sangyo Kankyo Kanri Kyokai (in Japanese).

Based on the emissions of SO_2 and NO_2 shown in Table 2.8, their damage costs are 1.6×10^{10} JPY and 5.2×10^{11} JPY, respectively, against the acidification of land soil, and other damage, including the influence of acid rain. The HR of the consumption of resources is 2.6×10^{12} JPY when the interest rate is 3%, based on the consumption of resources in Table 2.6. The HR of photochemical oxidant is 5.2×10^3 JPY because of the emissions of volatile organic compounds (VOCs) in Table 2.8.

The total HR consisting of global warming, soil acidification, resource consumption, and risks to human health and social resources caused by photochemical oxidants turns out to be -4.2×10^{13} JPY/year.

2.3.5 Cost and Benefit

The cost of COS was estimated by the Research Institute of Innovative Technology for the Earth (RITE) (2006) as 7959 JPY/t-CO_2. The benefit of COS is considered the profit earned from the right to sell sequestered CO_2, from which the emitted CO_2 is subtracted. Here the price of CO_2 was set at 2310 JPY/t-CO_2, which corresponds to 14 Euro/t-CO_2 based on the European Union market in 2006 and the average exchange rate in 2006. Therefore, C–B of global COS is 5649 JPY/t-CO_2 and, hence, 1.9×10^{14} JPY/year.

2.3.6 Ecological Risk

Risk is defined as the production of the damage of an endpoint and the probability of its occurrence. The extinction rate of species, or reduction in BD, is usually used as an endpoint in environmental risk assessments, and it was used in this study, too. At present, so far almost only mortality has been investigated as the effect of CO_2 on marine organisms. In this study, environmentally changed area versus species extinction was chosen to represent damage. The extinction probability was obtained as the occurrence probability of the reduction rates in the number of species caused by either OSA or COS using an expert questionnaire and statistical semi-quantification method.

2.3.6.1 Cause-Effect Relationships

Figure 2.3 shows a map of cause-effect relationships, in which the start point is the implementation of COS based on the two scenarios. Here, the endpoint, which is reduction in

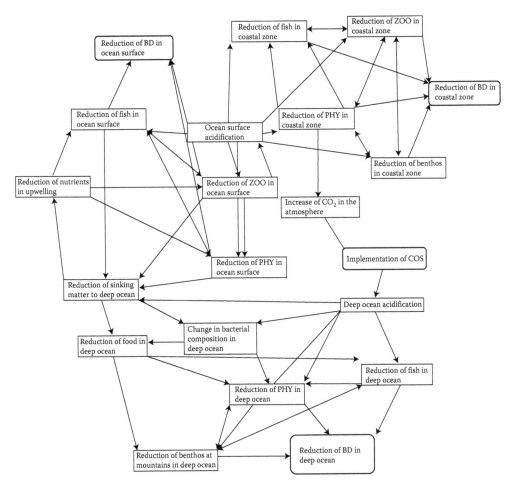

FIGURE 2.3
Cause-effect relationships of ocean surface acidification and CO_2 ocean sequestration.

BD, is categorized based on three spaces in the ocean: coastal zone, ocean surface, and deep ocean. Phytoplankton (PHY), benthos, zooplankton (ZOO), and fish were chosen as the marine organisms. In this study, it is assumed that PHY exist in the coastal zone and offshore surface and that benthos exists only in the coastal zone, and not in the middle-depth ocean far from the seafloor that necessary specialty fields are enough covered.

2.3.6.2 Occurrence Probability Using Cross-Impact Method

Because it is hard for average people to answer a questionnaire that requires specialized knowledge on marine ecosystems and COS, and because COS is seldom known to even experts on marine ecosystems, purposive sampling was adopted in this study, although it has been claimed that its results are not always appropriate for statistical analysis. However, it is believed that it is better to consult experts who are familiar with COS for accurate analysis than a large number of people who may not know the mechanism in marine ecosystems related to COS, despite the possibility of statistical errors. The number of expert respondents was 11, and their specialties are given in Table 2.9; we believe that the necessary fields of specialty were sufficiently covered.

TABLE 2.9

Specialties of respondents of Expert Questionnaire

Academic Field	Number of Experts
Marine biology	6
Marine biochemical cycle	2
Marine ecosystem	2
Marine chemistry	1
Total	11

TABLE 2.10

Choices of Event P_i and Corresponding Tentative Probabilities

Scales	Wording	Tentative Probability Range
1	Definitely yes	0.9–1.0
2	Yes	0.7–0.9
3	Probably yes	0.5–0.7
4	Probably no	0.3–0.5
5	No	0.1–0.3
6	Definitely no	0.0–0.1

TABLE 2.11

Choices of Impact Event P_j to Event P_i and Corresponding Tentative Impact Factors $\alpha_{j \to i}$

Scales	Wording	Tentative Impact Factor
−4	Heavily Suppress	−0.4
−3	Suppress moderately	−0.3
−2	Suppress to some extent	−0.2
−1	Suppress a little	−0.1
0	No effect	0
1	Promote a little	0.1
2	Promote to some extent	0.2
3	Promote moderately	0.3
4	Promote very much	0.4

The respondents were asked to select one of the six alternatives shown in Table 2.10 when indicating the probability that Event P_i would occur. The right column shows the range of occurrence probabilities, which were not shown in the question sheet that was used. Direct answers of the respondents were tentatively quantified using random numbers within the probability range shown in Table 2.11.

The tentative impact factor $\alpha_{j \to i}$ of the occurrence of Event i when Event j occurred is similarly expressed as shown in Table 2.11.

The resultant tentative occurrence probabilities and impact factors from the expert questionnaires were quantified to adjust general probability theory using a semi-quantification method, called the cross-impact method (e.g., Ishitani and Ishikawa, 1992).

The method substitutes quantitative data for expert opinions when occurrence probabilities of events and impact factors between events do not exist. It can be used to correct the tentative probability presumed by experts to meet the mathematical requirements of probability theory.

Using tentative P_i and $\alpha_{j \to i}$ obtained from the expert questionnaires, the occurrence probability $P(i)$ of Event i and the impact probability $P(j \to i)$, at which Event i occurs after Event j occurred, are given by

$$P(i) = P_i \times \prod_m (1 + \alpha_{m \to i}) \quad (0.1 \leq |\alpha_{m \to i}|), \tag{2.6}$$

$$P(j \to i) = P(i) + \alpha_{m \to i} \quad (0 \leq |P(j \to i)| \leq 1), \tag{2.7}$$

where m is an event resulting from the cross-impact analysis of groups in positions that were more subordinate to those on which the study focused.

Next, an impact probability $P(j \to i)$ is converted to a conditional probability $P(j,i)$ by the cross-impact method. Then, the resultant $P(i)$ and $P(j,i)$ are modified to $P^*(i)$ and $P^*(i,j)$, which are consistent with probability theory, by solving the following nonlinear optimization problem with a variable πk, which is the Nth-order bond probability for a certain State k:

$$\left[\sum_i^n (P^*(i) - P(i))^2 + \sum_{i<j}^n (P^*(i,j) - P(i,j))^2 \right] \to \text{minimize}, \tag{2.8}$$

$$P^*(i) = \sum_{k=1}^N \theta_k^i \pi_k, \tag{2.9}$$

$$P^*(i,j) = \sum_{k=1}^N \theta_k^i \theta_k^j \pi_k, \tag{2.10}$$

$$\theta_k^i = \begin{cases} 1 & \text{Event i occurs at State k} \\ 0 & \text{Event i does not occur at State k} \end{cases} \quad (i,k = 1,2,....,21), \tag{2.11}$$

$$\sum_{k=1}^N \pi_k = 1 \quad (\pi_k \geq 0). \tag{2.12}$$

Finally, the resultant probabilities at which BD reduction (species extinction) occurs are listed in Table 2.12. Scenario A has higher probabilities than Scenario B in all three sea areas. The reason why the deep ocean, where CO_2 is directly injected, in Scenario B has a smaller probability than that in Scenario A may be that the experts were worried about the reduction of food in the deep ocean that would result from the reduction in sinking organic matter due to the OSA in Scenario A and because the maximum additional CO_2 concentration in the deep ocean in Scenario B was limited to 500 ppm, which the experts did not consider to be very serious.

TABLE 2.12

Occurrence Probabilities of Endpoint Resulting from
Cross-Impact Method

Scenario	Coastal Zone	Ocean Surface	Deep Ocean
A	0.40	0.35	0.38
B	0.33	0.29	0.31

2.3.6.3 Quantifying Endpoint using Environmentally Changed Area

It is known that the following species-area relationship (SAR) between the number of species, S, and their habitat area, A, from the observation results in a land region (Rosenzweig, 1995),

$$S = cA^Z, \tag{2.13}$$

where Z is empirically set at 0.25, following Rosenzweig (1995). Let the variables prior to the environmental change have a subscript 0, with A given by

$$A = A_0(S/S_0)^{1/Z}. \tag{2.14}$$

Therefore, when the number of species changes from S0 to S, the changed area, ΔA, is given by

$$\Delta A = A_0 - A = A_0\left[1 - (S/S_0)^{1/Z}\right]. \tag{2.15}$$

The SAR can convert the value of the ecosystem, which provides human beings with no services, to a change in an area of land or sea. Here, the sea area was divided into three, namely, the target deep area for COS, pelagic sea surface, and coastal zones all over the world, and the reduction rate of the number of species is given as in Figure 2.4.

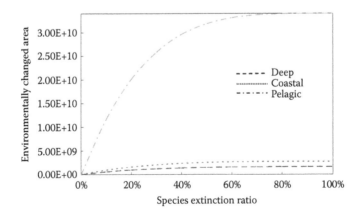

FIGURE 2.4
Relationship of environmentally changed area to species extinction ratio.

2.3.6.4 Calculation of ER

ER was obtained by taking a product of the probability, at which the endpoint event occurs, and the quantified damage, which is an environmentally changed area. The total areas were composed of the coastal zone at 2.7 billion ha, which is 7.6% of the global sea surface, the remaining surface, which was the pelagic surface area, and the area of the deep ocean, which were calculated considering that a 100×300 km site could accommodate 50 million t-CO_2/year (Research Institute of Innovative Technology for the Earth (RITE), 2006) and that the global sequestered CO_2 was 3.3×10^{10} t-CO_2/year, based on the assumption that only the sea area in which CO_2 is injected is affected by COS.

Although the present questionnaire analysis is able to specify the occurrence probability of species extinction, it cannot say how many kinds of species are extinct. In this study, the species extinction ratio, $1-S/S_0$, was set to be 40% as a case study. This means the sea area of 87% was changed according to Equation 2.15, and therefore it is possible to say that this study takes a rather conservative standpoint.

Although the equivalence factor of both coastal zones and continental shelves is assumed to be 0.36 in the EF, that of the ocean surface and deep ocean is 0. Since EF targets an economic activity to the end, the latter is not appropriate, and in this study, it is considered that there is no difference among the three sea areas in terms of species extinction, and the equivalence factor is 1.0 in all sea areas. The resultant ER is shown in Table 2.13. To calculate ΔER, the difference between both scenarios was taken.

The sensitivity of the species extinction ratio was also checked, and in the cases of 10 and 16%, the total ΔER are −823 and −1188 million gha, respectively.

2.3.7 Triple I

The resultant Triple I is shown in Table 2.14 in the case where the species extinction ratio is 40%, where ΣEF/ΣGDP is 2.8×10^{-6} gha/JPY, based on the EF and world GDP (EFN, 2006), and the currency exchange rate is the one on January 29, 2007. The contents of the Triple I are also shown in Figure 2.5. It is seen that the largest portion of the Triple I is occupied by EF and the total Triple I is negative, which means that the COS is determined to be effective.

As a result, the Triple I is negative and supports the scenario of keeping atmospheric CO_2 concentrations at a maximum of 550 ppm, with COS implemented globally, rather than maintaining a concentration of 1000 ppm, which would be accompanied by OSA. It is interesting to note that in this process the experts think that the risk of species extinction in the deep ocean caused by OSA is greater than the risk of deep ocean acidification caused by COS.

TABLE 2.13

Calculation of ΔER (million-ha)

Seas	Total Area	Changed Area	Scenario	Occurrence Probability	ER	ΔER
Coastal zone	2900	2520	A	0.40	1010	−180
			B	0.33	830	
Pelagic surface	36,200	31,510	A	0.35	11,030	−1890
			B	0.29	9140	
Deep ocean	1600	1390	A	0.38	530	−100
			B	0.31	430	
Total						−2170

TABLE 2.14

Triple I of CO_2 Ocean Sequestration

	JPY/y	gha
ΔEF		-6.9×10^9
ΔER		-2.2×10^9
ΔHR	-4.2×10^{13}	-1.2×10^8
Δ(C-B)	1.9×10^{14}	5.2×10^8
Triple I		-8.7×10^9

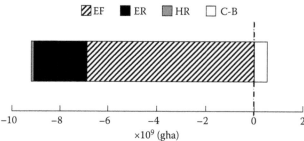

FIGURE 2.5
Contents of Triple I for CO_2 ocean sequestration.

2.3.8 Conclusions

In this study, a method for calculating the Triple I was proposed by choosing COS as an example of a large-scale technology that uses oceanic space, compared with OSA as BaU. In the process, ER was obtained by analyzing the occurrence probability of the endpoint of COS, that is, species extinction, based on an expert questionnaire and by quantifying the damage of the endpoint using SAR. It should be noted that the expert questionnaire was purposive sampling and may have statistical errors. Moreover, the ER was calculated using an assumed species extinction ratio of 40% as a case study, and therefore the resultant value of the Triple I varied depending on the ratio.

In the future, it will be necessary to consider more realistic scenarios and to prove the SAR in sea area to improve the accuracy of the Triple I and ER of COS.

2.4 Scaling of Ecological and Economic Values in the Inclusive Impact Index

Application to Artificial Upwelling

2.4.1 Introduction

The value of the Triple I can be significantly affected by the parameter γ, which converts economic value (monetary cost) to environmental value (EF). Thus, the correlation between γ and EF is reanalyzed to propose a method of calculating γ (Yoshimoto and

Tabeta, 2011). In the analysis, the industries contributing to GDP are grouped into three categories based on the correlation between EF and the scale of economic activities. In the proposed method, γ is calculated using the marginal abatement costs of CO_2 emissions and the scale of economic activity of the industrial category.

2.4.2 Scaling of Ecological and Economic Values

1. *Standard method*: In the standard method, γ is defined by EF and GDP in the target countries or regions as follows (Otsuka and Ouchi, 2008):

$$\gamma = \gamma_0 \equiv \frac{EF_{region}}{GDP_{region}} \tag{2.16}$$

When the target region is Japan, national EF (WWF, 2008) in Japan is used for EF_{region} and GDP in Japan is used for GDP_{region}. Then γ_0 is calculated as follows:

$$\gamma_0 = \frac{EF_{Japan}}{GDP_{Japan}} = \frac{6.3[gha]}{5.0 \times 10^{12}[\$ \cdot year^{-1}]} = 1.3 \times 10^{-4} \left[gha \cdot year \cdot \$^{-1} \right] \tag{2.17}$$

2. *Scaling by value of CO_2 emission*: Since 75% of total EF in Japan is accounted for by EF due to CO_2 emissions, γ should strongly correlate with the economic value of CO_2 emissions. The marginal abatement cost (MAC) in Japan is estimated to become about US\$25/t-CO_2 when the target of GHG emissions in 2020 is set to be at the same levels as in 2005 (Research Institute of Innovative Technology for the Earth (RITE), 2009). Let γ_1 be defined as the value of γ determined by MAC; it is calculated as follows:

$$\gamma_1 = \frac{1}{(MAC) \cdot (CO_2 \text{ absorption rate by forest})}$$
$$= \frac{1}{25[\$ \cdot (t\text{-}CO_2)^{-1}] \cdot 5.2[t\text{-}CO_2 \cdot ha^{-1} \cdot year^{-1}]} = 1.0 \times 10^{-2} \left[gha \cdot year \cdot \$^{-1} \right]. \tag{2.18}$$

3. *Proposed Scaling method*: The γ has the function of converting economic value to environmental value (EF) in the calculation of Triple I. The essential meaning of γ is 'additional EF gained (reduced) by an additional unit of economic activity.' Using the notation that γ is obtained from this definition by γ_t, it can be calculated using the following equation:

$$\gamma_t = \frac{d(EF)}{d(GDP)} = \frac{1}{(d(GDP)/d(EF))}. \tag{2.19}$$

In general, the reduction of GDP due to the unit reduction of EF gets larger as EF decreases. Hence, GDP as a function of EF is a convex function. Additionally, because GDP contains the economic value produced by industrial sectors with little EF (e.g., hospitality

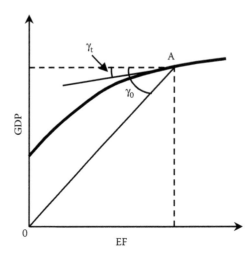

FIGURE 2.6
Assumed relationship between EF and GDP.

industry), GDP is positive when EF $= 0$. Therefore, GDP as a function of EF can be given by the curve in Figure 2.6. Now, let Point A denote the GDP and EF for the target region. The slope of the tangent at Point A is equivalent to $1/\gamma_t$. On the other hand, the tangent of the line OA is equivalent to $1/\gamma_0$. As shown in Figure 2.6, γ_t is usually larger than γ_0.

To calculate γ_t, the industrial sectors are divided into three categories, as shown in Table 2.15. The contribution of each industrial category to GDP is shown in Figure 2.7(a). The areas for Industries I and II in Figure 2.7(b) corresponds to GDP produced by the economic activity of Industries I and II, respectively.

The magnitude of $1/\gamma_1$ can be approximately expressed by the scale of Area I, as shown in Figure 2.7(b), because change in GDP caused by CO_2 emission is almost derived from Industry I. The magnitude of $1/\gamma_t$ can also be expressed by the scale of Areas I and II. The area for Industry C is added in the figure to show the magnitude of $1/\gamma_0$. Therefore, $1/\gamma_t$ can be calculated by summing $1/\gamma_1$ and Area II (scale of Industry II) over EF_{Japan}. The statistics from the Cabinet Office provides a breakdown of GDP for industrial sectors, which shows that GDP by Industry II accounts for 1.4% of the total GDP in Japan. Therefore, γ_t can be calculated as follows:

$$1/\gamma_t = 1/\gamma_1 + GDP_{Japan} \times 0.014/EF_{Japan} = 2.1 \times 10^2 [\$ \, gha^{-1} \, year^{-1}],$$
$$\gamma_t = 4.8 \times 10^{-3} [gha \, year \, \$^{-1}]. \tag{2.20}$$

TABLE 2.15

Categorization of Industrial Sectors

Category	Definition	Typical Sectors
I	High correlation with both EF and CO_2 emissions	Manufacturing, construction, mining, transport
II	High correlation with EF and low correlation with CO_2 emissions	Agriculture, forestry, fisheries
III	Less correlation with EF	Finance and insurance, information, service

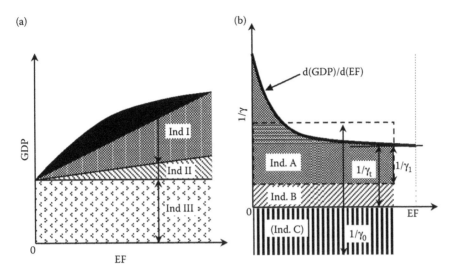

FIGURE 2.7
Contribution of industrial categories to GDP or γ. (a) Breakdown of GDP for each industrial category. (b) Relationships between γ and GDP for each industrial category.

2.4.3 Application to Artificial Upwelling Technology

For an application example, the Triple I of an artificial upwelling using a seabed mound is calculated. Artificial upwellings are a technology that is used to increase marine productivity by uplifting bottom seawater with rich nutrients to the euphotic zone using artificial structures. The target project is an artificial seabed mound located off Nagasaki Prefecture, which is made of 4800 blocks of fly ash concrete. In the calculation of the project's EF, the environmental impact due to CO_2 emissions was estimated in two ways. For the estimation of the direct influence, EF was estimated using the component method. EF including indirect influences is estimated using environmental input-output (I/O) analysis.

1. *EF and BC due to direct influences*: CO_2 emission due to the construction of the seabed mound is estimated as shown in Table 2.16. It is assumed that there are no CO_2 emissions after the construction since the seabed mound does not require maintenance. Let the CO_2 absorption rate of a forest be 4.2[t-CO_2 ha^{-1} year^{-1}], the equivalence factor be 1.3 [gha ha^{-1}], and the life of the seabed mound be 30 years; then the EF of the artificial seabed mound is

$$EF_{direct} = \frac{4.1\times10^3[\text{t-CO}_2]}{30[\text{year}]\cdot 5.2[\text{t-CO}_2\,\text{ha}^{-1}\,\text{year}^{-1}]\cdot 1.3[\text{gha ha}^{-1}]} = 3.5\times10[\text{gha}]. \qquad (2.21)$$

It is reported that the fish catch increased by 1.3×10^3 t/year owing to the artificial seabed mound, almost all of which is anchovy. Assume that the ratio of organic carbon to the wet weight of anchovy is 0.045, and the trophic level is 2.5; the increase in primary production corresponding to the fish catch is

$$1.3\times10^3[\text{t/year}]\times0.045[\text{t-C/t}]\times10^{(2.5-1)} = 1.8\times10^3[\text{t-C/year}]. \qquad (2.22)$$

TABLE 2.16

CO$_2$ Emissions for Construction of Seabed Mound

	Material	Weight [t]	CO$_2$ Emissions [t]		
			Hauling	Production	Total
Materials	Fly Ash	19,537	9	30	39
	Cement	3256	27	2722	2749
	Steel Reinforcement	137	1	64	65
	Frames	77	1	117	118
	Others		0	724	724
	Subtotal				3657
Work type	Install				23
	Hauling, Construction				371
	Subtotal				394
	Total				4089

Source: Research Institute of Innovative Technology for the Earth (RITE) (2005). Carbon dioxide fixation, effective utilization technologies 2004.

Since the average primary production rate for the productive ocean is 100 [g-C/m^2/year] and the equivalence factor is 0.40[gha/ha], the increase in biocapacity is

$$\text{BC}_{\text{direct}} = \frac{1.8 \times 10^9 \, [\text{g-C year}^{-1}]}{100[\text{g-C m}^{-2}\,\text{year}^{-1}] \cdot 0.4[\text{gha m}^{-2}]} = 7.1 \times 10^2 [\text{gha}]. \tag{2.23}$$

2. *EF and BC due to indirect influences*: EF including indirect influences is calculated based on CO$_2$ emissions estimated using environmental I/O analysis. Since an I/O table describes the entire economy (flows of goods and services) in the target nation or region, it can be applied to environmental analysis including indirect ripple impacts. In environmental I/O analysis, the inputs of various natural resources or outputs of various emissions and wastes by sector are additionally described. In the present case, CO$_2$ emissions are used as the environmental impact. When evaluating environmental impacts due to the introduction of a new technology, two categories of environmental impacts are considered, CO$_2$ emissions from building new plants, for example (C_p), and environmental impact by operation (C_u); these are calculated as follows:

$$C_p = i'_n \hat{E}(I-A)^{-1} \hat{D} f_p, \tag{2.24}$$

$$C_u = i'_n \left\{ \hat{E}(I-A)^{-1} + \hat{E}_f \right\} f_u, \tag{2.25}$$

where $\hat{D}$ is a diagonal matrix whose elements are the annual depreciation rate for each good, f_p is the plant vector denoting construction materials, machinery and equipment, and other materials necessary for plant construction, i'_n is the emission coefficient vector, $\hat{E}$ is the CO$_2$ emission matrix, $(I-A)^{-1}$ is the Leontief inverse matrix, $\hat{E}_f$ is the diagonal matrix of the CO$_2$ emission factor due to the

TABLE 2.17

CO$_2$ Emissions for Construction Estimated by I/O Analysis

Material	Amount of Use	Price Per Unit	CO$_2$ Emission [t year^{-1}]
Cement	3.3×10^3 ton	6.0×10^{-2} \$/kg	2.9×10
Steal	2.1×10^2 ton	8.0×10^3 \$/ton	4.8×10
Electricity	2.4×10^4 kWh	9.0×10^{-2} \$/kWh	2.0×10^{-1}
Water transport	4 years	8.5×10^4 \$/month	3.7×10^2
Total			4.5×10^2

consumption of fossil fuel as final demand, and f_u is the final demand vector denoting the materials and energy necessary for the operation of the plants. Here, I/O tables and basic unit data in year 2000 for matrix A and $\hat{E}$ are applied (National Institute for Environmental Studies (NIES), 2002). The materials, prices used for the construction of the seabed mound, and the calculated CO$_2$ emissions are shown in Table 2.17. Then EF due to the construction of the seabed mound including indirect effects is calculated as follows:

$$\text{EF1}_{\text{indirect}} = \frac{4.5 \times 10^2 \, [\text{t-CO}_2 \, \text{year}^{-1}]}{5.2 [\text{t-CO}_2 \, \text{ha}^{-1} \, \text{year}^{-1}] \cdot 1.3 [\text{gha ha}^{-1}]} = 1.1 \times 10^2 [\text{gha}]. \qquad (2.26)$$

Furthermore, the reduction of environmental impacts is estimated assuming that the increased production of fish due to the artificial upwelling will replace meat as protein. Using environmental I/O analysis, the reduction of CO$_2$ due to the reduction of meat demand is estimated to be 1.4×10^4 t/year (Yoshimoto and Tabeta, 2011), whose EF is calculated as follows:

$$\text{EF2}_{\text{indirect}} = \frac{1.4 \times 10^4 \, [\text{t-CO}_2 \, \text{year}^{-1}]}{5.2 [\text{t-CO}_2 \, \text{ha}^{-1} \, \text{year}^{-1}] \cdot 1.3 [\text{gha ha}^{-1}]} = 3.5 \times 10^3 [\text{gha}]. \qquad (2.27)$$

3. *Triple I Light of Artificial Upwelling Technology*: Table 2.18 shows the estimated EF for the artificial upwelling technology using a seabed mound. It is revealed that EF including indirect influences could be several times larger than direct effects. It is also found that an increase in BC (direct influence) and reduction of EF (indirect influence) due to the enhancement of primary production will be much greater than an EF increase due to the construction of a seabed mound.

   Triple I light is calculated under the assumptions that the construction cost of the artificial seabed is 1.2×10^7 USD and its economic life will be 30 years. Table 2.19 shows Triple I light in the case of $\gamma = \gamma_0$ (Case 1) and $\gamma = \gamma_t$ (Case 2). In Case 1,

TABLE 2.18

Ecological Footprint of Artificial Upwelling Technology Using Seabed Mound

	Direct Effects only	Including Indirect Effects
EF	3.5×10 gha	(CO$_2$ emission) 1.1×10^2 gha (reduction of meat demand) -3.5×10^3 gha
BC		7.1×10^2 gha
EF-BC	-6.8×10^2 gha	-4.1×10^3 gha

TABLE 2.19

Estimation of III Light for Artificial Upwelling Technology
Using Seabed Mound

	Case 1	Case 2
γ [gha year^{-1} \$$^{-1}$]	1.3×10^{-4}	4.8×10^{-3}
EF-BC [gha]	$-0.68 \sim -4.1 \times 10^{-3}$	
C-B[\$ year^{-1}]	4.0×10^{5}	
III_{light} [gha]	$-0.63 \sim -4.0 \times 10^{-3}$	$-2.2 \sim 1.2 \times 10^{-3}$

Triple I light becomes negative (which means sustainable), regardless of whether indirect or direct effects are considered. In this case, however, the economic value of the project is undervalued compared to the environmental value, as discussed in the previous section. In Case 2, Triple I light becomes positive (unsustainable) when considering only direct influences, and Triple I light becomes negative (sustainable) when including indirect influences.

2.5 Spatial Analysis of Inclusive Impact Index

Application to Offshore Wind Power Stations

2.5.1 Introduction

Currently, the continued depletion of fossil fuels and global warming are driving energy innovation to a new level around the world. Wind energy is one of the most popular green energy sources in this new stage. A large-scale wind farm that was under construction in this decade is extending to ocean spaces to capture stable and strong wind despite the necessity of advanced technology and high costs. Of course, electricity generation from green energy will decrease CO_2 emissions directly. However, if the costs incurred or energy consumed by the time the wind power stations are installed is too great, we will not be able to say that it is sustainable or acceptable technology. The Triple I (*III*) is a very suitable tool for discussing the balance of the cost and effects of the reduction in CO_2 emissions. So we should try to apply Triple I light (*III*$_{light}$) to the offshore wind turbine problem for discussion.

2.5.2 Object

There are a lot of offshore wind turbine types: bottom-mounted type, floating TLP type, floating SPAR type, floating semi-submersible type, and so on. In addition, the turbine system is roughly categorized into a horizontal axis type and cross-flow type. The size and capacity of the turbine of every category is increasing in recent years. Currently (2017) the largest floating offshore wind turbine is belong to the floating semi-submersible and the horizontal axis type, 7 MW capacity, 160 m in diameter and installed in Japan.

In this section, we treat seven sizes of a typical SPAR-type offshore wind turbine, as shown in Figure 2.8. The assumed sizes are from 250 kW to 4320 kW. The mass distributions of all wind turbines are shown in Table 2.20.

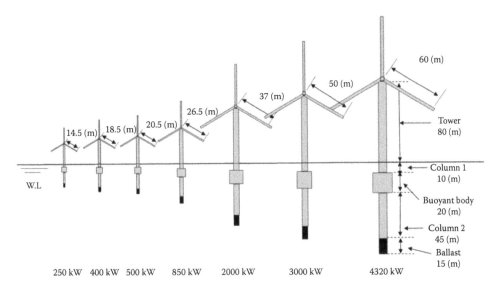

FIGURE 2.8
Sketch of floating wind power turbines of SPAR type.

2.5.2.1 The EF in a Life of SPAR-Type Wind Turbines

Ecological footprint (EF) is one of the important factors for obtaining Triple I light (III_{light}). The EF of a SPAR-type wind turbine mainly depends on the CO_2 emissions during its lifetime. Thus, we first obtain the level of CO_2 emissions during its lifetime. We simply assume that the life is divided into five stages: manufacture, transportation, instruction, operation/maintenance, and disposal stage. However, CO_2 emissions in the disposal stage cannot be determined unless the manner of re-use of the wind turbine is determined, so we consider the CO_2 emissions in just the first four stages and assume an operating period of 20 years. In the manufacturing stage, the CO_2 emissions can be calculated using Equation 1:

$$CO_{2-emission}\left[kg-CO_2\right] = Material[kg] \cdot Factor_{CO_2-emission}\left[\frac{kg-CO_2}{kg}\right]. \qquad (2.28)$$

TABLE 2.20

Mass Distribution of Components of All Wind Power Turbines

Class	250 kW	400 kW	500 kW	850 kW	1650 kW	3000 kW	4320 kW
Blade	12.6	16.0	17.8	25.5	32.1	43.4	52.0
Nacelle	20.7	26.5	29.3	42.0	52.9	71.5	85.8
Generator	46.1	58.8	65.1	93.3	117.6	158.9	190.6
Hub	137.0	174.8	193.7	250.4	349.6	472.4	566.9
Column 1	19.8	25.3	28.0	36.2	50.5	68.3	82.0
Buoyant body	337.4	430.4	477.0	616.6	860.9	1163.4	1396.1
Column 2	89.1	113.7	126.0	162.9	227.4	307.4	368.8
Ballast	1743.0	2223.8	2464.3	3185.5	4447.7	6010.4	7212.5
Total	2405.7	3069.4	3401.2	4412.4	6138.7	8295.6	9954.7

Unit: ton

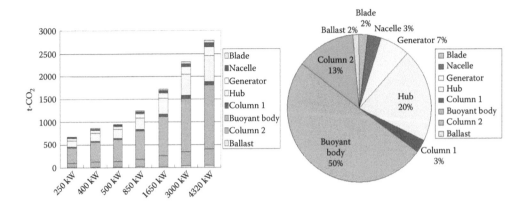

FIGURE 2.9
CO_2 emissions in manufacturing by generating capacity and ratio in manufacturing stage (4320 kW).

The breakdown of the amount of CO_2 emissions in a given size of energy generation systems at the manufacturing stage is shown in Figure 2.9 (left). The right panel shows the typical CO_2 emission ratio (4320 kW case) at the manufacturing stage. As shown, we understand most levels of CO_2 emissions were accounted for by the tower and floating structure parts.

The transportation stage means towing the constructed wind turbine from where the structure was made to the operating site. The CO_2 emissions in the transportation stage can be calculated using Equation 2.29:

$$CO_{2-emission}[kg-CO_2] = Material[kg] \cdot Distance[km] \cdot Factor_{CO_{2-emission}} \left[\frac{kg-CO_2}{kg} \right]. \quad (2.29)$$

The factor is changed by the towing vessel.

When the turbine system is constructed, a large amount of electrical power that is generated by burning coal, gas, and petroleum is used. The amount of CO_2 emissions in the construction stage of a 300 kW turbine was estimated by Hondo et al. (2004), as shown in Table 2.21. In this section, it is assumed that the emission amount is proportional to the diameter of a blade.

In the operation stage, which is assumed to last 20 years, the internal consumption rate is assumed to be 10%, as per Uchiyama (1995). The amount of CO_2 emissions for operations and repair is assumed to be 2% of the amount of the material and the fuel in manufacturing and transportation, based on Uchiyama (1995). Once the amount of CO_2 emissions is obtained, we can obtain the EF using the equivalence factor of forest and the area occupied by the structure.

TABLE 2.21

Fuels Required for Construction of Wind Power Plant

	Required	Conversion Factor	CO_2 Emission [t-CO_2]
Coal and Gas	2.46 t	2.327(t-CO_2/t)	5.72
Petroleum	3.00 t	2.698(t-CO_2/t)	8.11
Electrical power	1.414 MWh	0.355(t-CO_2/MWh)	0.50

Source: Hondo, H, Uchiyama, Y, and Moriizumi, Y (2004). *Evaluation of power genera-tion technologies based on life cycle CO_2 Emissions,* Socio-economic Research Center, Reo, No, Y99009.

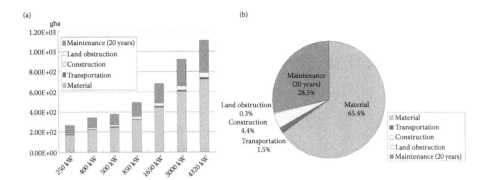

FIGURE 2.10
EF by wind power (a) and rate for 4320 kW (b).

The EF depends on the amount of CO_2 emissions during the lifetime of each wind turbine and on the occupied area and the scale of the EF in the stages for a 4320 kW turbine; typical rates are shown in Figure 2.10.

Then, as shown in Figure 2.10, EF does not have linear relation to the capacity of a wind turbine, and the relative EF per kW decreases with increases in generation capacity. It is understood that the largest part of the EF is in the manufacturing (from material production) and operating (maintenance) stages.

2.5.2.2 Comparison of III_{light} by LCA Among Power Generating Systems

As for the amount of CO_2, a comparison between wind power and other power generating systems is shown in Figure 2.11.

Here we assume that the operating period is 20 years, an offshore location is 10 km from a coast, the sea depth at the location is 200 m, the average wind velocity at the location is 7 m/s, and electricity production is calculated using a Rayleigh distribution. The 300 kW wind turbine in Figure 2.11 is assumed to be on land. The CO_2 emissions of renewable energy such as wind power and solar photovoltaic power is much less than that of fossil fuel power such as thermal power. As for offshore wind power generators, the amount of CO_2 emissions decreases with increasing capacity because of the improvement in generating efficiency. Figure 2.12 shows the EF per GWh and the cost per MWh in various kinds of electricity generating system and various sizes of the offshore wind power generator.

Readers interested in the details of the assumptions may consult Murai (2009). As shown in Figure 2.12, the power generating cost to use fossil fuels is lower than the lowest cost of offshore wind power generation. Among renewable energy sources, the generating cost of mega-size offshore wind power is not relatively high.

The III_{light} of each system is shown in Figure 2.13. To estimate the benefit, it is assumed that the exchange rate from US$ to yen is 100 yen = US$1, the sales price of the generated electricity is US$110/MWh, and the conversion factor of GDP_{Japan} and EF_{Japan} is 1.25×10^{-4} [gha/US$].

Figure 2.13 shows that the III_{light} of every generating system becomes positive, that is, they all are unsustainable technology systems. On the other hand, in fact, fuel-burning electricity generation systems are now accepted in society, which means that we unconsciously recognize that electricity is more important than conversion factors. At any rate, comparing III_{light} among the various choices makes us think about which one is better for society.

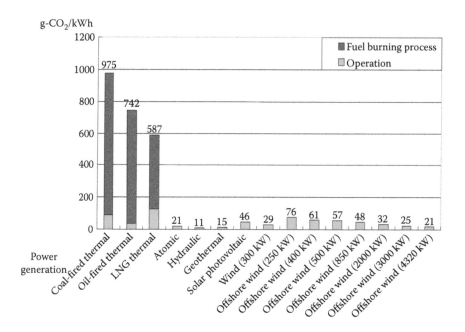

FIGURE 2.11
CO_2 emissions by power plant (numerical values other than offshore wind power generation are quoted from Hondo et al. (2004)).

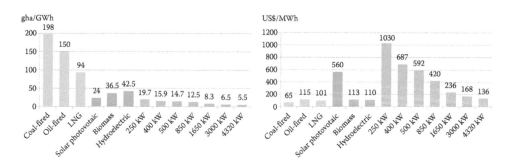

FIGURE 2.12
Their EF per GWh (numerical values of EF other than offshore wind power generation are quoted from Wackernagel and Rees (2004)) and the cost per MWh (cost of construction other than offshore wind power generation according to Japan Ocean Industries Association (2005)).

2.5.3 Spatial Analysis of III of Floating Offshore Wind Turbine

In the preceding section, we treat the III_{light} of the floating offshore wind turbine at a certain location. Here we combine the III_{light} and GIS to look for the best installation site for offshore wind power generation. The distance from the coast, the depth of the sea, and the average wind speed change the III_{light} with the location. The example result is shown in Figure 2.14.

The difference between the left panel (a) and the right one (b) is caused by the difference in the sale price of the generated electricity. The blue area is negative III_{light}, that is, it is sustainable, and the red area is positive III_{light}, meaning it is unsustainable in society. We understand easily and visually where the suitable area is for the wind farm business

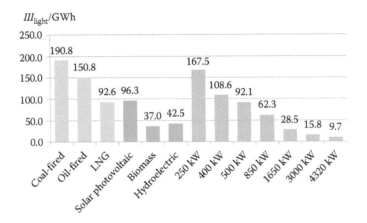

FIGURE 2.13
Comparisons of III_{light} among power generation systems.

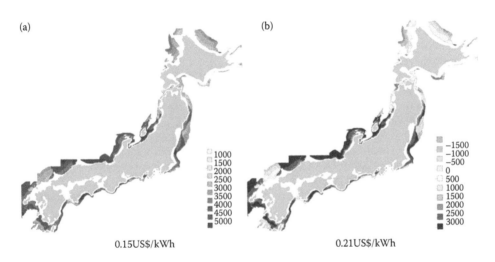

FIGURE 2.14
III_{light} map obtained in combination with GIS. (a) 0.15 US$/kWh. (b) 0.21 US$/kWh.

by looking at the map. Additionally, the map indicates that the higher acceptable price expands the area in which it is possible to operate a commercial wind farm. If we hope to reduce CO_2 emissions, acceptable electricity prices should be higher. On the other hand, when we choose economic efficiency, keeping the price of electricity down, the wind farm business will be driven out of society.

References

Arrhenius, O (1921). Species and Area, *Journal of Ecology*, Vol 9, pp 95–99.
Artificial Tideland Investigation Committee (1998). *Artificial Tideland Investigation Report* (in Japanese).

CEIS (Center for Environmental Information Science) (2005). CO$_2$ Emission Unit of Product ver. 1.2, *Website of Ministry of the Environment*, http://www.lifecycle.jp/manual/lca_unit_1.2.pdf.

Central Environment Council, Japan (2005). *Future action on the international movements of climate change - on long-term goal. 2nd Report* (in Japanese).

Chambers, N, Simmons, C and Wackernagel, M (2000). *Sharing Nature's Interest: Ecological footprints as an indicator of sustainability*. Earthscan, London.

Dodson, SI, Arnott, SE and Cottingham, KL (2006). The Relationship in Lake Communities between Primary Productivity and Species Richness, *Ecology*, Vol 80, No 10, pp 2662–2679.

Duan, F, Yamaguchi, H and Kawabuchi, M (2011). Inclusive Environmental Impact Assessment for Water Purification Technologies, *Proceedings of the Twenty-first International Offshore and Polar Engineering Conference*, pp 814–825.

Duarte, CM (1990). Seagrass Nutrient Content, *Marine Ecology Process Series*, Vol 67, pp 201–207.

Ecological Footprint Network (EFN) (2006). *Ecological Footprint and Biocapacity Technical Notes 2006 Edition*, Oakland.

Fukuoka City (2006). *The Eelgrass Field Restoration in Goto Area* (in Japanese).

Ewing, B, Reed, A, Galli, A, Kitzes, J and Wackernagel, M (2010). Calculation Methodology for National Footprint Accounts. 2010 Edition, Global Footprint Network, Oakland.

Hiraoka, K and Kameyama, M (2005). *Emission factors of gaseous exhaust emissions based on analysis of actual voyage logs of oceangoing cargo ships for life cycle assessment of seaborne transportation*. Technical Report of National Marine Research Institute 5: 25–90 (in Japanese).

Hondo, H, Uchiyama, Y and Moriizumi, Y (2004). *Evaluation of power generation technologies based on life cycle CO$_2$ Emissions*, Socio-economic Research Center, Reo, No, Y99009.

Horie, T (1987). Modeling of Ocean Material Circulation and Method on Predicting Effects of Cleaning Measures, *Report of National Institute of Port*, Vol 26, No 4, pp 57–123 (in Japanese).

Intergovernmental Panel on Climate Change (IPCC) (2001). Climate change 2001. IPCC.

Intergovernmental Panel on Climate Change (IPCC) (2005). IPCC Special Report on Carbon Dioxide Capture and Storage. IPCC.

Ishitani, H and Ishikawa, M (1992). *Social Systems Engineering*. Asakura Pub, Tokyo. (in Japanese).

Itsubo, T and Inaba, A (2005). Life Cycle Environmental Assessment Method – LIME. Sangyo Kankyo Kanri Kyokai (in Japanese).

Kamishiro, N and Sato, T (2007). Public acceptance of the oceanic carbon sequestration, *Marine Policy*, Vol 33, 466–471.

Kanto Regional Development Bureau, MLIT (2005). *The Plan of Environmental Restoration in Tokyo Bay* (in Japanese).

Kimura, K, Ichimura, Y, Sakamaki, T, Nishimura, O, Inamori, Y, Kohata, K and Sudo R (2002). Analysis of the Effect of Artificial Tidal Flat on Water Purification Capacity, *Proceedings of Coastal Engineering*, Vol 49, pp 1306–1310 (in Japanese).

Kitazawa, D (2001). Research on the Impact of the Mega Float Construction on the Ecosystem Using Numerical Simulation, *Doctoral thesis*, University of Tokyo (in Japanese).

Kitazawa, D, Tabeta, S and Fujino, M (2003). Response of the Water Quality to the Mass Load from Land Area in Tokyo Bay, *Coastal Ocean Research*, Vol 40, No 2, pp 159–169 (in Japanese).

Kozuki, Y, Nakanishi, T, Shigematsu, T and Ohtsuka, K (2001). To Establishment of Selection Method of Environmental Restoration Technology, *Ecosystem Engineering*, Vol 6, pp 53–89 (in Japanese).

Kuwae, T, Hosokawa, Y and Ozasa, H (2000). Mesocosm Experiments on the Recruitment of Benthic Communities into Created Intertidal Flats, *Proceedings of Coastal Engineering*, Vol 47, pp 1101–1105 (in Japanese).

Matsumura, T and Ishimaru, T (2004). Freshwater Discharge and Inflow Loads of Nitrogen and Phosphorus to Tokyo Bay (from April 1997 to March 1999), *Umi no Kenkyu (Oceanography in Japan)* Vol 13, No 1, pp 25–36 (in Japanese).

Ministry of Environment (2005). *Environmental Statistics Report* (in Japanese).

Ministry of Environment (2006). *The Basic Principle of the Total Reduction on the COD, Nitrogen and Phosphorus in the Tokyo Bay* (in Japanese).

Murai, M and Aono, T (2009). Inclusive environmental assessment for offshore wind power stations, *Proc. of ISOPE*, pp 406–413.

Nakajima, T and Nakajima, E (2003). Research on the LCA assessment of Sewage Treatment, *Proc. of 40th Workshop of Sewage Research*, pp 283–285 (in Japanese).

Nakanishi, J (1995). *Environmental Risk*. Iwanami Pub, Tokyo. (in Japanese).

National Institute for Environmental Studies (NIES) (2002). *Embodied Energy and Emission Intensity Data for Japan Using Input-Output Tables (3EID)*, 71 pp.

Okura, N (1993). *Tokyo Bay – Environmental Changes in 100 years*. Kouseisha Press, Tokyo. (in Japanese).

Orr, JC et al. (2005). Anthropogenic acidification over the twenty-first century and its impact on calcifying organisms, *Nature*, Vol 437, pp 681–686.

Otsuka, K (2006). A proposal of comprehensive environmental impact assessment index for large-scale utilization of the ocean. *Proc. Techno-Ocean 2006/19th JASNAOE Ocean Eng. Symp.*, CD-ROM (in Japanese).

Otsuka, K (2007). Ecological Footprint Accounting of TAKUMI, *Proc, of 7th ISOPE Ocean Mining Symposium (2007)*, pp 33–38.

Otsuka, K (2009). Inclusive Impact Index Triple I and Its Application for Ocean Technology. *Conf. Proc. The Japan Society of Naval Architects and Ocean Engineers*, Vol 8, pp 45–48 (in Japanese).

Otsuka, K and Ouchi, K (2008). Inclusive Environmental Impact Assessment for Ocean Nutrient Enhancer, *Journal of the Japan Society of Naval Architects and Ocean Engineers*, Vol 8, pp 17–25.

Otsuka, K and Sato, T (2005). Summary of Inclusive Marine Pressure Assessment and Classification Technology Committee, *Conf. Proc. The Society of Naval Architects of Japan*, Vol 5, pp 31–32 (in Japanese).

Research Institute of Innovative Technology for the Earth (RITE) (2005). Carbon dioxide fixation, effective utilization technologies 2004.

Research Institute of Innovative Technology for the Earth (RITE) (2006). www.rite.or.jp (in Japanese).

Research Institute of Innovative Technology for the Earth (RITE) (2009). Analyses on Mid-term Targets of Major Countries, 5p.

Rosenzweig, ML (1995). *Species Diversity in Space and Time*. Cambridge Univ. Press, Cambridge.

Sato, T (2004). Numerical simulation of biological impact caused by direct injection of carbon dioxide in the ocean, *Journal of Oceanography*, Vol 60, pp 807–816.

Sato, T, Watanabe, Y, Toyota, K and Ishizaka, J (2005). Extended probit mortality model for zooplankton against transient change of PCO2, *Marine Pollution Bulletin*, Vol 50, pp 975–979.

Tamaki, T (2003). Evaluation of the fishing village activation effect by fishing ground creation and urban interchange, *Bulletin of Fisheries Research Agency*, No 8, pp 22–111 (in Japanese).

Tokyo Bureau of Sewerage (2009). *Technical statistics* (in Japanese). http://www.gesui.metro.tokyo.jp

Uchiyama, Y (1995). Evaluation of power generation based on life cycle CO_2 Emissions, Socio-economic Research Center, Reo, No, Y99015.

Wackernagel, M and Rees, W (1996). *Our Ecological Footprint*. New Society Press, Gabriola Island.

Wackernagel, M and Rees, W (2004). *Our Ecological Footprint*. Godo-shuppan, Tokyo.

World Wide Fund For Nature (WWF) (2008). Living Planet Report 2008, WWF International, 44p.

World Wide Fund For Nature (WWF) (2010). Living Planet Report 2012, WWF International, Switzerland.

Yokoyama, T and Kudo, S (1995). Evaluation of chemical absorption process for CO_2 removal from fuel gases of LNG-fired power plant. *Technical Report of Central Research Institute of Electric Power Industry*, T94057 (in Japanese).

Yoshimoto, H and Tabeta, S (2011). A Study on the Scaling of Ecological and Economic Values in the Inclusive Impact Index, *Proceedings of the Twenty-first International Offshore and Polar Engineering Conference*, pp 804–809.

3

Biophysical Modelling of Marine Organisms: Fundamentals and Applications to Management of Coastal Waters

Dmitry Aleynik, Thomas Adams, Keith Davidson, Andrew Dale,
Marie Porter, Kenny Black and Michael Burrows

CONTENTS

3.1 Introduction 66
    3.1.1 Overview 66
    3.1.2 Management Challenges 66
    3.1.3 A Modelling Approach 67
    3.1.4 Summary 68
3.2 Model Framework 69
    3.2.1 Overview 69
    3.2.2 The Meteorological Model 69
    3.2.3 The Hydrodynamic Model 70
        3.2.3.1 Overview 70
        3.2.3.2 Parameterisations 71
        3.2.3.3 Boundary Forcing 71
        3.2.3.4 Validation Data 72
        3.2.3.5 Model Evaluation Tools 73
    3.2.4 Biological Models 74
        3.2.4.1 Harmful Algal Bloom Model 74
        3.2.4.2 General Larval Dispersal Model 75
3.3 Results 78
    3.3.1 Meteorological Model Validation 78
    3.3.2 Hydrodynamic Model Output and Validation 78
    3.3.3 HAB Model 82
        3.3.3.1 Drifter Tracking Simulations 82
        3.3.3.2 *Karenia mikimotoi* Simulations 83
        3.3.3.3 Biological Model Integration Results 83
    3.3.4 General Particle-Tracking Model 83
        3.3.4.1 Habitat Structure and Larval Duration 85
        3.3.4.2 Wind Forcing 87
        3.3.4.3 Diffusion and Velocity Computation 87
        3.3.4.4 Vertical Migration 88
        3.3.4.5 Salinity Driven Mortality 88
        3.3.4.6 Further Applications: Intertidal Organisms and Novel
                     Offshore Habitat 90

3.4 Discussion and Conclusions 91
Acknowledgements...93
References...93

3.1 Introduction

3.1.1 Overview

Fjordic coastlines are found around the world, in regions such as (but not limited to) the western edges of Scotland, Norway, Canada, USA and Chile. A defining feature of these areas is their long and convoluted coastline, consisting of many long and narrow inlets (fjords, or sea lochs) and numerous islands. These features create sections of coastline with varied aspects, from areas giving shelter from the open ocean to those fully exposed to wave action, and correspondingly varied environmental conditions within small geographic areas. Another key physical feature of these areas is their transition between freshwater and saltwater environments. These regions often experience high rainfall which, combined with steep-sided lochs and mountainous areas of land adjacent to the water's edge, results in high freshwater runoff and a rather different temperature to that of nearby ocean water. The dynamics of sea loch circulation are complicated, involving strong stratification, periodic overturning events, and a predominantly wind-influenced current (Stigebrandt, 2012).

The complex convoluted coastline allows these regions to support a diverse ecology and economy. Scotland has the second longest coastline in Europe with a geometric length of 18,588 km (Darkes and Spence, 2008), and its west coast is a characteristic example of a fjordic region. The coast includes around 800 major islands and over 100 sea lochs (Edwards and Sharples, 1986). These lochs provide sheltered locations for finfish and shellfish aquaculture (mainly salmon and mussels). Strongly tidal areas and more exposed patches of open water give ideal conditions for the development of renewable energy extraction activities (Davies and Watret, 2011; Davies et al., 2012). Commercial fishing activity is diverse, from demersal fish to crustaceans and benthic organisms such as scallops. These areas also support a diverse economy based upon recreational activities including water sports and wildlife viewing. Many of the unique habitats provided by these coastal features are protected by law in the form of marine protected areas (MPAs) or fishing restrictions.

3.1.2 Management Challenges

Each human activity interacts slightly differently with the marine environment. However, good management of biological flora and fauna in this environment requires an understanding of how organisms move or disperse in the sea, whether to help maintain sustainable populations or to guard against invasion and subsequent ecological and economic damage. In this chapter, we will present a computational modelling approach to investigating the dispersal process and understanding some of its implications for several of the most important contemporary challenges.

Our first example is that of harmful algal blooms (HABs), some of which develop offshore and are then advected onshore, affecting aquaculture. The Scottish west coast and islands suffer from regular HAB events, primarily related to the shellfish biotoxin-producing genera *Alexandrium*, *Pseudo-nitzschia* and *Dinophysis* and the ichthyotoxic species *Karenia mikimotoi* (Davidson et al., 2009, 2011; Touzet et al., 2010). These toxins do not harm the shellfish themselves but pose a health risk for the humans consuming them (Davidson and Bresnan, 2009). Blooms of *Pseudo-nitzschia* (Fehling et al., 2012) and *Dinophysis* (Farrell et al., 2012) are thought to develop offshore and are often advected toward the coast. Regulatory thresholds for *Pseudo-nitzschia* and *Dinophysis* are presently 50,000 cells L^{-1} and 100 cells L^{-1} respectively, and an early warning system must therefore be able to detect concentrations below these levels. The fish-killing dinoflagellate *Karenia mikimotoi* is a common member of the dinoflagellate community of the North-East Atlantic, and its blooms have reached potentially harmful densities in many recent years (Davidson et al., 2009). In contrast to the shellfish biotoxin genera mentioned previously, harmful *K. mikimotoi* events are characterised by rapid blooms that can reach concentrations of millions of cells per litre. The related species *K. brevis* contains a signature pigment (Kirkpatrick et al., 2000), which may be identified by satellite remote sensing (Shutler et al., 2011). Combining such an approach with numerical predictions of transport pathways may thus provide early warning for aquaculture.

Another issue affecting aquaculture is the management and control of diseases and parasites, which are transmitted between aquaculture sites by a combination of water movements and the passage of wild fish. A particularly topical example is sea lice, parasitic copepods that are a particular problem for salmonids in Scottish waters (principally the species *Lepeophtheirus salmonis* L.). Sea lice are endemic in the wild (even prior to widespread and intensive fish farming), but salmon farms are often cited as playing a key role in sea lice abundance (Costello, 2006, 2009) owing to the large numbers of potential host fish that are held in one location. Their impact on fish can include mortality of smolt (very young) fish (Heuch et al., 2005) or serious damage to individual adult fish in the case of heavy infections (Costello, 2006). Female adult sea lice give birth to non-infective nauplii that move passively in the water column and later develop into infective copepods. Control of lice at salmon aquaculture sites is most frequently carried out by chemical treatment of fish, coordinated within defined management areas. Biophysical models have several applications in this context. They may be used at the small scale of the loch/management area (50–60 km) to predict the expected density of lice at individual sites and throughout the loch (Salama et al., 2012) or the role of farm sites as the source or destination of lice for other sites (including reinfection at the site itself) (Adams et al., 2012), and they may be used to determine management schedules for the most effective chemical treatment regimes for particular site locations (potentially allowing reduced chemical use). The advent of larger-scale (100–500 km) coastal models opens up the possibility of more general investigations into the optimisation of management area boundaries (limiting transmission between separate areas).

For natural populations, the sustainability of coastal fisheries and MPAs can also depend on the recruitment of enough juvenile organisms, whether through retention or influx from other areas, to maintain levels of abundance of adults. Depending on how particular affected species disperse between these areas, the act of protecting or exploiting particular areas may have a profound effect on the broader population, or it may have no effect. It is therefore vital that the 'connectivity' of such areas (Roberts et al., 2010; Burgess et al., 2014) be taken into consideration in policy formulation.

3.1.3 A Modelling Approach

Difficulties in physically tracking minute biological organisms in the wild have led to a rapid and widespread surge in the popularity of simulation modelling, coupling hydrodynamic with biological models, for estimating larval dispersal processes and their impact on population dynamics. Some general results have been obtained using idealised hydrodynamic scenarios (Gaines et al., 2003; Siegel et al., 2003), but understanding processes in specific places requires bespoke model implementations (Aiken et al., 2007; North et al., 2008; Mitarai et al., 2009). Such models incorporate topography, freshwater influx and prevailing meteorological conditions, amongst other factors. Tidal, temperature and salinity conditions at model boundaries are generally derived from broader-scale models or observational data. However, accurate modelling of coastal regions is complicated and involves a balance between computational expense and spatial resolution.

The islands and fjords that are used for aquaculture, and their distinct ecology, present a significant challenge to modelling, which is compounded by complex bathymetry and local weather patterns. Existing structured grid models do not provide the resolution needed to represent these intricate coastlines in their wider shelf context (a regular square grid cannot efficiently represent both narrow channels and broad expanses of water). The advent and widespread adoption of finite-element methods (computation using a variable-resolution triangular grid) may therefore be seen as a response to this challenge. Finite-element methods provide both accuracy and efficiency in a wide range of situations (Chen et al., 2003; Zhao et al., 2006; Huang et al., 2008).

The representation of cell or larval biology in coupled biophysical models varies widely, and from general to highly species specific. In the simplest models, biological entities are treated as passive particles, with movement driven by advective and diffusive processes in the hydrodynamic model. Mortality is a common addition. Behaviour may also be incorporated, such as a switch in the ability of larvae to settle or changes in swimming behaviour. In addition to the possibility of encoding particular 'species' or habitat arrangements, model configuration inevitably involves a number of more fundamental choices, such as turbulent diffusion level, velocity interpolation procedure and so on. However, comparisons between specific scenarios and the influence of such choices on results are seldom documented and we aim to address this here.

3.1.4 Summary

Here we present a biophysical model framework of coastal waters that uses an unstructured finite-volume methodology, allowing simulation of biological transport within complex nearshore waters. We cover all aspects of development, forcing and validation of the hydrodynamic model. We also deal with implementation and application of biological models driven by the hydrodynamic model to give insight into real-world management challenges, using the specific context of the west coast of Scotland as an example system. Model-observation comparisons reveal close correspondence of tidal elevations for major semi-diurnal and diurnal tidal constituents. The thermohaline structure of the model and its current fields are also in good agreement with a number of existing observational data sets. We simulate the transport of Lagrangian drifting buoys and a bloom of *K. mikimotoi* (with the incorporation of an individual-based biological model), demonstrating that such an approach holds considerable promise for dispersal predictions.

We then present the results and applications of a generic particle-tracking model to some other contemporary management issues (sea lice dispersal and invasive species control). Most biophysical model studies have used parameterisations based upon previous work, sometimes for consistency, but often without clear justification. The use of biophysical models has now become rather widespread, and we felt it timely to discuss various choices that may be made in model configuration and their influence on model predictions such as distance dispersed, probability of successful return to habitat, and more general geographic descriptors of movement. Finally, more general ecological implications of model predictions and their role in predicting the fate of communities are discussed.

3.2 Model Framework

3.2.1 Overview

To tackle the challenges outlined previously, we use a hierarchical biophysical model framework consisting of several separate components (Figure 3.1). This allows for testing and validation of each model component in isolation. Furthermore, running models separately increases computational efficiency; once validated, there is no requirement to recompute the meteorological model for each hydrodynamic model run, for example. Likewise, each instance of the particle-tracking or algal bloom model is computed using stored output from the hydrodynamic model. These biological models are much simpler than the hydrodynamic model that drives them, and they therefore require information from only a portion of the full domain at any particular moment. In what follows, we describe each component of the model in detail.

3.2.2 The Meteorological Model

Any attempt to accurately model coastal hydrodynamics must be underpinned by high-quality meteorological forcing (Davies and Hall, 2002). Local meteorology was modelled using the open-source Weather Research and Forecasting (WRF) model v. 3.5.1 (Skamarock et al., 2008). This is a non-hydrostatic atmospheric model, nested within the NOAA National Centers for Environmental Prediction (NCEP) FNL operational forecast model with 1° spatial resolution (NCEP, 2000). Our WRF model domain covers Scotland and its neighbouring seas with a grid of 140 × 240 points. The finest resolution is around 2 km in the central part. Additional forcing for the WRF model was provided by daily real-time global sea surface temperature (SST) fields at a resolution of 1/12° (Gemmill et al., 2007).

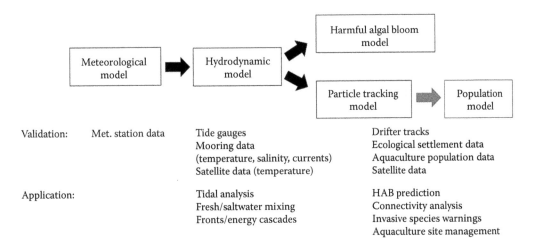

FIGURE 3.1
Schematic of approach showing hierarchical model framework. Each component of the model is validated in different ways. Output from each may be considered in isolation to answer particular questions, contingent on the development of dependent components. We do not discuss biological population models in detail here as these tend to focus on specific organisms. However, later sections do give some insight into how outputs may be utilised in driving them.

3.2.3 The Hydrodynamic Model

3.2.3.1 Overview

Our study of the region's hydrodynamics used the open source Finite Volume Community Ocean Model (FVCOM) v. 3.1.6 (Chen et al., 2011), a primitive-equation, free-surface, hydrostatic model. Modelled parameters include surface elevation, temperature, salinity, velocity and turbulence. The model domain extends from the Isle of Man to the northern tip of Scotland and from the Scottish coast to the Outer Hebrides archipelago Figure 3.2a and b). Model computation uses an unstructured mesh consisting of non-overlapping triangular prism elements of variable size, allowing increased resolution in areas of complex coastline or bathymetry. Horizontally, there are 46,878 nodes, giving 79,244 triangular elements. Node spacing ranges between 4.6 km at the open boundary to around 100 m in constricted areas, such as narrow straits and the heads of fjords. The model has 11 vertical sigma-coordinate (terrain-following) layers, with increased resolution close to the surface.

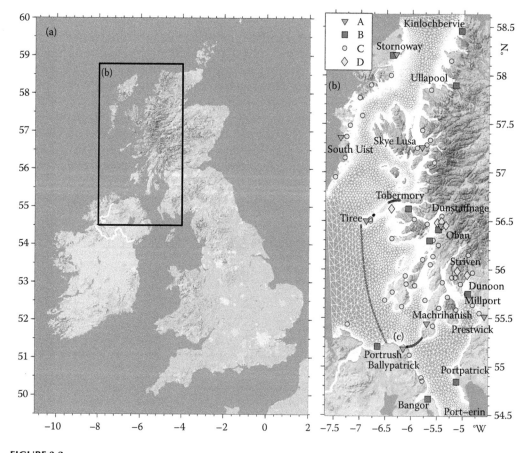

FIGURE 3.2
West coast of Scotland FVCOM model mesh and observation/validation locations. Legend: symbols A (▼) – UK MetOffice/Midas weather stations; B (■) – pressure-gauge sites from BODC; C (●) – International Hydrographic Office (IHO) port sites; D (◆) – moorings (SBE-37 or thermistor loggers). White rectangles indicate the parts of the domain that are shown with zoom at (b) and those with the line (c) indicate the ASIMUTH-FVCOM model domain open boundaries.

Initial development focused on the smaller Applied Simulations and Integrated Modelling for the Understanding of Toxic and Harmful Algal Blooms (ASIMUTH)-FVCOM model domain that has higher horizontal resolution (shortest element length 75 m), studying hydrodynamics in the Firth of Lorn area (Figure 3.2b and c), for which HAB transport simulations are presented subsequently. The final hydrodynamic model run for this domain covered the period 20 June to 31 October 2011. The domain was subsequently extended to the wider West Scottish Coastal Ocean Modelling System 'WeStCOMS-FVCOM' at approximately half the horizontal resolution (up to 130 m). In time, this will allow more spatially extensive HAB modelling to be undertaken. The hydrodynamic model run for the larger domain extended from 20 June 2013 till now. Presented particle-tracking simulations are based on this larger domain.

Model bathymetry was based on gridded data from SeaZone (2007) and refined in certain key areas using Admiralty charts and a number of multi-beam surveys. Depths of less than 5 m are found in fewer than 7% of mesh elements, and this was therefore used as the minimum model depth (long-line shellfish production and salmon aquaculture are conducted in areas deeper than this). To minimise hydrostatic inconsistencies, volume-preserving bathymetric smoothing was applied (Foreman et al., 2009), limiting the bottom slope gradient to 0.3 within mesh elements.

3.2.3.2 Parameterisations

Vertical eddy viscosity and diffusivity was computed using the Mellor-Yamada 2.5 scheme (Mellor and Yamada, 1982). No significant difference in performance was found between this and several more complex schemes available in the GOTM (General Ocean Turbulence Model). Horizontal diffusion was represented using a Smagorinsky (1963) eddy parameterisation with a coefficient of $C = 0.2$.

The bottom boundary layer was parameterised with a logarithmic wall-layer law using a drag coefficient $C_d = \max\{\kappa^2 / \ln[(Z / Z_0)^2], C_{d0}\}$, with $\kappa = 0.4$ and Z the vertical distance from the nearest element centroid to the seabed. The bottom drag coefficient was set to $C_{d0} = 0.0025$ across the entire domain, while the roughness length scale parameter Z_0 was set to $Z_0 = 4 \times 10^{-3}$ m, except in shallow regions and coastal areas, where it increased gradually to $Z_0 = 8 \times 10^{-3}$ m.

The hydrodynamic model was integrated numerically using a time-split method with an external time step of 0.6 s. Computation on the ARCHER Cray XC30 system used 9 hours of wall time for a 6-month simulation run, using 1,024 CPU cores based on Intel-Xeon E5v2 2.7 GHz architecture. WRF meteorological model runs with the same period required twice the computation time.

3.2.3.3 Boundary Forcing

Water movement within hydrodynamic models is predominantly driven by tidal forcing applied at the 'open' (water) boundaries. The tidal forcing was derived from the $1/30°$ degree implementation of the Oregon State University Tidal Prediction Software (OTIS) for the European shelf (Egbert et al., 2010). At each of the domain's open boundary nodes, the 11 primary tidal constituents were used to derive sea surface elevation time series. Open boundaries also included a 6-km-wide 'sponge layer' in order to filter high-frequency numerical wave reflection noise using a Blumberg-Kantha implicit gravity wave radiation condition (Chen et al., 2011).

One goal of the ASIMUTH project was the integration of multiple Western European modelling systems for HAB prediction, and this involved the configuration of a system of nested models (meaning that smaller-scale models derive some boundary conditions – in our case linearly interpolated 3-hourly values for temperature, salinity and velocity – from larger-scale models). The Scottish west coast model was one-way nested within the Irish Marine Institute's NE Atlantic Model (NEA-ROMS), which has a resolution in our boundary area of around 2 km and is itself nested within the operational $1/12°$ Mercator ocean model, which assimilates sea surface height and temperature over the North Atlantic (Dabrowski et al., 2014). Each model run was initialised from a state of rest for currents and three-dimensional (3D) temperature and salinity fields interpolated from the nearest temporal snapshot from an external model (NEA-ROMS), augmented by information from Conductivity, Temperature, Depth profiler (CTD) casts in several side lochs (areas of known low salinity).

Discharge from the 91 largest rivers in the domain was included via freshwater input at nearest mesh node points computed from a daily average of 3-hourly rainfall data over respective catchment areas (Edwards and Sharples, 1986), with a lag of 1–2 days (lag proportional to catchment area). This method was calibrated against available regulated discharge rate data. Evapotranspiration was estimated using a seasonally varying scaling applied to rainfall totals. This simplified approach incorporates significant (more than 50%) inter-annual variations in precipitation (Parry et al., 2014), which would be lost using 'climatological' river discharge. River temperature was calculated from a combination of night-time air and sea surface temperatures, with data sources including local and satellite-derived observations.

3.2.3.4 Validation Data

Tides: Eight tide gauges, maintained by the UK Tide Gauge Network, are sparsely distributed in the region. Data from these were augmented by International Hydrographic Office (IHO) historical tidal analyses from 59 local ports and harbours (Figure 3.2).

Temperature, salinity and currents: Temperature and salinity data sources included frequent (weekly or fortnightly) summer CTD transects and thermistor loggers in the central Firth of Lorn area (Fehling et al., 2006). In the southern part of the domain, the Scottish Environment Protection Agency (SEPA) provided CTD time series from two mooring sites in the Clyde estuary system. Finally, a mooring in the Tiree Passage provided an 18-year time series of currents and subsurface CTD measurements (Inall et al., 2009). All these locations are indicated by diamonds in Figure 3.2c.

Further validation was made using daily satellite-derived Multi-scale Ultra-high Resolution Sea Surface Temperature (MUR-SST) at 1 km horizontal resolution (Armstrong et al., 2012) and real-time global sea surface temperatures from the $1/12°$ NCEP operational analysis (Gemmill et al., 2007). The latter product was also used as an additional forcing source of the WRF meteorological modelling system.

Drifter tracks: In summer 2013, 30 Surface Velocity Program (SVP) drifters were deployed on the Malin Shelf slope as part of the Fluxes Across Sloping Topography in the North East Atlantic (FASTNEt) project, to the south-west of the WeStCOMS-FVCOM domain (Inall et al., 2013). Half of these had drogues centred at 15 m and half at 70 m, with each drifter reporting surface temperature and GPS-derived position every 3 hours. All drifters that were drogued at 15 m and 10 of those drogued at 70 m moved north-eastward from their release positions onto the shelf and entered the model domain. However, most of these did so either outside the simulation period or for only a brief amount of time, remaining close to the domain boundary. Those that remained near the western edge of the domain and were quickly deflected out of it tended to remain on the ocean-ward side of the Islay Front, a tidal mixing feature (Simpson et al., 1979), which effectively blocked their further advection into the region. This front is not well resolved spatially in the model, limiting the effectiveness of validation using drifters in this area of the domain.

Four of the drifters drogued at 15 m made a sufficient (and appropriately timed) incursion into the model domain, allowing comparison of passive transport processes without the complicating effects of biological growth and food-web interactions. Observed drifter track positions were resampled at regular time intervals (3 and 24 hours) and compared with virtual drifters. Model-drifter separation time and spatial scales were estimated, together with the contribution of different factors using realistic model simulations under the best available weather and tidal forcing.

Drifter 23 was used as an example throughout this study, as it was the drifter that spent the most time within the model domain and the only one that was advected across its full length. This drifter entered the model domain from the south-west and was advected towards Mull and Tiree before being deflected by the Islay Front away from the Tiree Passage into the Minch and northwards to Skye. In this southern region the path of the drifter was smooth, suggesting a poleward current, further north, north of Coll/Tiree and south of Skye the mean flow was interrupted by numerous eddies, which north of Skye subsided to once again give a smooth, poleward flow. The different circulation regimes highlighted by the drifter are an ideal validation challenge for the model.

3.2.3.5 Model Evaluation Tools

Comparisons between the model and the observations may be represented by Taylor diagrams (Taylor, 2001), which depict geometrically the relationship between fundamental statistical measures:

$$D^2 = \sigma_m^2 + \sigma_o^2 - 2\sigma_m\sigma_o R, \tag{3.1}$$

where (σ_m, σ_o) are standard deviations of the model (m) and observational (o) data sets, and R is their correlation coefficient over N samples with means $(\bar{m}, \bar{o})$:

$$R = \frac{(1/N)\sum_{i=1}^{N}(M_i - \bar{m})(O_i - \bar{o})}{\sigma_m\sigma_o}. \tag{3.2}$$

The Taylor skill score (S) is defined as

$$S = \frac{1}{4} \cdot \frac{(1-R)^2}{(\sigma_m/\sigma_o + \sigma_o/\sigma_m)^2}. \tag{3.3}$$

In addition, we used a measure of model skill based on regression analysis (Wilmott, 1982) in which the Wilmott index of agreement (d) is defined in terms of the root-mean-square (RMS) error (E) and centred RMS difference (D):

$$E = \frac{\sum_{i=1}^{N}(M_i - O_i)^2}{N}, D = \left[\frac{1}{N}\sum_{i=1}^{N}[(M_i - \bar{m}) - (O_i - \bar{o})]^2\right]^{1/2}, \tag{3.4}$$

$$d = 1 - \frac{\sum_{i=1}^{N} (M_i - O_i)^2}{\sum_{i=1}^{N} \left(|M_i - \bar{o}| + |O_i - \bar{o}|\right)^2} . \tag{3.5}$$

The measures d and S approach one when agreement is excellent and zero when agreement is poor.

3.2.4 Biological Models

The physical model produced hourly 3D velocity and scalar (temperature and salinity) fields. We implemented two different biological models: one for the transport of HABs and another for general particle dispersal.

These models were run 'off-line': using output data files from the hydrodynamic model rather than as part of a combined model run. This has the advantage that the generally simpler and less computationally expensive biological simulations can be run many times using different configurations without the need for further hydrodynamic simulation. We describe two models: a HAB model and a general larval dispersal model.

3.2.4.1 Harmful Algal Bloom Model

Individual biological models (IBMs) for phytoplankton are an important and useful tool in predicting the spread of dangerous species in areas of developing aquaculture. The central part of the west coast of Scotland has been a major area of shellfish harvesting for decades and, therefore, was naturally selected as the ASIMUTH model domain.

3.2.4.1.1 Population Growth and Mortality

To simulate the population dynamics of one of the most harmful algae (*Karenia mikimotoi*), biological variables (i.e., cell concentrations) and processes (i.e., growth and mortality) were calculated at the same unstructured mesh locations as the scalars in the hydrodynamic model. We used the biological model of Gentien et al. (2007), which includes growth, mortality, predation and simplified behavioural algorithms. Physical processes (advection and diffusion) affecting cell concentration were also included. Some applications have used other biogeochemical models for *K. mikimotoi*, e.g., Vanhoutte-Brunier et al. (2008), but they require the parameterisation of complex and poorly understood food-web interactions, making them difficult to use for forecasting.

In the model, growth (division) and mortality of cells (removal) are expressed by

$$\frac{\partial C}{\partial t} = g(T)C - \mu\gamma C^2, \tag{3.6}$$

where C is the cell density, $g(T)$ is the growth rate as a function of water temperature T,

$$g(T) = 2.5 \cdot 10^{-3} \cdot T^3 - 0.15 \cdot T^2 + 2.8775 \cdot T - 17.25, \tag{3.7}$$

and $\mu = 0.3 \cdot 10^{-5}$ was an empirical mortality coefficient obtained from experimental observations (Gentien et al., 2007). The shear term γ was defined by the turbulent kinetic energy (TKE) dissipation rate defined within the physical model ε (m^2 s^{-3}) and kinematic viscosity of sea water $\nu = 1.15 \cdot 10^{-6}$ m^2 s^{-1}:

$$\gamma = \left(\varepsilon \Big/ 7.5 \cdot \upsilon \right)^{0.5}. \tag{3.8}$$

Shear increases the cell encounter rate, which can lead to aggregation and sinking.

3.2.4.1.2 HAB Transport

The horizontal movement of *K. mikimotoi* cells was defined by advection. For every time interval, particle position X_P was governed by the formula

$$X_P^t(x,y,z) = X_P^{t-\Delta t}(x,y,z) + \Delta t \cdot \left[U_P(x,y,z) + w_p(z) \right] + \delta_h(x,y) + \delta_z(z), \tag{3.9}$$

where Δt is the model time step, U_P is the 3D advection velocity from the hydrodynamic model, w_p is the vertical swimming motion of the particle and δ_h and δ_z are displacements due to horizontal and vertical diffusion respectively, defined by

$$\delta_h(x,y) = R\left[6 \cdot K_h \cdot \Delta t\right]^{0.5}, \qquad \delta_z(z) = R\left[6 \cdot K_v \cdot \Delta t\right]^{0.5}. \tag{3.10}$$

$K_h = 1$ and $K_v = 0.001$ are horizontal and vertical eddy diffusivities (m$^2 \cdot$ s^{-1}) and R is a uniform random number in the interval $[-1, 1]$. Diel vertical plankton migration was not included in the simulation for simplicity.

3.2.4.1.3 Scenarios and Simulations

We used the movement (transport) module of the HAB model to instantiate virtual particles (VPs) for the purpose of hydrodynamic model validation, via comparison with the movement patterns of physical drifters (described in Section 3.2.3.4). Releases of VPs, centred at resampled drifter positions within an uncertainty radius (UR), were made. The UR was defined as half the distance to the nearest sequential drifter GPS position and therefore varied over time and space from 0 to 2 km. VP trajectories were calculated at the drogue depth using the model velocity field. Eleven VPs were initiated around the point at which a drifter entered the domain interior (having passed the model sponge layer). Calculated VP trajectories associated with a real drifter were traced for as long as they remained within the domain (1–10 weeks). We also made repeated 'daisy chain' VP releases along their trajectory (Robel et al., 2011), starting at midnight each day during the drifter's residence time in the domain.

The HAB model was used to simulate short-term development and advection of algal populations developing within the ASIMUTH-FVCOM domain. In particular, a HAB event occurring in the waters west of Scotland from 27 August 2011 was investigated. Firstly, we studied the movement of VPs driven by the movement module of the HAB model alone. In this chapter we focus on a bloom originating in the Tiree Passage (approximately −6.5 W, 56.5 N). Forty neutrally buoyant VPs were released at the centre of the HAB maximum, both at the surface and close to the seabed. Their movement was tracked for 5 days, covering the full duration of the observed bloom. Secondly, the full HAB model (including growth and mortality) was implemented to track the fate of the same bloom, for comparison with observed dynamics from *MODIS-MERIS* satellite data from MyOcean (Saulquin et al., 2011).

3.2.4.2 General Larval Dispersal Model

Another general particle-tracking model was implemented based on those of Amundrud and Murray (2009) and Adams et al. (2014a). It was used to study general and specific aspects of dispersion of particles representing microorganisms such as larvae of sea lice or intertidal organisms from several different habitat site configurations within the model domain and the connectivity of the habitat sites.

3.2.4.2.1 Horizontal Movement

The horizontal movement of each particle in the model used horizontal components of the movement process defined earlier in Equation 3.9, that is

$$X_P^t(x,y) = X_P^{t-\Delta t}(x,y) + \Delta t \cdot [U_P(x,y)] + \delta_h(x,y). \tag{3.11}$$

A time interval of $\Delta t = 0.005$ hours was used, linearly interpolating between hourly FVCOM output values. Velocity U is that in the closest depth layer and either the nearest FVCOM element centroid or a weighted average of the nearest element centroids. We investigated the role of the methodological choice on particular dispersal scenarios.

Most recent biological studies have used a fixed value of $K_h = 0.1$ for diffusivity (Amundrud and Murray, 2009; Stucchi et al., 2010), though this is known to vary over time and space (Turrell, 1990). We investigated the role of variation over several orders of magnitude.

3.2.4.2.2 Vertical Movement

Three types of modelled larvae were considered: (1) particles that remain in the surface layer all the time (e.g., sea lice) (Heuch et al., 1995), (2) particles that remain in the bottom layer all the time, and (3) particles that alter their depth based on the flooding or ebbing of the tide ('selective tidal stream transport') (Criales et al., 2011). For the third type, increasing surface elevation at a particle's location (element) was taken to indicate local flooding of the tide, in which case the particle moves to the surface layer. Decreasing surface elevation indicates local ebbing of the tide, and the particle moves into the lowest depth layer. Vertical migration due to active swimming and diffusion was not considered.

3.2.4.2.3 Development (Settlement Competency)

Marine organisms generally undergo some form of development during their pelagic phase. Sea lice, for example, are only able to attach to suitable hosts after an initial stage lasting around 4 days (Stien et al., 2005). This is also true of other species such as barnacles. We assumed that the model larvae spent the first half of their pelagic duration unable to settle, after which they could settle in a suitable habitat, if available.

3.2.4.2.4 Mortality

Most previous larval modelling included a constant mortality rate (Amundrud and Murray, 2009; Stucchi et al., 2010), and we used a default value of $\mu = 0.01$ h^{-1} in our simulations.

However, Bricknell et al. (2006) suggest that the mortality of sea lice may be related to salinity. We investigated how this may affect dispersal between sites, using the following functional form:

$$\mu(S) = 0.0011S^2 - 0.07S + 1.144, \tag{3.12}$$

where S is the hydrodynamic model salinity (psu) at the nearest element centroid (this form is the best fitting second-order polynomial to mean mortality rates over Bricknell's presented data range, not shown). This gives $\mu > 0.01$ h^{-1} in most situations encountered.

3.2.4.2.5 *Larval Duration*

Our previous work considered particles with dispersal durations encompassing those observed common pelagic dispersing rocky shore species in the U.K. (1–28 days; summarised by Burrows et al., 2009, Appendix C). Sea lice generally have a larval duration of around 14 days (Costelloe et al., 1998). For this study, we restricted our comparative examples to 2- and 14-day larval duration.

3.2.4.2.6 *Dispersal Measures*

We recorded the number of particle time steps spent in each mesh element, the start and end locations, and the dispersal duration (or calculated mortality) of all particle trajectories, in addition to whether they had been 'successful' or 'unsuccessful' in reaching a habitat site at which to settle.

We used start and end locations to estimate *dispersal kernels* (probability distributions of dispersal distances) for the particles. Kernels were summarised over all habitat sites within the network, subdivided between successful and unsuccessful dispersers.

Information on the start and end locations of successful particle trajectories allows for the construction of *connectivity matrices* (*n*-by-*n* matrices containing pairwise bidirectional dispersal probabilities between sites) (Adams et al., 2012). Summing this matrix and dividing by the number of sites (*n*) gives the overall probability that particles released from the sites composing that habitat type will be retained within that habitat type.

3.2.4.2.7 *Habitat Sites and Wind Scenarios*

In the general larval dispersal study, we considered three particular habitat scenarios (with example organisms):

- a. Salmon farm sites (parasites such as sea lice, viruses);
- b. Evenly spread coastal habitat (intertidal organisms such as barnacles, limpets and algae);
- c. Subtidal soft sediment (benthic organisms such as nephrops).

These habitats are displayed in Figure 3.3 and differ greatly in spatial distribution. Fish farms tend to be located in somewhat sheltered areas across the west coast of Scotland. We used location information on active salmon farm sites (from http://aquaculture.scotland.gov.uk/data/data.aspx). Evenly spaced coastal sites were obtained from the hydrodynamic model mesh, starting at one end of each connected set of coastal points (section of coastline) and selecting the next point in the set that was greater than 1 km from all existing selected points. The soft sediment habitat was chosen by randomly selecting sites with at least 1 km separation from within areas defined from UKSeaMap (McBreen et al., 2011).

To illustrate the role of wind forcing on particle fate, we considered dispersal commencing at two different start dates: early April 2014, during which the wind was strong and predominantly south-south-westerly; and late April 2014, during which winds were generally lower and south-easterly.

Particles were released from the nearest element centroid to each habitat site in the selected configuration, and settlement was deemed to have occurred once a particle moved within 500 m of a habitat site (once development, detailed earlier, had occurred). Each model run (given habitat configuration, larval duration, model assumptions) consisted of 100 particles being released from each on the *n* habitat sites.

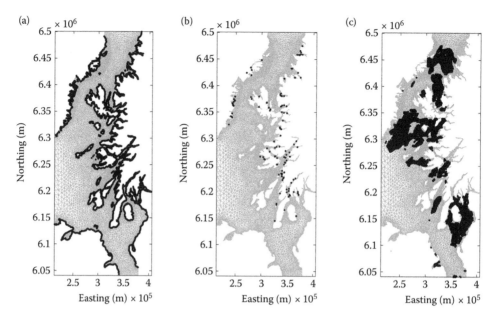

FIGURE 3.3
Habitat configurations used for particle-tracking investigations: (a) equally spaced (1 km) coastal sites; (b) salmon aquaculture sites; (c) subtidal soft sediment habitat (sites with min 1 km separation).

3.3 Results

3.3.1 Meteorological Model Validation

Wind velocity (u, v) predictions of a WRF model run were compared with eight U.K. Met Office weather stations, subsampled at 3-hourly intervals from January 2011 until September 2013. Excellent agreement was achieved in the more exposed locations such as Stornoway Airport (Figure 3.4a and b).

Long-term mean wind direction in the region measured at Met Office weather stations was 219°; the average value from the WRF model was 214°. Performance metrics (Equations 3.1–3.5) confirmed the high quality of this reproduction of wind components with an average correlation coefficient for the two velocity components R = [0.92, 0.90], Taylor Score S = [0.80, 0.75], and Wilmott index of agreement d = [0.94, 0.91]. The WRF model was capable of resolving the steering and enhancement of prevailing winds along fjordic inlets within the model domain (Figure 3.4c shows an example of North Minch).

3.3.2 Hydrodynamic Model Output and Validation

Tides: Tidal model validation is commonly made by comparing the amplitude and phase of key harmonic constituents. In western Scottish waters, most tidal energy is accounted for by two major constituents, M_2 and S_2, responsible together for between 68 and 77% (south to north) of total amplitude variation. Comparison of the WeStCOMS-FVCOM tidal predictions against pressure-gauge sites demonstrated very small differences in amplitude and phase of major tidal constituents, with average relative errors for M_2 (S_2) of −1.5 (+1.3) cm and 3.3° (5.8°), respectively (Aleynik et al., 2016).

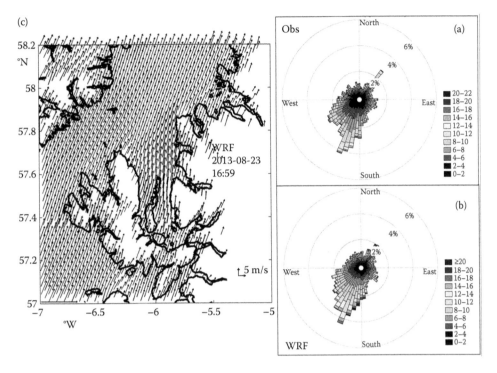

FIGURE 3.4
Wind rose diagrams for Stornoway Airport weather station subsampled at 3-hour intervals from UKMO-Midas hourly data set (a) and from Scottish-WRF model output (b). The left panel shows every fourth vector in the wind field at 10 m above sea level averaged over the period 16–30 September 2013 at the Northern part of the domain (Minch).

Combined elevation differences between observed (A_o, P_o) and modelled (A_m, P_m) values was calculated following Foreman et al. (2006):

$$DF = [(A_o cos P_o - A_m cos P_m)^2 + (A_o sin P_o - A_m sin P_m)^2]^{0.5}. \tag{3.13}$$

Average DF values for the nine gauged sites for M_2 (S_2) are 8.6 (8.1) cm, respectively.

The amplitude and phase of the modelled M_2 are plotted on a cotidal chart in Figure 3.5a. The figure also includes M_2 amplitudes from IHO tables. The overall distribution of the main tidal harmonics matched the IHO values with minimal deviation: for M_2 and S_2 it was −0.1 and +3.1 cm, respectively (Aleynik et al., 2016). The WeStCOMS–FVCOM model also correctly predicted the location of the tidal amphidrome near the North Channel and increasing K_1 amplitude towards the Outer Hebrides and the North Minch.

Observed sea surface elevation time series at sites along the west coast of Scotland mostly demonstrate very good agreement with our model predictions for tides (after the removal of surge signals). Full analysis is detailed for M_2 and S_2 harmonics (with Taylor diagrams) in Aleynik et al. (2016).

Currents: Weekly averaged residual surface currents (run summer–autumn 2013) give a generally northward flow with speed around 6–7 cm · s^{-1} in the open parts of the basin. Residual currents during a spring tidal period for August 2013 are shown in Figure 3.5b. These results are in general agreement with estimates of radio-nuclide tracer spread, originating from Sellafield (McKinley et al., 1981; McKay et al., 1986). In these studies,

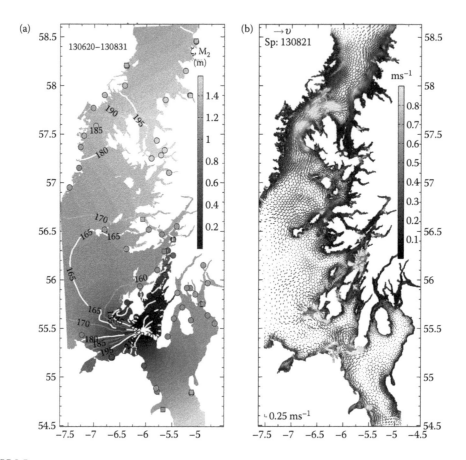

FIGURE 3.5

(a) Cotidal charts of sea surface elevation for M_2 tidal harmonic from WeStCOMS-FVCOM for summer 2013 run. Phase in degrees (white lines) and amplitude in metres (grey). Circles: corresponding amplitudes from International Hydrographic Office (IHO) Port harmonic analyses; squares: BODC pressure-gauge sites. The grey colour scheme for sites of both types is matched with the model's to aid comparison. (b) Residual surface currents calculated for duration of 7 days centred at spring tide period in August 2013.

flow through the Sea of the Hebrides was 2–3 km $\cdot$ day^{-1} (2.3–3.5 cm $\cdot$ s^{-1}) and around 4–5 km $\cdot$ day^{-1} (4.6–5.7 cm $\cdot$ s^{-1}) in the Minches. WeStCOMS-FVCOM residual currents in spring tides are slightly (1–2 cm $\cdot$ s^{-1}) higher than during neap tides, while the difference of the averaged maximum residual currents (in narrow sounds, in the Strait of Corryvreckan, for example) exceeded 10–20 cm $\cdot$ s^{-1}.

Direct current measurements are available for the layer 34 m below the surface from the SAMS mooring station at the Tiree Passage, where the RCM-9 was set at a depth of 11 m above the seabed. Model tidal parameters generally slightly overestimate observed eastern (northern) U (V) current speeds, e.g., the M_2 model exceeded the observed values on 5.9 (1.3) cm $\cdot$ s^{-1} and 6° (11°) for the tidal phase. Currents fluctuate along the channel and the major (M_2 component) axis of observed (modelled) tidal currents is equal to 59.5° (54.4°), which is in good agreement with longer-term estimates (57°) (Inall et al., 2009). The daily averaged time series of the residual currents are shown in Figure 3.5a. The correlation coefficients between observed and model U(V) components are high and equal 0.80 (0.86), which confirm the good response of the model to the influence of the lower-frequency atmospheric (wind) signal.

Temperature and salinity: Seasonal temperature cycles are well captured by the model, evident through a comparison of modelled and observed time series from several mooring sites (Figure 3.6c and d). The best model performance (averaged Wilmott index of agreement $d = 0.90$) was obtained for the higher-resolution ASIMUTH domain at coastal stations in the central part of the domain (temperature loggers at Foram and Saulmore) and

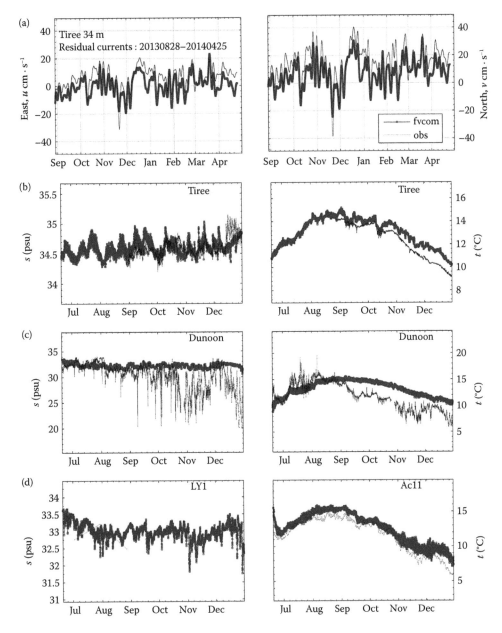

FIGURE 3.6

Time series of observed data (thin lines) and model (thick lines): (a) current velocity (u, v) components at Tiree, depth 34 m in September 2013–April 2014; (b) salinity (left) and temperature (right) at Tiree, 33 m; (c) Dunoon, 0 m; (d) CTD casts at LY1 station (salinity, left) and Access-11 0.5 m moorings (temperature, right), for June–December 2013. Thick lines: FVCOM model output for the same locations.

for the Tiree Passage mooring. In the wider, and relatively coarse, WeStCOMS-FVCOM domain, the average correlation with measured temperatures decreased from 0.91 to 0.77, and Wilmott's index of agreement decreased from 0.89 to 0.64.

3.3.3 HAB Model

3.3.3.1 Drifter Tracking Simulations

Full residence time simulations: Overall northward displacement of virtual particles was close to the observed track of Drifter 23, drogued at 15 m (Figure 3.7). In conjunction with spatially varying wind fields, the strong tidal and turbulent flows found in our model domain mean one would expect deviations between real and virtual drifter tracks. However, the location of the VP cloud centre should remain close to observed drifter locations within a short time.

Daisy chain simulations: Figure 3.7b shows contours encircling 50% (inner contour) and 95% (outer contour) of the VPs released each day over a fortnightly period along the track of Drifter 23. The real drifter trajectory usually exits the VP trajectory cloud within a few days in calm conditions, or more rapidly (1–1.5 days) in more dynamic environments.

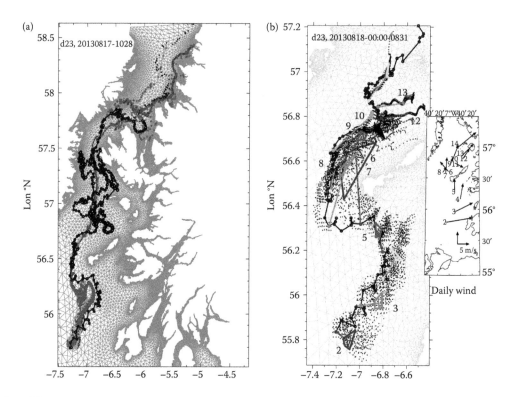

FIGURE 3.7

(a) Observed track for FASTNEt Drifter 23 (black dots) and 11 virtual particle (VP) trajectories released near the 'entrance' location and traced for 71 days within WeStCOMS FVCOM model. Light grey dots refer to all VPs initiated from a cloud within one-third distance to the next observed drifter fixing (1 hour 'radius'), connected dark grey points refer to VPs released from centre position within a cloud. (b) Daisy chain experiment for Drifter 23 (black line) and 11 VP trajectories (dots) are shown for the first 2 weeks of its tracking. The contours are shown, encircling 50% (inner) and 95% (outer) of the final positions for synthetic VPs after a day from restart, their spreading is grey-colour coded daily. (c) The average daily wind is shown with vectors for the location of synthetic drifters' centre of mass.

In our experiments the separation distance between real drifter fixes and the centre of the probability cloud remained in a range of 3–9 km. Adjustments to model drogues (to account for wind response and friction due to geometrical characteristics) were proposed by Edwards et al. (2006). This produced a limited improvement in the numerical drifter performance (separation distance reduced to 2.7–8 km).

There is good general agreement between the model VPs and observed drifter tracks. One would not expect a real drifter inhabiting a chaotic environment to perfectly match the trajectories of modelled particles over long timescales, with any deviations in trajectory being amplified by differences in oceanographic conditions encountered. Nonetheless, drifters provide a useful way to ground-truth the general patterns of the movement component of the physical model.

3.3.3.2 Karenia mikimotoi *Simulations*

High concentrations of (chlorophyll a CHL-A) were evident in the waters west of Scotland in late summer 2011. The extent of this is shown in combined *MODIS-MERIS* satellite data from MyOcean (Saulquin et al., 2011), presented in Figure 3.8a. Such conditions can indicate a high probability of the presence of a harmful bloom. Local sampling confirmed a bloom of *K. mikimotoi* with cell densities in samples from lochs on the south-east coast of the Isle of Mull in excess of 100,000 cells L^{-1}.

In this context, we evaluated the ability of the model to simulate a harmful bloom event using (1) particle tracking only and (2) the full *K. mikimotoi* movement, growth and mortality model. Simulations were restricted to the ASIMUTH-FVCOM model domain because of its higher resolution in the region of interest and the availability of directly applicable boundary forcing from the Irish NEA-ROMS model.

Particle tracking model: The Lagrangian tracking module of the HAB model was used, releasing 40 neutrally buoyant particles at the centre of the CHL-A maximum, both at the surface and near the seabed, at the beginning of the HAB event (Figure 3.8b). Surface particles experienced tidal oscillations and propagated towards the north-east, almost completely leaving the model domain within the same time scale (4–5 days) as the *K. mikimotoi* patch disappeared in the same direction. Near-bottom particles remained in the area for much longer (a few weeks). There is general agreement with the observed pattern of the August–September 2011 HAB event in the Tiree Passage and off the west coast of the Isle of Mull. However, the transport model alone cannot describe the density of cells over time.

3.3.3.3 *Biological Model Integration Results*

CHL-A satellite data derived from the www.MyOcean.eu website were also used for forcing the *K. mikimotoi* biological model. Daily data, converted into cell $\cdot$ L^{-1} in a similar fashion to Gillibrand et al. (2014), provided an initial condition on 26 August 2011, a day before the main *K. mikimotoi* event within the model subdomain. The evolution of cell concentrations in the surface layer over the subsequent 3 days is shown in Figure 3.8c. This HAB event was well captured by the model, albeit with small biases in intensity and duration. Again, we note a general agreement with the observed horizontal pattern of spreading of the HAB in the model domain.

3.3.4 General Particle-Tracking Model

We now turn to our general investigation of the factors influencing particle spread more generally within the model domain.

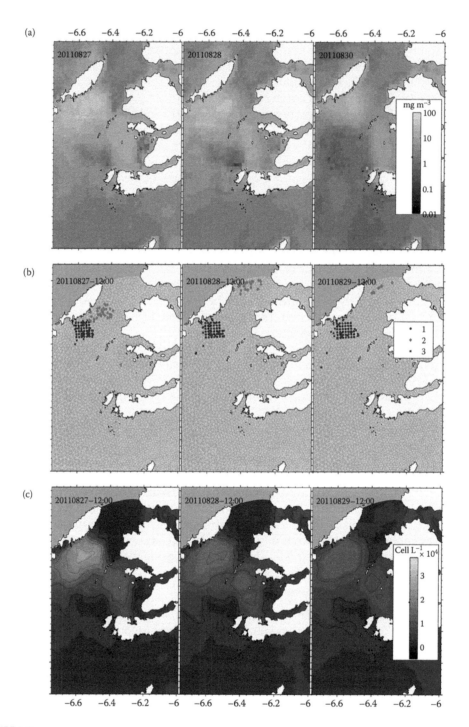

FIGURE 3.8
(a) Satellite observations of chlorophyll a concentrations (CHL-A_MODIS_MERIS_L4-RAN-OBS merged MyOcean product, 0.01mg m^{-3}) during the primary HAB event in the Tiree Passage, 27–29 August 2011. (b) Three snapshots of Lagrangian tracking of 40 neutral buoyant virtual particles released from surface (light dots) and near bottom (dark dots) during HAB event. Initial VP positions are indicated by circles. (c) *Karenia mikimotoi* bloom modelling results (concentration of species cells L^{-1} in surface layer).

3.3.4.1 Habitat Structure and Larval Duration

Organisms may have similar biological dispersal characteristics but occupy different environments (whether by 'choice' or circumstance). The structure of available habitat for attached adult stages differs greatly between habitat types. Consider the case of an organism with fixed dispersal characteristics (2-day larval lifespan, surface dwelling, constant mortality rate of 0.01 h^{-1}). Dispersing from the three different environments gives a rather different picture. The average probability of successful dispersal is 0.33 in the coastal habitat case, 0.07 in the fish farm case and 0.44 in the subtidal soft sediment case, reflecting the relative abundance of each habitat type. The spread of particles is distinctly different in the three cases, with more ubiquitous habitat leading to a more even spread of particles. However, common features are evident (2-day duration case: Figure 3.9). In particular, there are areas of high particle density in all cases in principal current systems, such as the south coast of Mull and the northward current around Skye and into the Northern Minch area.

Larval duration plays a very important role in determining how far particles migrate. A long larval duration gives more time for prevailing currents to transport larvae, but it may also give a longer window for meteorological variation to allow escape from the confines of embayments and narrow sea lochs. As a consequence, the total distance travelled from natal habitat is generally longer for particles with longer larval durations (Shanks, 2009). Long-range connectivity between habitat sites is higher. This is at the expense of short-range connectivity (larvae adapted to long dispersal are less likely to be fully developed for settlement at early times). Average distance travelled by all particles, including both successful and unsuccessful dispersers, in each habitat was broadly similar. However, successfully settled particles in the fish farm scenario typically travelled shorter distances than settled particles in other habitats, particularly in the case of 14-day dispersal (the

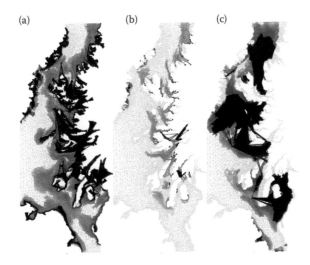

(a)   (b)   (c)

FIGURE 3.9

Spatial distribution and connection networks of dispersing particles with a 2-day larval duration from three different habitat scenarios: (a) coastal habitat, (b) salmon aquaculture sites and (c) subtidal soft sediment. Relative density (particle-hours) over the duration of a simulation is indicated by the shading of hydrodynamic model mesh elements. Black arrows indicate connections between sites; thicker lines/larger wedges indicate higher connection probability. In some cases wedges are longer than the connection line; their tip is located at the destination site.

separation of connected sites was lower, not shown). This is probably because long-distance dispersers locate sparse fish farm habitat sites with very low probability.

Larger areas of the sea are covered by particles dispersing for long durations. However, for a given level of reproductive output, this also implies dilution of a particular source's larvae over many destinations, resulting in a well-mixed population. Mortality may also affect long dispersing particles more acutely due to predation risk, a factor that is rather difficult to measure in practice.

The resultant connection networks vary distinctly between the habitat scenarios. In the fish farm scenario, sites are generally weakly connected, particularly at long ranges (Figure 3.9b, but also in the 14-day dispersal case). However, inter-connectivity of more ubiquitous habitat is much more extensive. For short dispersal, coastal habitat sites are generally connected in a fairly continuous fashion (Figure 3.9a). For the subtidal mud sites, there are generally groups of strongly connected sites, with somewhat weaker links to other areas. In the case of 14-day dispersal, particles spread over much longer distances and sites in the coast and subtidal habitat cases become very strongly connected.

Connections between sites are directional, dependent on the region's hydrodynamics. This is illustrated by arranging the sites from north to south and plotting pairwise connectivity between sites in a matrix (Figure 3.10). In this case, movement is generally in a northerly direction, with a corresponding bias in the direction of connections.

Sites with high connectivity tend to occur in clusters, reflecting the geography of the region (separate lochs, for example). Long-range connectivity is relatively uncommon, though it may be noted that certain sites have a tendency to act as a destination for larvae from many sources (horizontal lines in figure) or to act as a source of larvae for many other sites (vertical lines in figure). Depending on specific objectives, such as ensuring linkages across the network of sites (MPAs), or restricting them (fish farms), these factors would influence management decisions relating to sites.

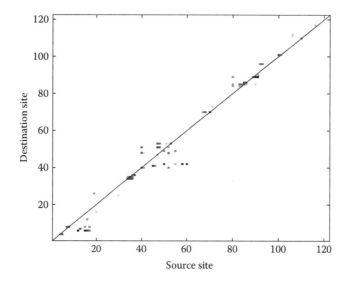

FIGURE 3.10
Pairwise connectivity between salmon aquaculture sites (numbered south to north) for particles with 2-day lifespan. The majority of connections are close to the leading diagonal, with highly connected regions of Loch Fyne and Loch Linnhe being notable.

3.3.4.2 Wind Forcing

Meteorological conditions at the time of initial release and onward throughout their dispersal can play a very important role in the fate of larvae. Identical particles released at different dispersal start dates have rather different fates. This is illustrated in Figure 3.11.

Areas moved through, distance travelled, width of 'plumes', and ultimate pattern of connectivity are all affected by wind conditions throughout the dispersal period. Note in particular the contrasting difference in areas of spread from sites in the central eastern shores of the domain between the two periods (increased range in the south-easterly wind case) and those in the north-western areas of the domain (increased range in south-westerly wind case). Dispersal distances and success rate may be increased or decreased by particular wind conditions, though the nature and magnitude of the effect will depend very much on the location of the site in question. In the example given in Figure 3.11, total connectivity between sites (probability of successful dispersal) decreases from 0.074 to 0.070 (that is, around 5%) between the SE and SW wind scenarios. However, in the full coastline scenarios, the same switch in wind periods instead gives an increase of around 10% (SE wind: 0.330, SW wind: 0.363). These differences underline the need for careful (and often detailed and specific) modelling when investigating dispersal in particular environments, especially complex coastal areas.

3.3.4.3 Diffusion and Velocity Computation

Turbulent diffusion in marine environments causes increased dispersal and broadening of plumes of tracer propagules such as larvae, dyes and effluents (Thorpe, 2012). In particle-tracking simulations, a fixed value for turbulent diffusion (over both space and time) is generally taken. However, the impact of this choice on model behaviour, or its reflection of reality, is rarely investigated. This is partly due to the difficulties involved in the tracking of larvae (or other particles) in the physical environment. However, it is possible to test model sensitivity to variations in the magnitude of the diffusion parameter.

(a)   (b)

FIGURE 3.11
Wind forcing. Spatial distribution and connectivity for surface-dwelling particles with 2-day lifespan under two contrasting wind scenarios: (a) light southerly/south-easterly wind and (b) moderate south-westerly wind. Black dots indicate the sites from which particles are released, grey shading indicates relative number of 'particle time steps', and black arrows/lines indicate connections between sites.

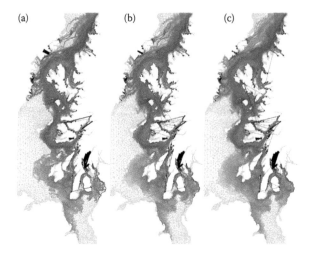

FIGURE 3.12
Spread of particles and connection network (fish farm sites, 14-day dispersal), with turbulent diffusion parameter: (a) $D_h = 0$, (b) $D_h = 0.1$, (c) $D_h = 10$.

It was found that varying this parameter had very little impact on model results. The distribution of particles over the domain remains very similar, and the connection network predicted is not altered dramatically by using either a 'low' value of $D_h = 0.1$ or a value 100 times higher (though some additional low-probability long-range connections do appear; see Figure 3.12). Total connection probabilities are unchanged, and dispersal kernels are not affected (not shown).

Computing particle velocity using an inverse-square weighted average of its containing hydrodynamic mesh element and neighbouring elements (sharing sides), as opposed to the velocity of the containing element alone, results in a reduction in overall connectivity. In the fish farm scenario, this is around 16% for a 2-day dispersal and 8% for a 14-day dispersal. In general, the relative connectivity of individual sites under each computation method is not altered. The pattern of connections or spread of particles also does not change notably (not shown). However, some sites (particularly in the shorter 2-day dispersal case) experienced altered outflux or influx of particles, as the average computation method has the potential to alter particle transport direction in complex inshore areas, creating or removing (generally short-range) connections in very few cases.

3.3.4.4 Vertical Migration

In general, surface currents are higher and change more in response to meteorological conditions (especially wind) than do bed currents. This is reflected by a general rule that particles spending more time in surface layers tend to travel further from their source location (Figure 3.13).

3.3.4.5 Salinity Driven Mortality

Introducing the salinity-driven mortality relationship derived from data presented by Bricknell et al. (2006) has the potential to alter both total and site-specific connectivity (Figure 3.14). In all cases, the variable mortality rate derived from the data gives much lower levels of connectivity than the fixed rate assumed in previous sea lice connectivity

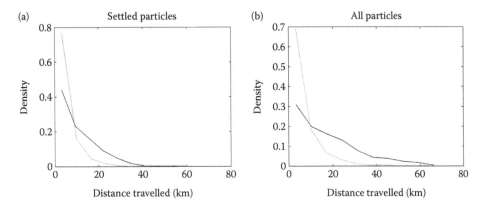

FIGURE 3.13
Dispersal kernels for particles occupying different depths (2-day dispersal, coastal habitat). Black lines: surface-dwelling particles; grey dashed lines: surface/bed alternating particles. (a) Distance travelled by particles successful at finding settlement habitat; (b) distances travelled by all particles. Particles remaining at the surface generally travel farther from their point of origin.

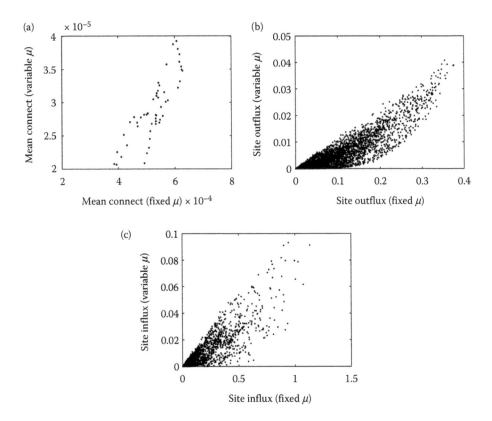

FIGURE 3.14
Variable mortality. The impact of varying mortality spatially according to local surface salinity on connectivity between salmon aquaculture sites (14-day dispersal duration), in comparison with equivalent fixed-rate mortality, for simulations beginning on each of 51 different days. (a) Average inter-site connectivity. (b) Outflux of individual sites. (c) Influx at each individual site.

studies. In particular, many sites that have moderate success in producing particles that later settle at other sites (outflux) have almost zero success in producing successful dispersers when salinity dependence is added. If larvae produced at low-salinity locations (close to the head of sea lochs, for example) do die more quickly as a result of exposure to freshwater, this would mean that these sites play a smaller role in the spread of lice larvae than existing published modelling studies would suggest.

3.3.4.6 Further Applications: Intertidal Organisms and Novel Offshore Habitat

In a previous study, we used the particle-tracking model in conjunction with a subdomain of the hydrodynamic model detailed in this chapter to assess the potential spread of intertidal organisms. This study had two key goals. Firstly, we sought to identify how generic coastal features related to dispersal potential (success of offspring produced, likely dispersal distance, and number of arriving larvae) (Adams et al., 2014a). Secondly, we investigated whether novel offshore structures (such as those associated with renewable energy

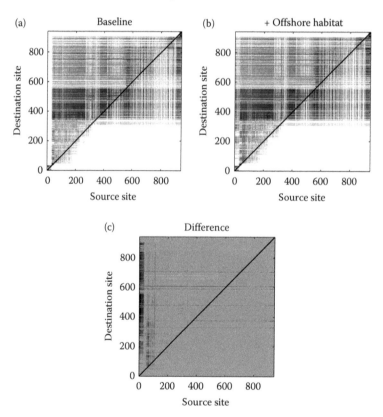

FIGURE 3.15
Multi-step shortest path connection probability of 940 equally spaced coastal habitat sites in ASIMUTH subdomain (arranged south to north; shading indicates relative path probability). (a) Baseline scenario (connections between coastal sites alone), (b) with addition of 352 habitable offshore sites (Adapted from Adams, T. P. et al. (2014b). *Journal of Applied Ecology*, 51, 330–338), and (c) difference between the two cases (black: positive change, grey: no change, white: negative change). Note in particular source sites in range [0, 40]. These represent sites on Ireland's northern coast; remaining sites are those on Scottish coastlines. In the baseline case, no connections between these sites and Scottish sites are made. But in the case including offshore habitat, these connections are some of the strongest in the network.

developments) were likely to alter dispersal pathways by providing 'stepping stones' across pre-existing biogeographical barriers (Adams et al., 2014b).

We found that open areas such as headlands and isolated islands produced larval particles that tended to travel further but were less likely to successfully 'locate' downstream habitats (these being effectively diluted by proximity to open water). Larger numbers of arriving larval particles are generally found in enclosed areas. However, we found that meteorological conditions defied definitions of precise relationships; the location of a particular site and prevailing wind conditions were very important in determining the number of arriving larvae in particular.

We identified a biogeographical barrier between the northern Irish coastline and Scottish coastlines, created by the presence of strong flows parallel to the coast (northward through the channel at the south end of the model domain here). The lack of various species endemic in Ireland on Scottish coastlines was noted many years ago by Lewis (1964); the model provides a possible explanation for this.

The model predicts that species currently resident on northern Irish coastlines but unable to reach Scotland owing to pelagic dispersal limitation would become able to cross to Scotland in the presence of novel hard substrate (particularly those with dispersal durations between 2 and 8 days) (Figure 3.15). The spatial distribution of stepping-stone sites and connecting coastal sites is possible using this approach and has obvious implications for the monitoring of novel infrastructure for the presence of particular non-native invasive species (some of which, e.g., *Crassostrea gigas* L., are found frequently in Ireland but rarely in Scotland).

3.4 Discussion and Conclusions

We have presented a hierarchical combined biophysical modelling approach for the study of organisms with pelagic dispersal in coastal marine environments. Our study region (the west coast of Scotland) has much in common with many other coastal regions in terms of geography, oceanography, ecology and local economy. As such, while the implementation uses a specific representation of the local topography, the methods that we presented here are applicable to any other region facing similar issues. Indeed, combined biophysical models have been developed for very varied applications in coastal environments the world over. Our hierarchical framework simplifies the analysis and validation of the model components and makes for more efficient computation. Using output from hydrodynamic models, biological models may be adapted to tackle a huge range of management challenges and issues involving dispersal. The framework may be extended and applied to develop spatio-temporal population models including both dispersing larvae and attached/sessile adult organisms (Bendtsen and Hansen, 2013; Adams et al., 2015). Such models can be parameterised directly from observational data and provide insights into the role of sites within networks of reproducing adult organisms (such as sea lice at host salmon farms) in an experimental context. In the HAB case, established models such as those detailed here describe fully the population dynamics of organisms solely inhabiting the water column. In doing so, the complete range of physical and biological processes may be taken into account to tackle in managing these issues.

The advent of unstructured grid finite-volume hydrodynamic models provides exciting opportunities for marine science. The advantages have already been mentioned, but they are worth reiterating: these models allow representation of fine-scale features and narrow channels without the expense of unnecessarily high-resolution computation in less complicated areas of open water. This makes the simulation of regions involving processes operating at a range of spatial scales (from metres to kilometres) much more realistic. For the first time, this means that the impact of large-scale processes (such as regional meteorological events and dominant broad-scale current patterns) on physical processes, ecological populations, and human activities in complex near and inshore environments may be investigated directly in a coherent modelling framework.

Modelling HABs involves many varied challenges. Model initiation is based upon spatially and temporally specific satellite data. As a result, it may be hampered by cloud cover, which renders data absent, incomplete, or difficult to analyse in the region/timeframe of interest. Should a clear view of ocean colour be available, there remains the challenge of discriminating between harmful and benign organisms. In spite of much empirical and modelling work on HABs, their life cycle remains poorly understood. There are several key areas in which understanding might usefully be advanced. The factors controlling cell death (internal 'programming' versus external factors) (Franklin et al., 2006) are not always clear, leading to difficulties in modelling the mortality process. Similarly, mixotrophy (ability to mix different sources of energy) (Burkholder et al., 2008) is poorly understood, while phototaxis (movement in response to light) (Brand et al., 2012), in common with many other marine organisms, is difficult to measure in the field.

Connectivity analysis methods are now well established in marine ecology, with studies having been carried out in diverse environments and with diverse study species. Assuming the adoption of a modelling framework sufficient to resolve detail in coastal and bathymetric topography, the key limiting factor in the study of particular species is our knowledge of larval behaviour. As demonstrated by our simple vertical placement example, larval behaviour can have profound implications for the fate of larvae. More complicated behaviour certainly takes place in reality. Behaviour of the models implemented is robust to many of the assumptions that are built in during development. However, it is clear that some assumptions have the potential to substantially modify results. We demonstrated here that, while approximately maintaining overall behaviour, salinity-dependent mortality might alter the role of particular sites within the connected network. The implications of such a finding may affect the interpretation of previous studies in systems involving environmentally sensitive organisms and act as a guide to future research investigations. Differences in how particles of varying size, mass and buoyancy are transported by moving water (McManus and Woodson, 2012) are also not often taken into account, which could also affect the results obtained. Nevertheless, modelling the movement of marine larvae via frameworks such as ours is a monumental step forward from simplistic distance-based assumptions (particularly in complex coastal areas) and is likely to be subject to refinement rather than fundamental alteration.

Having accounted correctly for potential larval connectivity between sites, there remains the question of what this means in practical terms for biological populations. What is the population density of particular organisms at particular locations? What is the probability of persistence at particular sites? To understand this, we must also consider the full life cycle of organisms, including the adult stages that produce larvae in the first place (and are created by settling larvae). Burgess et al. (2014) described a framework for analysing site-level persistence and population density in the context of a full connected network of populations, necessarily including three components: (1) an assessment of potential connectivity between sites (what we consider here), (2) the 'lifetime reproductive value' of successful recruits (does an average adult replace itself during its lifespan), and (3) the probability of successful establishment of arriving larvae. Field measurements are important for obtaining this information. Burgess et al. (2014) also discussed how inter-site connectivity could be analysed (or validated) using microchemistry techniques.

Validation of model output remains a challenge. We have described various approaches to validating the hydrodynamic model. Of these, assessing the model representation of currents is particularly difficult owing to small-scale fluctuations and subgrid-scale turbulent diffusion, though for the purpose of understanding dispersal, this is the feature of the model in which we are most interested. The validation of the biological components of such models is an ongoing challenge that requires both laboratory and extensive field studies involving genetic analysis (Dyhrman et al., 2006; Sammarco et al., 2012), cyst bed mapping (McGillicuddy Jr et al., 2011), the study of nutrient or environmental interactions (Gowen et al., 2012), or microchemistry techniques (Hogan et al., 2014), though no technique is necessarily applicable or useful in all species. Our previous work on this topic yielded some success in comparing predicted arrival levels with empirically observed population densities, though more complete population models should consider the full life cycle of organisms (as separate models driven by connectivity matrices; see earlier examples). As the field becomes more established, we expect that such studies will become more commonplace.

In summary, our model framework provides an adaptable and robust methodology for investigating how environmental processes affect the dispersal of marine organisms. It also provides a rapid and cost-effective approach to answering many of the questions faced by ecologists and local industries alike, even in situations where ecological data are sparse or unavailable.

Acknowledgements

The WeStCOMS Modelling System development was made possible by support from EU FP7 projects *ASIMUTH* (Grant 261860), *HYPOX* (226213) and EU HORIZON 2020 *Aquaspace* (633476); Marine Scotland Science and the European Fisheries Fund through the 'Sea Lice Dispersal in the North Minch' project; and NERC Programmes Shelf Seas Biogeochemistry (NE/K001884/1), MultiMARCAP (NE/L013029/1) and FASTNeT (NE/I030224/1). The WeStCOMS forecast system became operational in 2015 and distributes results via its portal http://www.habreports.org, which is supported by BBSRC/NERC WindyHABs (BB/M025934/1) and the European Maritime and Fisheries Fund. Model assessment observational data were provided by SAMS Physics and Technology Dept., NERC Facility for Scientific Diving (Dr M. Sayer) and SEPA. The Irish Marine Institute kindly supplied lateral open boundary conditions and RDA (NCEP, 2000) provided weather model forcing. SAMS Samhanach and ARCHER/Cray teams facilitated all required high-performance computing resources.

References

Adams, T. P., Black, K., Macintyre, C., Macintyre, I. and Dean, R. (2012). Connectivity modelling and network analysis of sea lice infection in Loch Fyne, west coast of Scotland. *Aquaculture Environment Interactions*, 3, 51–63.

Adams, T. P., Aleynik, D. and Burrows, M. T. (2014a). Larval dispersal of intertidal organisms and the influence of coastline geography. *Ecography*, 37, 698–710.

Adams, T. P., Miller, R. G., Aleynik, D. and Burrows, M. T. (2014b). Offshore marine renewable energy devices as stepping stones across biogeographical boundaries. *Journal of Applied Ecology*, 51, 330–338.

Adams, T. P., Proud, R., and Black, K. (2015). Connected networks of sea lice populations: Dynamics and implications for control. *Aquaculture Environment Interactions*, 6, 273–284. doi:10.3354/aei00133

Aiken, C. M., Navarrete, S. A., Castillo, M. I. and Castilla, J. C. (2007). Along-shore larval dispersal kernels in a numerical ocean model of the central Chilean coast. *Marine Ecology Progress Series*, 339, 13–24.

Aleynik, D., Davidson, K., Dale A. C. and Porter, M. (2016). A high resolution hydrodynamic model system suitable for novel harmful algal bloom modelling in areas of complex coastline and topography. *Harmful Algae*, 53(3), 102–117. doi:10.1016/j.hal.2015.11.012.

Amundrud, T. L. and Murray, A. G. (2009). Modelling sea lice dispersion under varying environmental forcing in a Scottish sea loch. *Journal of Fish Diseases*, 32, 27–44.

Armstrong, E. M., Wagner, G., Vazquez-Cuervo, J. and Chin, T. M. (2012). Comparisons of regional satellite sea surface temperature gradients derived from MODIS and AVHRR sensors. *International Journal of Remote Sensing*, 33, 6639–6651.

Bendtsen, J. and Hansen, J. L. S. (2013). A model of life cycle, connectivity and population stability of benthic macro-invertebrates in the North Sea/Baltic Sea transition zone. *Ecological Modelling*, 267, 54–65.

Brand, L. E., Campbell, L. and Bresnan, E. (2012). Karenia: The biology and ecology of a toxic genus. *Harmful Algae*, 14, 156–178.

Bricknell, I. R., Dalesman, S. J., O Shea, B., Pert, C. C. and Mordue Luntz, A. J. (2006). Effect of environmental salinity on sea lice Lepeophtheirus salmonis settlement success. *Diseases of Aquatic Organisms*, 71, 201–212.

Burgess, S. C., Nickols, K. J., Griesemer, C. D., Barnett, L. A. K., Dedrick, A. G., Satterthwaite, E. V., Yamane, L., Morgan, S. G., White, J. W. and Botsford, L. W. (2014). Beyond connectivity: how empirical methods can quantify population persistence to improve marine protected-area design. *Ecological Applications: A Publication of the Ecological Society of America*, 24, 257–270.

Burkholder, J. M., Glibert, P. M. and Skelton, H. M. (2008). Mixotrophy, a major mode of nutrition for harmful algal species in eutrophic waters. *Harmful Algae*, 8, 77–93.

Burrows, M. T., Harvey, R., Robb, L., Poloczanska, E. S., Mieszkowska, N., Moore, P., Leaper, R., Hawkins, S. J. and Benedetti-Cecchi, L. (2009). Spatial scales of variance in abundance of intertidal species: effects of region, dispersal mode, and trophic level. *Ecology*, 90, 1242–1254.

Chen, C., Beardsley, R. C. and Cowles, G. (2011). *An Unstructured Grid Finite-Volume Coastal Ocean Model: FVCOM User Manual*. USA, University of Massachusetts-Dartmouth.

Chen, C., Liu, H. and Beardsley, R. C. (2003). An unstructured grid, finite-volume, three-dimensional, primitive equations ocean model: application to coastal ocean and estuaries. *Journal of Atmospheric and Oceanic Technology*, 20, 159–186.

Costello, M. J. (2006). Ecology of sea lice parasitic on farmed and wild fish. *Trends in Parasitology*, 22, 475–483.

Costello, M. J. (2009). How sea lice from salmon farms may cause wild salmonid declines in Europe and North America and be a threat to fishes elsewhere. *Proceedings - Royal Society. Biological Sciences*, 276, 3385–3394.

Costelloe, M., Costelloe, J., O'donohoe, G., Coghlan, N. J., Oonk, M. and Van Der Heijden, Y. (1998). Planktonic distribution of sea lice larvae, Lepeophtheirus salmonis, in Killary Harbour, west coast of Ireland. *Journal of the Marine Biological Association of the United Kingdom*, 78, 853–874.

Criales, M. M., Robblee, M. B., Browder, J. A., Cardenas, H. and Jackson, T. L. (2011). Field observations on selective tidal-stream transport for postlarval and juvenile pink shrimp in florida bay. *Journal of Crustacean Biology*, 31, 26–33.

Dabrowski, T., Lyons, K., Berry, A., Cusack, C. and Nolan, G. D. (2014). An operational biogeochemical model of the North-East Atlantic: Model description and skill assessment. *Journal of Marine Systems*, 129, 350–367.

Darkes, G. and Spence, M. (2008). *Cartography - An Introduction*. London, The British Cartographic Society.

Davidson, K. and Bresnan, E. (2009). Shellfish toxicity in UK waters: a threat to human health? *Environmental Health*, 8, S12.

Davidson, K., Miller, P. I., Wilding, T. A., Shutler, J., Bresnan, E., Kennington, K. and Swan, S. (2009). A large and prolonged bloom of Karenia mikimotoi in Scottish waters in 2006. *Harmful Algae*, 8, 349–361.

Davidson, K., Tett, P. and Gowen, R. (2011). Harmful algal blooms. In Harrison, R. M. and Hester, R. E. (eds.), *Marine Pollution and Human Health*, London, RSC Publishing, 95–127.

Davies, A. M. and Hall, P. (2002). Numerical problems associated with coupling hydrodynamic models in shelf edge regions: the surge event of February 1994. *Applied Mathematical Modelling*, 26, 807–831.

Davies, I. M. and Watret, R. (2011). *Scoping Study for Offshore Wind Farm Development in Scottish Waters*. Marine Scotland, Aberdeen, UK.

Davies, I. M., Gubbins, M. and Watret, R. (2012). *Scoping Study for Tidal Stream Energy Development in Scottish Waters*. Marine Scotland, Aberdeen, UK.

Dyhrman, S. T., Erdner, D., La Du, J., Galac, M. and Anderson, D. M. (2006). Molecular quantification of toxic Alexandrium fundyense in the Gulf of Maine using real-time PCR. *Harmful Algae*, 5, 242–250.

Edwards, A. and Sharples, F. (1986). *Scottish Sea Lochs: A Catalogue*. Oban, Scottish Marine Biological Association.

Edwards, K. P., Werner, F. E. and Blanton, B. O. (2006). Comparison of observed and modeled drifter trajectories in coastal regions: an improvement through adjustments for observed drifter slip and errors in wind fields. *Journal of Atmospheric and Oceanic Technology*, 23, 1614–1620.

Egbert, G. D., Erofeeva, S. Y. and Ray, R. D. (2010). Assimilation of altimetry data for nonlinear shallow-water tides: quarter-diurnal tides of the Northwest European Shelf. *Continental Shelf Research*, 30, 668–679.

Farrell, H., Gentien, P., Fernand, L., Lunven, M., Reguera, B., González-GIL, S. and Raine, R. (2012). Scales characterising a high density thin layer of Dinophysis acuta Ehrenberg and its transport within a coastal jet. *Harmful Algae*, 15, 36–46.

Fehling, J., Davidson, K., Bolch, C. J. and Tett, P. (2006). Seasonality of Pseudo-nitzschia spp. (Bacillariophyceae) in western Scottish waters. *Marine Ecology Progress Series*, 323, 91–105.

Fehling, J., Davidson, K., Bolch, C. J. S., Brand, T. D. and Narayanaswamy, B. E. (2012). The relationship between phytoplankton distribution and water column characteristics in North West European shelf sea waters. *PLoS ONE*, 7, e34098.

Foreman, M. G. G., Stucchi, D. J., Zhang, Y. and Baptista, A. M. (2006). Estuarine and tidal currents in the Broughton Archipelago. *Atmosphere-Ocean*, 44, 47–63.

Foreman, M. G. G., Czajko, P., Stucchi, D. J. and Guo, M. (2009). A finite volume model simulation for the Broughton Archipelago, Canada. *Ocean Modelling*, 30, 29–47.

Franklin, D. J., Bussaard, C. P. D. and Berges, J. A. (2006). What is the role and nature of programmed cell death in phytoplankton ecology?. *European Journal of Phycology*, 41, 1–14.

Gaines, S. D., Gaylord, B. and Largier, J. L. (2003). Avoiding current oversights in marine reserve design. *Ecological Applications*, 13, 32–46.

Gemmill, W., Katz, B. and Li, X. (2007). Daily Real-Time, Global Sea Surface Temperature High-Resolution Analysis: RTG_SST_HR. *Technical Note Nr.* 260. [Online]. Available from: http://polar.ncep.noaa.gov/mmab/papers/tn260/MMAB260.pdf.

Gentien, P., Lunven, M., Lazure, P., Youenou, A. and Crassous, M. P. (2007). Motility and autotoxicity in Karenia mikimotoi (Dinophyceae). *Philosophical Transactions of the Royal Society B: Biological Sciences*, 362, 1937–1946.

Gillibrand, P. A., Siemering, B., Miller, P. I. and Davidson, K. (2014). Individual-based modelling of the development and transport of a Karenia mikimotoi bloom on the North-West European continental shelf. *Harmful Algae*, 53, 118–134.

Gowen, R. J., Tett, P., Bresnan, E., Davidson, K., Mckinney, A., Milligan, S., Mills, D. K., Silke, J., Gordon, A. and Crooks, A. M. (2012). Anthropogenic nutrient enrichment and blooms of harmful micro-algae. *Oceanography and Marine Biology: An Annual Review*, 50, 65–126.

Heuch, P. A., Parsons, A. and Boxaspen, K. (1995). Diel vertical migration: A possible host-finding mechanism in salmon louse (Lepeophtheirus salmonis) copepodids?. *Canadian Journal of Fisheries and Aquatic Sciences*, 52, 681–689.

Heuch, P. A., Bjørn, P. A., Finstad, B., Holst, J. C., Asplin, L. and Nilsen, F. (2005). A review of the Norwegian 'National Action Plan Against Salmon Lice on Salmonids': The effect on wild salmonids. *Aquaculture*, 246, 79–92.

Hogan, J. D., Mcintyre, P. B., Blum, M. J., Gilliam, J. F. and Bickford, N. (2014). Consequences of alternative dispersal strategies in a putatively amphidromous fish. *Ecology*, 95, 2398.

Huang, H., Chen, C., Cowles, G. W., Winant, C. D., Beardsley, R. C., Hedstrom, K. S. and Haidvogel, D. B. (2008). FVCOM validation experiments: Comparisons with ROMS for three idealized barotropic test problems. *Journal of Geophysical Research: Oceans*, 113(C7), C07042, 1–14.

Inall, M. E., Gillibrand, P. A., Griffiths, C. R., Macdougal, N. and Blackwell, K. (2009). Temperature, salinity and flow variability on the North-West European shelf. *Journal of Marine Systems*, 77, 210–226.

Inall, M. E. et al. (2013). RRS James Cook Cruise JC088. FASTNEt Cruise to the Malin Shelf Edge. Internal Report, SAMS, Oban, 213 p. https://www.bodc.ac.uk/resources/inventories/cruise_inventory/report/13391/

Kirkpatrick, G., Mille, D. F., Moline, M. A. and Schofield, O. (2000). Optical discrimination of a phytoplankton species in natural mixed populations. *Limnology and Oceanography*, 45, 467–471.

Lewis, J. R. 1964. *The Ecology of Rocky Shores*. London, Universities Press.

McBreen, F., Askew, N., Cameron, A., Connor, D., Ellwood, H., and Carter, A. (2011). UK SeaMap 2010: Predictive mapping of seabed habitats in UK waters (No. 446). Peterborough: Joint Nature Conservation Committee. Retrieved from http://jncc.defra.gov.uk/PDF/jncc446_web.pdf

Mcgillicuddy Jr, D. J., Townsend, D. W., He, R., Keafer, B. A., Kleindinst, J. L., Li, Y., Manning, J. P., Mountain, D. G., Thomas, M. A. and Anderson, D. M. (2011). Suppression of the 2010 *Alexandrium fundyense* bloom by changes in physical, biological, and chemical properties of the Gulf of Maine. *Limnology and Oceanography*, 56, 2411–2426.

Mckay, W. A., Baxter, J. M., Ellett, D. J. and Meldrum, D. T. (1986). Radiocaesium and circulation patterns West of Scotland. *Journal of Environmental Radioactivity*, 4, 205–232.

Mckinley, I. G., Baxter, M. S., Ellett, D. J. and Jack, W. (1981). Tracer applications of radiocesium in the sea of the hebrides. *Estuarine Coastal and Shelf Science*, 13, 69–82.

McManus, M. A. and Woodson, C. B. (2012). Plankton distribution and ocean dispersal. *The Journal of Experimental Biology*, 215, 1008–1016.

Mellor, G. and Yamada, T. (1982). Development of a turbulence closure model for geophysical fluid problems. *Reviews of Geophysics and Space Physics*, 20, 851–875.

Mitarai, S., Siegel, D. A., Watson, J. R., Dong, C. and Mcwilliams, J. C. (2009). Quantifying connectivity in the coastal ocean with application to the Southern California Bight. *Journal of Geophysical Research*, 114(10), C10026.

NCEP. (2000). FNL Operational Model Global Tropospheric Analyses, continuing from July 1999, National Centers for Environmental Prediction/National Weather Service/NOAA/U.S. Department of Commerce, updated daily. Research Data Archive at the National Center for Atmospheric Research, Computational and Information Systems Laboratory.

North, E. W., Schlag, Z., Hood, R. R., Li, M., Zhong, L., Gross, T. and Kennedy, V. S. (2008). Vertical swimming behavior influences the dispersal of simulated oyster larvae in a coupled particle-tracking and hydrodynamic model of Chesapeake Bay. *Marine Ecology Progress Series*, 359, 99–115.

Parry, S., Muchan, K., Lewis, M. and Clemas, S. (2014). 2014 Hydrological summary for the United Kingdom: December 2013. Available: http://www.ceh.ac.uk/data/nrfa/nhmp/monthly_hs.html [Accessed 5 April 2014].

Robel, A. A., Lozier, S. M., Gary, S. F., Shillinger, G. L., Bailey, H. and Bograd, S. J. (2011). Projecting uncertainty onto marine megafauna trajectories. *Deep Sea Research Part I: Oceanographic Research Papers*, 58, 915–921.

Roberts, C. M., Hawkins, J. P., Flectcher, J., Hands, S., Raab, K. and Ward, S. (2010). *Guidance on the size and spacing of Marine Protected Areas in England*. Natural England, Sheffield, UK.

Salama, N. K. G., Collins, C. M., Fraser, J. G., Dunn, J., Pert, C. C., Murray, A. G. and Rabe, B. (2012). Development and assessment of a biophysical dispersal model for sea lice. *Journal of Fish Diseases*, 36(3), 1–14.

Sammarco, P. W., Brazeau, D. A. and Sinclair, J. (2012). Genetic connectivity in scleractinian corals across the northern gulf of Mexico: oil/gas platforms, and relationship to the flower garden banks. *PLoS ONE*, 7, e30144.

Saulquin, B., Gohin, F. and Garrello, R. (2011). Regional objective analysis for merging high-resolution MERIS, MODIS/Aqua, and SeaWiFS chlorophyll- a data from 1998 to 2008 on the European atlantic shelf. *IEEE Transactions on Geoscience and Remote Sensing*, 49, 143–154.

SeaZone Ltd. (2007). Hydrospatial, Digital Survey Bathymetry, Charted Vector and Charted Raster, User Guide V.1.1e. P 43. Available from SeaZone Ltd, Bentley, Hampshire, UK.

Shanks, A. L. (2009). Pelagic larval duration and dispersal distance revisited. *The Biological Bulletin*, 216, 373–385.

Shutler, J. D., Davidson, K., Miller, P. I., Swan, S. C., Grant, M. G. and Bresnan, E. (2011). An adaptive approach to detect high-biomass algal blooms from EO chlorophyll-a data in support of harmful algal bloom monitoring. *Remote Sensing Letters*, 3, 101–110.

Siegel, D. A., Kinlan, B. P., Gaylord, B. and Gaines, S. D. (2003). Lagrangian descriptions of marine larval dispersion. *Marine Ecology Progress Series*, 260, 83–96.

Simpson, J. H., Edelsten, D. J., Edwards, A., Morris, N. C. G. and Tett, P. B. (1979). The Islay front: Physical structure and phytoplankton distribution. *Estuarine and Coastal Marine Science*, 9, 713–726.

Skamarock, W. C., Klemp, J. B., Dudhia, J., Gill, D. O., Barker, D. M., Duda, M. G., Huang, X.-Y., Wang, W. and Powers, J. G. (2008). A description of the Advanced Research WRF version 3. *NCAR Technical Note*, 125 p, http://dx.doi.org/10.5065/D68S4MVH (electronic book).

Smagorinsky, J. (1963). General circulation experiments with the primitive equations. *Monthly Weather Review*, 91, 99–164.

Stien, A., Bjørn, P. A., Heuch, P. A. and Elston, D. A. (2005). Population dynamics of salmon lice Lepeophtheirus salmonis on Atlantic salmon and sea trout. *Marine Ecology Progress Series*, 290, 263–275.

Stigebrandt, A. (2012). Hydrodynamics and circulation of fjords. *Encyclopedia of Lakes and Reservoirs*, ed. L. Bengtsson, RW Herschy and RW Fairbridge, 327–344, Dordrecht, Springer Netherlands.

Stucchi, D. J., Guo, M., Foreman, M. G. G., Czajko, P., Galbraith, M., Mackas, D. L. and Gillibrand, P. A. (2010). Modeling sea lice production and concentrations in the Broughton Archipelago, British Columbia. In S. Jones and R. Beamish, eds., *Salmon Lice: An Integrated Approach to Understanding Parasite Abundance and Distribution*, pp. 117–150. Wiley-Blackwell, Oxford, UK.

Taylor, K. (2001). Summarizing multiple aspects of model performance in a single diagram. *Journal of Geophysical Research*, 106, 7183–7192.

Thorpe, S. A. (2012). On the biological connectivity of oil and gas platforms in the North Sea. *Marine Pollution Bulletin*, 64, 2770–2781.

Touzet, N., Davidson, K., Pete, R., Flanagan, K., Mccoy, G. R., Amzil, Z., Maher, M., Chapelle, A. and Raine, R. (2010). Co-occurrence of the West European (Gr.III) and North American (Gr.I) ribotypes of alexandrium tamarense (Dinophyceae) in Shetland, Scotland. *Protist*, 161, 370–384.

Turrell, W. R. (1990). *Simulation of Advection and Diffusion of Released Treatments in Scottish sea Lochs*. Aberdeen, DAFS Marine Laboratory.

Vanhoutte-Brunier, A., Fernand, L., Ménesguen, A., Lyons, S., Gohin, F. and Cugier, P. (2008). Modelling the Karenia mikimotoi bloom that occurred in the western English Channel during summer 2003. *Ecological Modelling*, 210, 351–376.

Wilmott, C. J. (1982). Some comments on the evaluation of model performance. *Bulletin American Meteorology Society*, 63(11), 1309–1313.

Zhao, L., Chen, C. and Cowles, G. (2006). Tidal flushing and eddy shedding in Mount Hope Bay and Narragansett Bay: An application of FVCOM. *Journal of Geophysical Research: Oceans*, 111(C10), C10015, 1–16.

4

Oceanic and Fisheries Response in the Northwest Pacific Marginal Seas with Climate Variability

Chung Il Lee, S. M. Mustafizur Rahman, Hae Kun Jung,
Chang-Keun Kang and Hyun Je Park

CONTENTS

4.1 Introduction 99
4.2 Teleconnected Pattern of Climate Forcing in Korean Marine Environment 101
    4.2.1 How Are Climate Variabilities Teleconnected with Each Other in the
        Northwest Pacific Ocean? 101
    4.2.2 How Does Direct Climate Forcing Occur in the Korean Waters? 106
    4.2.3 How Does Lag Oceanic Forcing Occur in Korean Waters? 108
4.3 Fishery Productivity and Ecosystem Dynamics Response to Climate Forcing 110
4.4 Concluding Remarks 114
Acknowledgements 115
References 115

4.1 Introduction

The marginal seas in the Pacific Northwest make a 25% contribution to world marine fishery capture and have become one of the most productive regions (FAO, 2012). Three contiguous marginal seas, the East Sea/Sea of Japan, the East China Sea (ECS), and the Yellow Sea (YS), make the Korean marine waters, which are the main sources of fishery resources also (Figure 4.1). These seas are vulnerable to extreme fishing pressures, higher anthropogenic activities and frequent climate-ocean forcing. Korean marine waters lie between a dry continental air mass and high humid oceanic air mass. They are frequently affected by the Yangtze River air mass, Okhotsk air mass, North Pacific air mass, and Siberian air mass, though they all have their own seasonal variability.

Marginal seas are highly sensitive to climate forcing because of their long continental slope, varying topography, closed current system, and relatively lower water exchange rate. They are subject to climate forcing just as much as open seas are, but they take more time to recover. The three seas in the Korean marine waters have received greater attention for their large-scale anthropogenic activities and alarming warming trend. Several types of air mass can affect the Korean marine waters: the continental dry air mass, the Yangtze River humid air mass, and Siberian dry cool air mass. Along with these local atmospheric forces, fluctuations in atmospheric and oceanic forces in the Northern Pacific and even at the equator also play a major role either directly or in indirect teleconnected patterns.

As with other East Asian marginal seas, the Korean marginal seas have also been subject to rapid warming since the 1980s (Belkin, 2009). The Kuroshio Branch Current (KBC) large marine ecosystem (LME) in the ECS warmed most rapidly in the period 1981–1998, when the

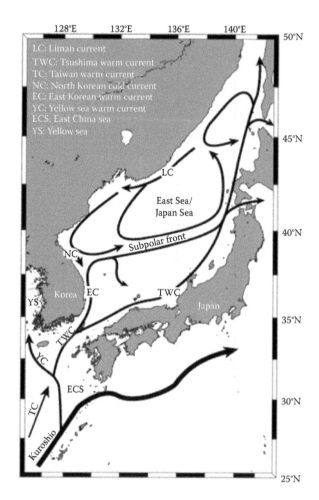

FIGURE 4.1
Current systems of three different marginal seas; East Sea/Sea of Japan, East China Sea and Yellow Sea of Korean marine waters.

sea surface temperature (SST) rose by 1.5°C. In the East Sea/Sea of Japan, the most rapid warming occurred in the period 1986–1998, when the SST rose by 2°C. Along with this long-term warming trend, seasonal and inter-annual variability in these seas' marine environments are also dominant features. However, decadal or long-term oscillation in relation to climate variability is relatively new to Korean researchers. Most importantly, the complex hydrodynamic features of the Kuroshio Current (KC), Taiwan Current (TC), and Tsushima Warm Current (TWC) in the ECS and the strong polar front in the East Sea/Sea of Japan with local and tele-connected atmospheric forcing make Korean marine waters highly in predictable.

Recent abrupt shifts have been detected both in Korean marine waters and fishery resources. Cold, warm, and cold regimes occurred respectively right after the 1976/1977, 1988/1989, and 1998 climate regime shifts (CRSs) (Rahman and Lee, 2012; Zhang et al., 2000; Kim et al., 2007; Jung et al., 2013; Kang et al., 2012). Zhang et al. (2000) talked about a regime shift in the surface and 50-m-depth temperature in the East Sea/Sea of Japan during mid-1976/1977 and 1988/1989 and for the mixed layer depth in the 1976/1977 CRS.

The East Sea/Sea of Japan is situated between subtropical and subpolar zones. It represents a miniature version of a large ocean because it has both wind-driven and buoyancy-driven boundary currents, a subpolar front, mesoscale eddies, intense air–sea interactions, subduction and deep convection, and topographic trapping (Mooers et al., 2006). Kim et al. (2002) discussed bottom water formation in the East Sea/Sea of Japan during the severe 2000–2001 winter, which can initiate a conveyor belt. The Tsushima Warm Current (TWC) in the southern region and the Liman Cold Current (LCC) in the northern region are two major currents in the East Sea/Sea of Japan, and they are divided into warm (southern) and cold (northern) regions, respectively, with the boundary (polar front) around 40°N (Figure 4.1). Inflow is primarily through the Korea/Tsushima Strait (KTS) in the south, and the outflow is primarily through the Tsugaru and Soya Straits in the east. The northern and southern regions of the East Sea/Sea of Japan are hydrographically and biologically distinct, with the southern region being more tropical/oligotrophic and the northern region being more boreal/eutrophic (Tian et al., 2008; Ashjian et al., 2005). The southern part of the sea is mainly influenced by the volume transport of the TWC and its branch current, the East Korea Warm Current (EKWC), which actually controls the upper-layer water properties (e.g., temperature), whereas the northern part is influenced by the LCC and its branch current, the North Korea Cold Current (NKCC). Spatial and temporal variability between the northern homogeneous cold deep water mass and the southern surface warm water are both associated largely with the dynamics of the ecosystem structure from lower to higher trophic levels (Tian et al., 2006, 2008).

Yeh et al. (2010) mentioned that atmospheric forcing was the primary cause of the low-frequency SST variability in the East Sea/Sea of Japan. They also mentioned that the large-scale circulation over the East Sea/Sea of Japan during winter is dominated by two atmospheric forcings, the Siberian High Pressure (SHP) system centred on the Asian continent and the Aleutian Low Pressure (ALP) system centred over the North Pacific. Changes in the intensity of these two pressure systems are related to changes in the magnitude of the wind blowing over the East Sea/Sea of Japan during winter. Therefore, the mean SLP gradient and total wind speed over the East Sea/Sea of Japan weakened, resulting in an increase in the East Sea/Sea of Japan SST due to a reduction in the sensible heat flux with weak entrainment of cold water beneath the mixed layer. In winter, strong northwesterly winds occurred by the East Asian winter monsoon (EAWM) pattern over the East Sea/Sea of Japan and weak southeasterly winds in summer (Kim et al., 2007).

The YS is characterized as a shallow waterbody (average depth of about 44 m) with mud flats and a low-saline, high-nutrient, semi-enclosed shelf sea. The average depth of the ECS is about 272 m and is mostly controlled by the TWC, the KBC, and the Changjiang River discharge. The ECS has only a weak cold water source (Kang, 1974) and is warmer than the East/Japan Sea and the YS (Kim and Kang, 2000). While warm and saline water enters from the ECS into the YS episodically in winter (Lie et al., 2001), cooler and fresher water originating from the rivers also enters the YS (Lee, 1998).

4.2 Teleconnected Pattern of Climate Forcing in Korean Marine Environment

4.2.1 How Are Climate Variabilities Teleconnected with Each Other in the Northwest Pacific Ocean?

Climate variabilities in the Pacific Ocean are linked with each other whether directly or in a teleconnected pattern. They transfer their signal to the marine environment only in

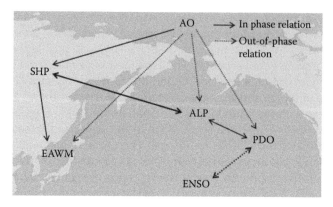

FIGURE 4.2
Schematic view of teleconnected pattern of major climate variabilities in the North Pacific Ocean.

coupled form. They can influence the upper-layer marine environment in two ways, by wind forcing, called atmospheric forcing, and by an upper-layer current, known as oceanic forcing. Typically the latter involves a time lag forcing. Both atmospheric and oceanic forcing is involved in Korean marine waters. Atmospheric forcing develops from the Arctic Oscillation (AO)–ALP-SHP-EAWM coupled system in all Korean waters (Figure 4.2), while oceanic forcing can develop from the ALP-Pacific Decadal Oscillation (PDO) pattern linked to the North Pacific gyre oscillation, which can regulate KC-TC-TWC dynamics and influence the ECS and the southern part of the East Sea/Sea of Japan (Gordon and Giulivi, 2004; Chiba et al., 2005; Rahman et al., 2014).

The SHP in wintertime plays an important role in climate at middle to high latitudes in Eurasia (Guo, 1996; Zhu et al., 1997; Gong and Wang, 1999; Miyazaki et al., 1999; Yin, 1999). It has a significant negative correlation with the mean surface temperature in mid- to high-latitude Asia (30°E–140°E and 30°N–70°N) (Gong and Ho, 2002). The AO, SHP, and EAWM are all highly teleconnected to each other in winter (Table 4.1). Though the AO has significant out-of-phase relationships with the EAWM, it influences the EAWM through its impact on the SHP (Gong et al., 2001). On the other hand, the SHP influences the Korean marine waters through its impact on the EAWM. There has been a significant negative correlation between the 10-m temperature of the East Sea/Sea of Japan marine waters and

TABLE 4.1

Correlation Statistics for Major Climate Variabilities in Pacific Ocean

	AO	SHP	PDO	ALP	EAWM	ENSO
AO	1	−.626**	.081	.070	−.549**	.269*
SHP	−.626**	1	−.255	−.455**	.192	−.290*
PDO	.081	−.255	1	.901**	.424**	.820**
ALP	.070	−.455**	.901**	1	.430**	.738**
EAWM	−.549**	.192	.424**	.430**	1	.069
ENSO	.269*	−.290*	.820**	.738**	.069	1

Note: ALP: Aleutian Low Pressure; AO: Arctic Oscillation; EAWM: East Asian winter monsoon; ENSO: El Nino Southern Oscillation; PDO: Pacific Decadal Oscillation; SHP: Siberian High Pressure.
* Significance at $P < 0.05$.
** Significance at $P < 0.01$.

TABLE 4.2

Correlation Statistics for Major North Pacific Climate
Variabilities and 10 m Sea Water Temperature in Different
Korean Waters

	SES	NES	ECS	YS
AO	.672**	.608**	.399**	.826**
SHP	−.467**	−.414**	.014	−.777**
PDO	−.396**	−.088	−.327*	−.019
ALP	−.300*	−.180	−.440**	.089
EAWM	−.671**	−.235	−.517**	−.477**
ENSO	−.069	−.105	−.102	.159

Source: Rahman, M.S.M. et al. (2014). Marine environment and fisheries
response to climate–ocean variability in Northwest Pacific marginal seas: a focus on Korean waters. Submitted to Fisheries
Oceanography. submitted to Fisheries Oceanography.

Note: ALP: Aleutian low pressure; AO: Arctic oscillation; EAWM: East
Asian winter monsoon; ECS: East China Sea; ENSO: El Nino
Southern Oscillation; NES: Northern part of the East Sea; PDO:
Pacific decadal oscillation; SES: Southern part of the East Sea;
SHP: Siberian high pressure; YS: Yellow Sea.

* Significance at $P < 0.05$.

** Significance at $P < 0.01$.

the SHP since the 1950s (Table 4.2). In the late 1980s, the AO changed to positive values,
the SHP changed to a negative phase, and, consequently, the EAWM decreased gradually (Figure 4.3). A negative EAWM indicates weak mid-latitude westerly winds and mild
winters in East Asia. The milder, calmer winters since the late 1980s may have resulted in
reduced wind mixing and convection in winter and thereby increased water temperatures
in the upper layer of the Korean marginal seas.

The positive phases of the ALP from the mid-1970s could have significant impacts on the
PDO through mid-latitude atmosphere–ocean interactions over the North Pacific region and
vice versa. Chiba et al. (2005) suggest that environmental variation at the surface layer in the
East Sea/Sea of Japan might be closely related to the PDO. Gordon and Giulivi (2004) investigated the long-term East Sea/Sea of Japan baroclinic sea level variability association with the
PDO. They found that sea surface height (SSH) relative to 200 dbar within the station (posi-
tioned south of the polar front at around 40°N in the East Sea/Sea of Japan) had a negative
correlation with the PDO. They also examined the close relationship between temperature
and salinity in association with the PDO – colder and saltier water when the PDO is positive, warmer and fresher water when the PDO is negative. They note that during the positive
phase of the PDO, the ALP becomes deeper and shifts to the south and the westerlies over
the North Pacific strengthen. This causes an increased southward Ekman transport, leading
to lower SST at mid-latitudes. They propose a negative correlation between the SSH the PDO
in the south East Sea/Sea of Japan owing to changes in the Kuroshio geostrophic transport,
which is weaker during a negative PDO (stronger during a positive PDO). They further point
out that the weaker phase of the KC would increase the amount of buoyant subtropical water
delivered to the East Sea/Sea of Japan, via the Korea Strait (KS), accounting for the higher
SSH and higher water temperature during a negative phase of the PDO and vice versa.

Overland and Wang (2007) investigated future PDO projections in the 12 models. The
PDO pattern is likely to continue to exist during the 21st century. However, the projection

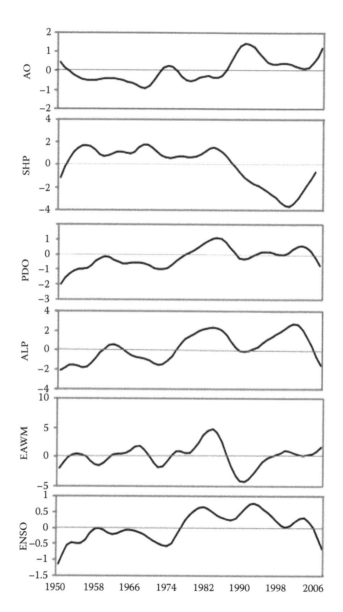

FIGURE 4.3
Time series of major climate variabilities in the Pacific Ocean since 1950s.

also indicated that the change in the mean background SST field under anthropogenic influences will surpass the magnitude of historical natural PDO variability in the North Pacific in less than 50 years. This result indicates that temperature changes associated with the PDO can exceed long-term global warming in certain places in the 21st century, but after that the warming trend will dominate the entire North Pacific. The PDO could dominate the volume transport of the KC. In the colder regime following the 1976/1977 CRS, the PDO became more intense and was in a positive phase. This positive phase of the PDO caused the KC to flow faster than average, so the volume transport to the Kuroshio-Oyashio region was higher than average, but the volume transport to the Tsushima Strait via the Kuroshio Branch Current was getting lower than average. In that case the Korean

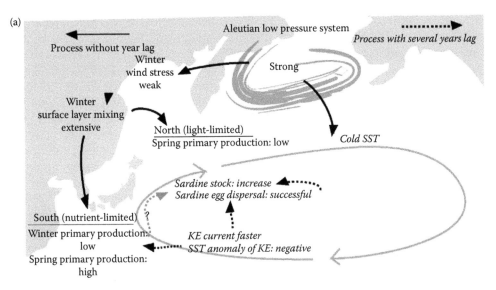

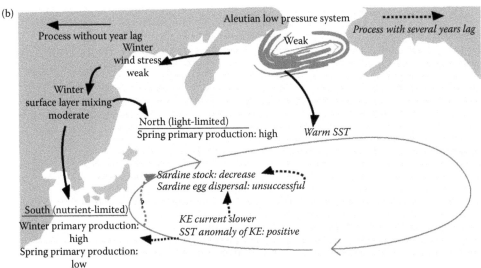

FIGURE 4.4
Schematic diagram of decadal-scale ecosystem changes in KOE region for (a) strong Aleutian Low (positive PDO index) and (b) weak Aleutian Low (negative PDO index). Processes occurring in phase with the Aleutian Low are shown in normal letters with solid arrows. Processes occurring with a several-year lag are shown in italics and with broken arrows. (Miller, A.J. et al. (2004). *Journal of Oceanography* 60, 163–188.)

part of the ECS and the East Sea/Sea of Japan was getting colder. After the 1988/1989 CRS, the exact opposite pattern was observed in Korean waters, which started becoming warmer than it previously had been.

The ALP is becoming coupled with the SHP and influences the EAWM (Jhun and Lee, 2004; Limsakul et al., 2001). Miller et al. (2004) illustrate that the strong ALP with a positive PDO system influenced winter wind stress and EAWM and, consequently, caused an increase in wintertime vertical mixing in both the Oyashio and Kuroshio regions after the 1976/1977 CRS through the 1988/1989 CRS (Figure 4.4). Chiba et al. (2005) found both direct and lagging influences from the PDO in the northern East Sea/Sea of Japan (NES) and the southern East Sea/Sea of Japan (SES), respectively. They discussed the higher spring stratification just after the 1976/1977 CRS in the NES region, whereas in the SES region, higher stratification was observed several years after the 1976/1977 CRS. However, there is no information about the direct or lagging influences in the ECS and YS regions.

Most of the atmosphere–ocean general circulation models (AOGCM) in the Intergover nmental Panel on Climate Change Fourth Assessment Report suggested that the El Niño Southern Oscillation (ENSO) would continue to exist in the 21st century. However, the effect of ENSO on the Pacific Northwest is unclear. Mantua (2001) mentions that the ENSO and PDO influence sea surface temperatures, sea level pressure, and surface winds in very similar ways. ENSO events tend to persist on the order of 1 year, while the PDO signature can last up to 30 years. A positive, or warm-phase, PDO produces climate and circulation patterns that are very similar to those of El Niño. Likewise, a negative, or cool-phase, PDO produces climate and circulation patterns similar to those of La Niña (Gershunov and Barnett, 1998). After the 1976 CRS, the frequency and amplitude of El Niño increased in the tropics. Lee et al. (2008) reported a strong correlation between water temperature and El Niño events in the East Sea/Sea of Japan with a time lag of 1.5 and 5.5 years for periods of 3 to 6 years and of decades, respectively. Park and Oh (2000) also report a 5- to 9-month lag phase of Korean waters' SST change with El Niño; a cold summer may occur in the East Sea/Sea of Japan following a winter El Niño. Kim and Kang (2000) also show a negative correlation between the Southern Oscillation Index (SOI) and the December SST in the ECS. Hong et al. (2001) indicate a higher correlation coefficient ($r = 0.55$) between sea surface temperature anomaly (SSTA) in the East Sea/Sea of Japan and El Niño events at the start of the year and the highest (0.81) in summer when El Niño events develop. However, the effect of tropical El Niño into Korean waters is somewhat ambiguous at the present and suggests a need for future research.

4.2.2 How Does Direct Climate Forcing Occur in the Korean Waters?

The changes in the patterns of the SST with respect to climate variability in the Pacific Northwest seem to be in the phase opposite that of the Northeast Pacific. The upper-layer water (10 m) temperature of all three seas around the Korean peninsula was in a warmer regime from the 1960s to the mid-1970s, a shorter cold regime from the mid-1970s to the late 1980s, and again a warmer regime that began right after the 1988/1989 CRS and persisted until the late 1990s (Figure 4.5). Upper-layer water temperatures in the different seas are significantly correlated with each other (Table 4.3); however, their fluctuation patterns regarding climate forcing seem to be different. Rebstock and Kang (2003) compared the temperature and salinity of the three seas in Korean waters. They found that temperature anomalies at depths of 10 and 50 m changed from negative to positive in the late 1980s in all three seas, but salinity anomalies at both depths changed from negative to positive only in the late 1970s. They confirmed that there was no evidence of a response to the 1976/1977 regime shift in the temperature of the three different seas. Kang et al. (2012), on the other hand, suggest that early winter (December) warming occurred as a response to the 1977 CRS, whereas there was winter warming (February) in response to the 1989 CRS in the four regions of the Korean waters; the eastern YS, the northern ECS, and the south-western and north-western East Sea/Sea of Japan. Kang et al. (2012) also confirmed an increasing pattern of surface salinity (10 m) after the 1977 CRS and a decreasing patterns after the

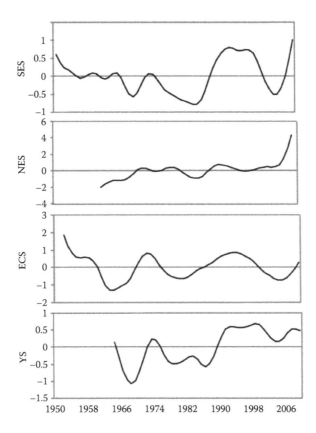

FIGURE 4.5

Decadal scale of time series in 10 m sea water temperature anomaly of southern East Sea/Sea of Japan (SES), northern East Sea/Sea of Japan (NES), East China Sea (ECS), and Yellow Sea (YS).

1998 CRS during summertime, whereas they did not confirm any response to the 1989 CRS in the eastern YS, northern ECS and south-western East Sea/Sea of Japan. They proposed that the salinity of the northern ECS and south-western East Sea/Sea of Japan is mostly controlled by the TWC, while the salinity of the eastern YS is controlled by Chinese river discharges.

TABLE 4.3

Correlation Statistics for 10 m Sea Water Temperature in Various Korean Waters

	SES	NES	ECS	YS
SES	1	.258	.506**	.781**
NES	.258	1	.317*	.498**
ECS	.506**	.317*	1	.464**
YS	.781**	.498**	.464**	1

Note: ECS: East China Sea; NES: Northern part of the East Sea; SES: Southern part of the East Sea; YS: Yellow Sea.

* Significance at $P < 0.05$.

** Significance at $P < 0.01$.

4.2.3 How Does Lag Oceanic Forcing Occur in Korean Waters?

Major oceanic changes cause variations in volume transport of the TWC by the KS, upper-layer seawater temperature, salinity, and seasonal primary productivity. Also, a lag correlation between the PDO and ALP with all sea water temperatures has been identified. For the PDO, we identified a 6-year positive lag for all of the regions' sea water temperature in wintertime, whereby, when the PDO leads sea water temperatures by 6 years, the lagging correlation coefficient between them becomes strongest (Figure 4.6, left panel). For the ALP, we found a 6-year positive lag for NES and SES regions and a 5-year positive lag for the ECS and YS regions (Figure 4.6, right panel). However, the cause(s) behind this lag correlation is (are) still unknown.

Seasonal volume transport through KS is an important parameter that defines the upper-layer environment (i.e., SST and salinity) of the East Sea/Sea of Japan (Zhang et al., 2000; Kang et al., 2012). It has a direct influence on the onset of the primary and secondary productivity in both seas, which are extremely important for some pelagic fishes' recruitment patterns. The decadal pattern of the PDO has a closer link with the TWC flow via KC flow. After the 1976/1977 CRS, when the PDO was in a positive phase, the intensification of the subtropical gyre was observed. The intensification of the subtropical gyre caused the KC to flow faster than average. At the same time, for the higher northwesterly winds, the KC started meandering higher on the southern Japanese coast, so the current flow from the KC to the TWC became weak (Gordon and Giulivi, 2004). Consequently, the upper layer became colder in that period in all of the Korean waters. After the 1988/1989 CRS, the exact opposite pattern occurred in all of the phenomena. Zhang et al. (2000) noted that the sharp increase in the KC strengthened the TWC and the EKWC following the 1976/1977 CRS through the 1988/1989 CRS, which in turn delayed the onset of primary productivity in the spring and increased the intensity of autumn blooms in the East Sea/Sea of Japan. The collapse of the saury following the 1976 CRS was demonstrated for the mismatch with the onset of spring primary productivity and migration of the saury (Zhang et al., 2000; Zhang and Gong 2005) in the NES. On the other hand, a higher abundance of sardine and filefish after the 1976 CRS was demonstrated by the increased intensity of autumn blooms. Volume transport through the KS from 1966 to 2001 indicates that total volume transport increased from the middle of the 1970s to the middle of the 1980s, which supports the higher volume transport of the KC in the ECS in general. However, those authors did not mention the different sources and seasonal variability of TWC volume via the KS.

There are different concepts regarding the origin of the TWC. Some researchers claim that the TWC is a continuation of the KBC to the west of Kyushu (KBCWK) that splits from the KBC south-west of Kyushu other researchers say that the source of the TWC is the TC, which flows northwards through the Taiwan Strait between Taiwan and China (Isobe, 1999). Isobe (1999) reanalysed the origin of the volume transport of the TWC and its seasonality and found that the Taiwan–Tsushima Warm Current (TTWC) is continuous throughout the year except during autumn. Isobe (1999) noted that about 66% of the volume transport of the TSWC comes directly from the Kuroshio region in autumn, crossing the shelf edge of the ECS. Lin et al. (2001) concluded from their surface current data that in the cold half-year, the KBCWK is the sole source of the TWC and in the warm half-year, the TWC has multiple sources, such as the KBCWK, the TC including the KBC to the north of Taiwan (KBCNT), and the currents transporting Yangtze River Diluted Water (YRDW) and mixed water in the northern ECS. Guo et al. (2001) found that, without tidal forcing and Yangtze River discharge, the TWC is supplied by three sources: the TC, KBCNT, and KBCWK. In the upper layer (0–50 m), the TC prevails over the KC (KBCNT and KBCWK

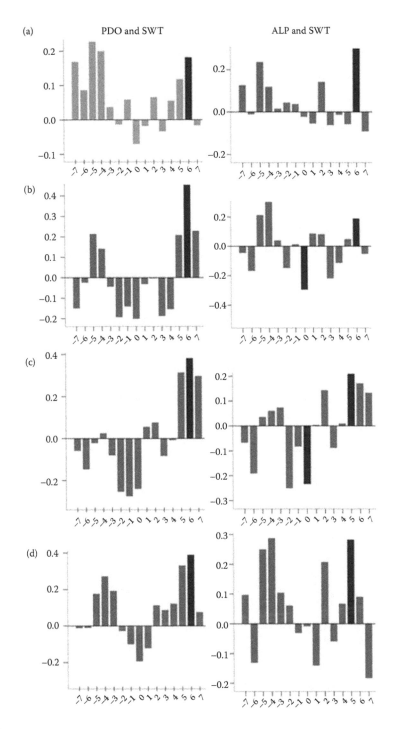

FIGURE 4.6
Lag phase between PDO and SWT (left panel) and lag phase between ALP and SWT (right panel) in wintertime. (a) NES, (b) SES, (c) ECS, and (d) YS.

from the East Taiwan Strait) in summer, but the KC prevails over the TC in winter. In the middle layer (50–100 m), the TC constitutes a small proportion while the KBCNT is the main source of the TWC. In the lower layer (100–150 m), the KBCWK is the main source of the TWC. Thus, from different retrospective analyses it can be concluded that the seasonal TWC flow through the KS is related to the seasonal variability of its different water sources. In winter, the TWC has its minimum water flow through the KS because the water source is single, the KBC, whereas in summer, it has maximum water flow with its multiple water sources, the TWC, KBC, and YRDW.

4.3 Fishery Productivity and Ecosystem Dynamics Response to Climate Forcing

The southern part of the East Sea/Sea of Japan and the ECS are more tropical and oligotrophic than the northern part of the East Sea/Sea of Japan, which seems to be more boreal and eutrophic (Ashjian et al. 2005; Tian et al. 2008). The YS is the largest shallow continental shelf area in the world. The East Sea/Sea of Japan has a moderately productive ecosystem (150–300 gm $Cm^{-2}Y^{-1}$) (Heiliman and Belkin, 2009) as a whole, whereas both the ECS (Heiliman and Tang, 2009) and the YS (Heiliman and Jiang, 2009) have higher productive ecosystems (>300 gm $Cm^{-2}Y^{-1}$). The upper layer of the ECS and the southern East Sea/Sea of Japan are relatively lower in nutrient availability for primary production than the northern part of the East Sea/Sea of Japan and the YS. The well-developed mixed-layer depth in the northern part of the East Sea/Sea of Japan make nutrients available for primary production, whereas the southern part of the East Sea/Sea of Japan is highly stratified water with a low level of nutrients in the upper layer, which makes it less productive. The YS is highly affected by run-off from numerous rivers, especially in fall, which makes it vulnerable to frequent blooming. The ECS is forced by mixed volume transport from several sources, which makes for more complex dynamics with respect to phytoplankton productivity.

The most common large predatory warm-water species showed a remarkable increasing pattern in the mid-1980s, and from the mid-2000s it increased continuously every year (Figure 4.7). Warm-water small pelagic species have also been increasing in the catch since the early 1980s. But cold-water demersal species yielded higher catches between the early 1970s and the late 1980s. These pattern changes or shifts show good coincidence with the 1976/1977 CRS and the late 1988/1989 CRS.

Kim et al. (2007) noted that the fishing area of walleye pollock in the 1990s was restricted to the coastal areas in the northern part of the East Sea/Sea of Japan because of a higher SST. Most East Sea/Sea of Japan fish species spawn in the ECS and TWC region, whereas they prefer to go farther north to the East Sea/Sea of Japan for feeding. Walleye pollock mostly inhabits the northern part of the East Sea/Sea of Japan, whereas pelagic fish like the chub mackerel, anchovy, common squid, and sardine mostly inhabit the TWC region. Highly migratory fish like Pacific saury inhabit both regions, the southern part for their spawning ground and the northern region for their feeding ground. The abundance of walleye pollock fisheries is highly dependent on cold-water intrusions in the southern part of the East Sea/Sea of Japan; after the 1976 CRS, this fishery abundance increased, whereas the warm-water pelagic fisheries were in greater abundance following the 1988 CRS when the EKWC became strong and the sub polar front went more northwards.

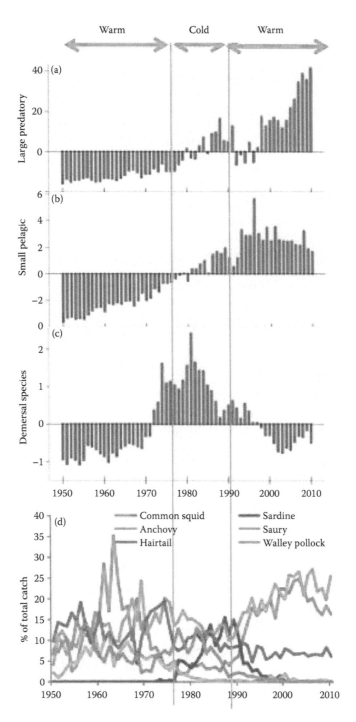

FIGURE 4.7
Time series of total catch of different fishery since 1950s. (a), (b) and (c) for total catch of large predatory, small pelagic and demersal fishery respectively. (d) for contribution of major six fish species in total catch respectively. The solid blue vertical line indicates the warm-cold-warm regimes.

Warming temperature in the ECS and the Oyashio region generally favours the winter cohort of common squid (Yatsu et al., 2013). When the East Sea/Sea of Japan warmed up, the major spawning migration route of the squid from the Hokkaido to the ECS will likely delay the onset of spawning migration and subsequently would delay their spawning season from winter to spring. This delay of spawning and subsequent hatching will cause a mismatch with the spring bloom, which is projected to occur earlier under warming scenarios. Also the body size of squid in a particular season will become smaller in response to delayed hatching. It is uncertain whether the winter cohort of common squid will increase or not because of opposing factors: warm temperature vs. more offshore transport and temporal mismatch between their spawning and spring bloom.

In many ecosystems all over the world, the population dynamics of the stock of anchovy is a mirror image of that of sardine (Chavez et al., 2003); however, in Korean waters, this phenomenon is not common. Warming may generally favour this stock through temperature-dependent growth rates (Yatsu et al., 2013; Zhang et al., 2008). Kim and Kang (2000) mentioned a positive relation with El Niño when temperature is increasing in December. They also mentioned a high correlation with chlorophyll-a concentrations in summer and with large zooplankton such as chaetognaths, euphausiids, and amphipods in fall and winter.

The optimum temperature for food consumption and specific growth rates of juvenile pollock were 12.3°C and 11.5°C, respectively (Yatsu et al., 2013). Most pollock in Korean waters reside in the northern part of the East Sea/Sea of Japan. After the 1988/1989 CRS, the fishing area was restricted to the coastal areas in the northern East Sea/Sea of Japan because of higher SST intrusion (Kim et al., 2007). Kang et al. 2000 mentioned a significant correlation between Pacific pressure and pollock abundance in the East Sea/Sea of Japan.

Peak fishing season for the major fisheries has been influenced by climate forcing as well. Before the 1988/1989 CRS, walleye pollock was caught for a longer time (October to March), but after that CRS, it was caught for a relatively shorter period of time (November to January). Sandfish changed their peak fishing time around the 1976/1977 CRS from autumn (October–December) to early summer (June–August). Pacific herring catch increased after the 1988/1989 CRS and also changed its peak fishing time. Before the 1988/1989 CRS, the main fishing time was from October to December, whereas after the 1988/1989 CRS, it changed to the January–March period. Hairtail in the YS also changed its peak fishing time after the 1988/1989 CRS from October–November to August–October. Zhang et al. (2004) reported on the changes in habitat of three major small pelagic species, chub mackerel, horse mackerel, and sardine, after the 1988 CRS. A change was observed in overlapping habitat for these major small pelagic species; after the 1988 CRS, sardine shifted away from mackerel habitats and there was a decrease in the distributional overlap of horse mackerel and chub mackerel. The researchers also noted that the continuous competition for prey and space among the major small pelagic fishes was the main factor in the change in distributional pattern. Changes in fishing season also have been identified in both cold and warm water species. The highest catches of walleye pollock were recorded in January–March during the 2000s compared to November–December during the 1980s (Kim et al., 2008).

The biomass diversity and mean trophic level (MTL) of each ecosystem have decreased during the last 3–4 decades (Figure 4.8). Zhang et al. (2004) identified the impact of CRS both in the late 1970s and late 1980s on the trophic level and relative biomass of species in the south-west East Sea/Sea of Japan ecosystem. Kim et al. (2008) mentioned a significantly decreasing trend in the trophic level of YS and ECS fishery resources during the period 1967–2000. They also discussed the higher declining trend in MTL for YS fishery than for

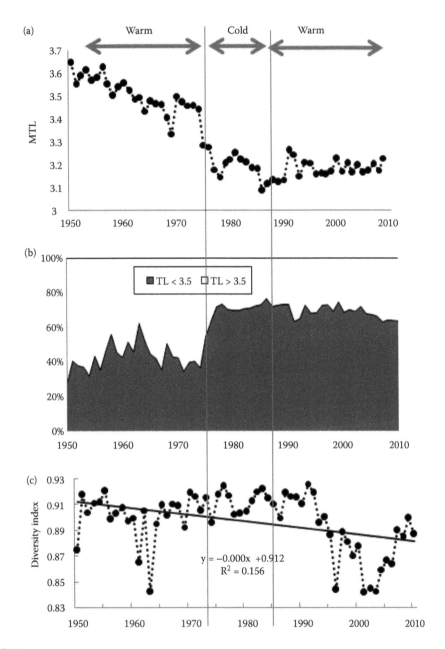

FIGURE 4.8
Ecosystem structure change pattern of Korean marine waters. (a) Mean Trophic Level (MTL) changes pattern, (b) change pattern of lower and higher trophic-level species, (c) Diversity Index (DI) changes pattern. The solid black vertical line indicates the warm-cold-warm regimes.

the ECS fishery. They found about a 62% reduction in the number of demersal species and a shift in major species in the Korean side of the YS in 1980/81 bottom trawl surveys compared to the 1967 survey. On the other hand, Zhang et al. (2004) reported on the increasing trend of MTL (from 3.09 to 3.24) in the East Sea/Sea of Japan during the 1976/1977 regime shift period. However, three different regimes in the MTL of the entire Korean marine fishery have been identified over the last six decades: (1) a relatively slowly decreasing trend from the 1950s to the 1970s; (2) a sharp decline within the entire 1970s period; and (3) a continuation of the lower MTL during the period from the early 1980s to 2010, with some variation. The average MTL during recent years (1995–2010) for Korean waters is 3.18. At around the time of the 1976/1977 CRS, species number (TL < 3.5) started increasing in amounts in the total catch, which correlates with and apparently may have directly contributed to a sharp decline in MTL in the 1970s (Figure 4.8).

Simpson's biodiversity index (DI) (Simpson, 1949) for all Korean marine fisheries indicates three major different patterns since the 1950s (Figure 4.8). From the 1950s to the late 1960s, DI tended to decrease and in 1963 reached its lowest value of 0.84. From the early 1970s to the late 1990s its value gradually increased up to around 0.92. From the early 1990s, its value fell until the early 2000s, when it again moved back up. Lower biomass diversity in Korean waters from around the early 1990s till the mid-2000s led to a higher abundance of common squid, chub mackerel, and Japanese anchovy. The fluctuation pattern in DI related to cold and warm regimes indicated that Korean marine fisheries might be more stable during colder than warmer regimes.

4.4 Concluding Remarks

Based on the Special Report of Emissions Scenarios (SRES) A2 CO_2 emission scenario for East Asia, air temperature is expected to increase 5.5°C by the end of the century and annual precipitation will also increase by 2.6% (2.68 mm/day); however, more precipitation on fewer rainy days will be observed by 2100 (Zhang et al., 2008). The effect of a 3°C increase in the SST is equivalent to a 500 km northward shift of isotherms on the Pacific side of the Japanese islands (Kim, 1995). Also, thermal expansion of the sea surface due to warming might cause a different circulation pattern in the Pacific Northwest (Zhang et al., 2008).

The present state of knowledge on the future intensity of the ALP, PDO, and El Niño is limited, and it is very difficult to talk about future periodical patterns and their intensity in a warming climate with respect to the present. It is also very difficult to determine the future teleconnection pattern of these large-scale climate variabilities. But it can be presumed that if there are very frequent changes in patterns of any climate variability under warming conditions, the regime shift might occur frequently. If the more frequent and intense Aleutian Lows observed in the 1980s occur in the future, periods of higher sardine abundance may be more frequent in Korean waters (Zhang et al., 2008). If regime shifts in the Korean marine environment occur more frequently in future conditions, the alternation between demersal cold water fishery and pelagic warm water fishery will occur very frequently. The invasion of warm-water species that is already evident in Korean waters (Jung et al., 2013) might increase and the extinction of cold-water species might accelerate. The timing and intensity of spring and autumn primary productivity will change in such a way that nutrient availability and proper vertical mixing will play a major role. Spring blooms might come earlier than before in the northern part of the East Sea/Sea of Japan, where light and temperature are the main factors in primary productivity. In the southern East Sea/Sea of Japan, the spring bloom might be disrupted for limited vertical mixing in winter and spring in lower-wind-stress conditions. The productivity dynamics of the ECS might alternate for the upper-layer current dynamics. The YS might experience more frequent blooming under future warming conditions. The ecosystem structure will also be changed in such a way that small pelagic fishery might be the dominant fishery in all seas. In particular, the yellow croaker in the YS and the pollock and cod in the East Sea/Sea of Japan might be lost or their habitats might change, whereas sardine, anchovy, and squid would benefit from the warmer conditions. Higher-trophic-level species like spotted mackerel (*Scomber australasicus*), yellowtail (*Seriola quinqueradiata*), and skipjack tuna (*Katsuwonus pelamis*) increased remarkably in the early 1990s when higher air temperatures, weak wind speeds, and warmer sea water temperatures prevailed on the southern coast of the Korean peninsula (Lee and Go, 2006). Thus, a large predatory fishery might be the major fishery in the future commercial catch of Korean waters; however, there will always be uncertainty about the future.

Acknowledgements

This research was a part of the project titled 'Walleye Pollock stock management based on Marine Information & Communication Technology', funded by the Ministry of Oceans and Fisheries, Korea. SM Mustafizur Rahman was supported by the Basic Science Research Program throughthe National Research Foundation of Korea (NRF) funded by the Ministry of Education (2016R1A6A1A05011910).

References

Ashjian, C.J., Davis, C.S., Gallager, S.M. and Alatalo, P. (2005). Characterization of the zooplankton community, size composition, and distribution in relation to hydrography in the Japan/East Sea. *Deep-Sea Res.- II* 52, 1363–1392.

Belkin, I.M. (2009). Rapid warming of large marine ecosystems. *Progress in Oceanography* 81, 207–213.

Chavez, F.P., Ryan, J., Lluch-Cota, E. and Niquen, M. (2003). From anchovies to sardines and back: Multidecadal change in the Pacific Ocean. *Science* 299, 217–221.

Chiba, S., Hirota, Y., Hasegawa, S. and Saino, T. (2005). North–south contrast in decadal scale variations in lower trophic ecosystems in the Japan Sea. *Fisheries Oceanography* 14, 401–412.

Food and Agriculture Organization of the United Nations (2012). *The State of World Fisheries and Aquaculture*. FAO, Rome.

Gershunov, A. and Barnett, T. (1998). Inter-decadal modulation of ENSO teleconnections. *Bulletin of the American Meteorological Society* 79, 2715–2725.

Gong, D.Y. and Ho, C.H. (2002). The Siberian high and climate change over middle to high latitude Asia, *Theoretical and Applied Climatology* 72, 1–9.

Gong, D.Y. and Wang, S.W. (1999). Long–term variability of the Siberian High and the possible influence of global warming. *Acta Geographica Sinica* 54, 125–133. [In Chinese.]

Gong, D.Y., Wang, S.W. and Zhu, J.H. (2001). East Asian winter monsoon and Arctic oscillation, *Geophysical Research Letters* 20, 2073–2076.

Gordon, A.L. and Giulivi, C.F. (2004). Pacific decadal oscillation and sea level in the Japan/East Sea. *Deep-Sea Research Part I-Oceanographic Research Papers* 51, 653–663.

Guo, Q.Y. (1996). Climate change in China and East Asian monsoon. In Yafeng, S. (ed). *Historical Climate Change in China*. Shandong Science and Technology Press, Jinan, pp. 68–483. [In Chinese.]

Guo, X., Miyazawa, Y., Hukuda, H., and Yamagata, T. (2001). Tracer experiments on the origin of the Tsushima Warm Current. *Proceedings, 11th PAMS/JECSS Workshop*, April 11–13, Cheju, Korea, p. 45–47.

Heiliman, S. and Belkin, I. (2009). Sea of Japan/East Sea: LME #50. In Sherman, K. and Hempel, G. (Eds.). *The UNEP Large Marine Ecosystem Report: A Perspective on Changing Conditions in LMEs of the World's Regional Seas- UNEP Regional Seas Report and Studies*, 182. United Nations Environment Programme, Nairobi, Kenya, 413–422.

Heiliman, S. and Jiang, Y. (2009). Yellow Sea: LME# 48. In Sherman, K. and Hempel, G. (Eds.). *The UNEP Large Marine Ecosystem Report: A Perspective on Changing Conditions in LMEs of the World's Regional Seas- UNEP Regional Seas Report and Studies*, 182. United Nations Environment Programme, Nairobi, Kenya, 441–452.

Heiliman, S. and Tang, Q. (2009). East China Sea: LME# 47. In Sherman, K. and Hempel, G. (Eds.). *The UNEP Large Marine Ecosystem Report: A Perspective on Changing Conditions in LMEs of the World's Regional Seas- UNEP Regional Seas Report and Studies*, 182. United Nations Environment Programme, Nairobi, Kenya, 381–392.

Hong, C.H., Cho, K.D. and Kim H.J. (2001). The relationship between ENSO events and sea surface temperature in the East (Japan) Sea, *Progress in Oceanography* 49, 21–40.

Isobe, A. (1999). On the origin of the Tsushima Warm Current andits seasonality. *Cont. Shelf Res.* 19, 117–133.

Jhun, J.G. and Lee, E.J. (2004). A new east asian winter monsoon index and associated characteristics of the winter monsoon. *J. Climate* 17, 711–726.

Jung, S., Pang, I.C., Lee, J-H., Choi, I. and Cha, H.K. (2013). Latitudinal shifts in the distribution of exploited fishes in Korean waters during the last 30 years: a consequence of climate change. *Reviews in Fish Biology and Fisheries.* doi: 10.1007/s11160-013-9310-1.

Kang, C.J. (1974). A study on the seasonal variation of the water masses in the southern sea of Korea. *Bulletin of Fisheries Research and Development Agency* 12, 107–121. [In Korean.]

Kang, S., Kim, S. and Bae, S.W. (2000). Changes in ecosystem components induced by climate variability off the eastern coast of the Korean Peninsula during 1960–1990. *Progress in Oceanography* 47, 205–222.

Kang, Y.S., Jung, S., Zuenku, Y., Choi, I. and Dolganova, N. (2012). Regional differences in the response of the mesozooplankton to oceanographic regime shifts in the northeast Asian marginal seas. *Progress in Oceanography* Vols. 97–100, 120–134.

Kim, S. (1995). Climate change and the fluctuation of fishery resources in the North Pacific. *Ocean Policy Res.* 10, 107–142. (In Korean with English abstract.)

Kim, S. and Kang, S. (2000). Ecological variations and El Niño effects off the southern coast of the Korean Peninsula during the last three decades. *Fisheries Oceanography* 9, 239–247.

Kim, K.R., Kim, G. and Kim, H. (2002). A sudden bottom-water formation during the severe winter 2000-2001: The case of the East/Japan Sea. *Geophysical Research Letters* 29 (8), 75(1)–75(4).

Kim, S., Zhang, C.I., Kim, J.Y., Oh, J.H., Kang, S. and Lee, J.B. (2007). Climate variability and its effects on major fisheries in Korea. *Ocean Science Journal* 42, 179–192.

Kim, S., Zhang, C.I., Kim, J.Y., Kang, S. and Lee, J.B. (2008). Country Reports, Republic of Korea, Impacts of Climate and Climate Change on the Key Species in the Fisheries in the North Pacific. *PICES Scientific Reports* 35, 101–135.

Lee, J.H. (1998). Hydrographic observations in the Yellow Sea. In: Hong, G.H., Zhang, J. and Park, B.K. (eds). *Health of the Yellow Sea*. The Earth Love Publication Association, Seoul, pp. 13–42.

Lee, S.J. and Go, Y.B. (2006). Winter warming and long term variation in catch of yellowtail (*Seriola quinqueradiata*) in the South Sea, Korea. *Korean J. Ichthyology* 18: 319–328.

Lee, C.I., Lee, J-Y., Choi, K-H. and Park, S.E. (2008). Long-term trends in pelagic environments of the East/Japan Sea Ecosystem. *Ocean Science Journal* 43, 1–7.

Lie, H.J., Cho, C.H., Lee, J.H., Lee, S., Tang, Y. and Zou, E. (2001). Does the Yellow Sea Warm Current really exist as a persistent mean flow? *Journal of Geophysical Research* 106, 22199–22210.

Limsakul, A., Saino, T., Midorikawa, T. and Goes, J.I. (2001). Temporal variations in lower trophic level biological environments in the northwestern North Pacific Subtropical Gyre from 1950 to 1997. *Progress in Oceanography* 49, 129–149.

Mantua, N.J. (2001). The Pacific Decadal Oscillation. In Mac Cracken, M.C. and Perry, J.S. (eds). *The Encyclopedia of Global Environmental Change, Vol. 1. The Earth System: Physical and Chemical Dimension of Global Environmental Change.* John Wiley & Sons, Ltd, pp. 592–594.

Miller, A.J., Chai, F., Chiba, S., Moisan, J.R. and Neilson, D.J. (2004). Decadal-scale climate and ecosystem interactions in the North Pacific Ocean. *Journal of Oceanography* 60, 163–188.

Miyazaki, S., Yasunari, T. and Adyasuren, T. (1999). Abrupt seasonal changes of surface climate observed in North Mongolia by an automatic weather station. *Journal of Meteorological Society Japan* 77, 583–593.

Mooers, C., Kang, H., Bang, I. and Snowden, D. (2006). Some lessons learned from Comparisons of numerical simulations and observations of the JES circulation. *Oceanography* 19(3): 86–95.

Overland, J.E. and Wang, M. (2007). Future climate of the North Pacific Ocean, *Eos* 88, 178–182.

Park, W.S. and Oh, I.S. (2000). Interannual and interdecadal variations of sea surface temperature in the East Asian Marginal Seas. *Progress in Oceanography* 47, 191–204.

Rahman, M.S.M. and Lee C.I. (2012). Long term changes pattern in Marine Ecosystem of Korean waters. *Journal of the Korean Society of Marine Environment & Safety* 18, 193–198.

Rebstock, G.A. and Kang, Y.S. (2003). A comparison of three marine ecosystems surrounding the Korean peninsula: Responses to climate change. *Progress in Oceanography* 59, 357–379.

Simpson, E.H. (1949). Measurement of diversity. *Nature* 163, 688.

Tian, Y., Kidokoro, H. and Watanabe, T. (2006). Long-term changes in the fish community structure from the Tsushima warm current region of the Japan/East Sea with an emphasis on the impacts of fishing and climate regime shift over the last four decades. *Progress in Oceanography* 68, 217–237.

Tian, Y., Kidokoro, H., Watanabe, T. and Iguchi, N. (2008). The late 1980s regime shift in the ecosystem of Tsushima Warm Current in the Japan/

Yatsu, A., Chiba, S., Yamanaka, Y., Ito, S.I., Shimizu, Y., Kaeriyama, M. and Watanabe, Y. (2013). Climate forcing and the Kuroshio/Oyashio ecosystem. *ICES Journal of Marine Science* 70: 922–933.

Yeh, S.-W., Park, Y.-G., Min, H.-S., Kim, C.-H. and Lee, J.-H. (2010). Analysis of characteristics in the sea surface temperature variability in the East/Japan Sea, *Progress in Oceanography* 85, 213–223.

Yin, Z.Y. (1999). Winter temperature anomalies of the North China Plain and macro scale extra extra-tropical circulation. *International Journal of Climatology* 19, 291–308.

Zhang, C.I. and Gong, Y. (2005). Effect of ocean climate changes on the Korean stocks of Pacific saury, *Cololabis saira* (BREVOORT). *Journal of Oceanography* 61, 313–325.

Zhang, C.I., Lee, J.B., Kim, S. and Oh, J.H. (2000). Climatic regime shifts and their impacts on marine ecosystem and fisheries resources in Korean waters. *Progress in Oceanography* 47, 171–190.

Zhang, C.I., Lee, J.B., Seo, Y.I., Yoon, S.C. and Kim, S. (2004). Variations in the abundance of fisheries resources and ecosystem structure in the Japan/East Sea. *Progress in Oceanography* 61, 245–265.

Zhang, C.I., Kim, S., Oh, J.H. and Gunderson, D.R. (2008). Impacts of climate change on marine ecosystems and fisheries productivity in marginal seas of the northwest Pacific. *In Climate Change Research Progress.*

Zhu, Q.G., Shi, N. and Wu, Z.H. (1997). Low frequency variation of winter ACAs in north hemisphere and climate change in China during the past century. *Acta Meteorologica Sinica* 55, 750–758.

5

Environmental Effects and Management of Oil Spills on Marine Ecosystems

Piers Chapman, Terry L. Wade and Anthony H. Knap

CONTENTS

5.1   Introduction.. 119
5.2   Physics and Chemistry of Crude and Refined Oils 121
5.3   Toxicity of Oil to Marine Life... 126
     5.3.1   Acute Effects... 127
     5.3.2   Sublethal Effects... 129
5.4   Basic Methodology of Mitigation 131
     5.4.1   Booms and Skimmers 132
     5.4.2   Dispersants 133
     5.4.3   Gelling Agents and Sorbents 134
     5.4.4   Burning... 134
     5.4.5   Bioremediation 134
     5.4.6   Doing Nothing 135
5.5   Management Plans for Different Ecosystems... 135
     5.5.1   Contingency Planning 135
     5.5.2   Environmental Sensitivity Index... 136
     5.5.3   Specific Environments... 137
           5.5.3.1   Beaches 137
           5.5.3.2   Exposed Tidal Flats.. 138
           5.5.3.3   Rocky Shores... 138
           5.5.3.4   Marshes 139
           5.5.3.5   Mangroves.. 139
           5.5.3.6   Coral Reefs 140
5.6   Conclusions and Recommendations.. 141
Acknowledgements... 141
References.. 141

5.1 Introduction

Petroleum, formed from the organic remains of dead organisms over geological time (Tissot and Welte, 1984), contains thousands of individual compounds, from methane with a molecular weight of 16 to asphaltenes with molecular weights in the thousands. More than 90% of the mass of most crude oils consists of hydrocarbons, compounds containing only the elements carbon and hydrogen, and a general definition of petroleum is any hydrocarbon mixture that can be produced through a drill pipe (Hunt, 1979). Petroleum can be produced as natural gas (which does not condense at atmospheric pressure and 15.6°C), condensate, which exists in the gas state in the reservoir but condenses to a liquid at the surface, crude oil produced as a liquid, and crude oil produced as a liquid that becomes solid at the surface (Hunt, 1979). The various biogeochemical processes that affect the global distribution of petroleum reservoirs have been summarized by McElroy et al. (1989).

Petroleum is extremely important for energy production and as a chemical feedstock (e.g., in the manufacture of plastic). However, the extraction, transport, and use of petroleum and its distilled products have led to both accidental and chronic inputs of petroleum into the environment. Estimates of the relative importance of various inputs of petroleum to the marine environment annually (NRC, 2003) are summarized in Table 5.1. It should be noted that while spills are the most dramatic and most frequently reported input source for petroleum, they represent only 8% of the estimated total annual global input, and their contribution has been decreasing steadily since the 1960s (ITOPF, 2014). Despite this, petroleum inputs into the marine environment have led to concerns regarding the hazards of petroleum contamination and its potential impact on ecosystems including fishery resources and human health.

Despite a general concern over the effects of large, isolated spills of petroleum and its products, hydrocarbons in the environment generally comprise a complex mixture of compounds derived from many sources. For example, coastal subtidal sediments near urban or industrialized areas contain a significant background of anthropogenic and biological

TABLE 5.1

Average Annual Releases (1990–1999) of Petroleum by Source (in thousands of tonnes)

	North America		Global	
Source	Range	Best Estimate	Range	Best Estimate
Natural seeps	80–240	160	200–2000	600
Extraction of petroleum	2.3–4.3	3.0	20–62	38
Platforms	0.1 5–0.18	0.16	0.29–1.4	0.86
Atmospheric deposition	0.07–0.45	0.12	0.38–2.6	1.3
Produced water	2.1–3.7	2.7	19–58	36
Transportation of petroleum	7.4–11.0	9.1	120–260	150
Pipeline spills	1.7–2.1	1.9	6.1–37	12
Tank vessel spills	4.0–6.4	5.3	93–130	100
Operational discharges	na[a]	na	18–72	36
Coastal facility spills	1.7–2.2	1.9	2.4–15	4.9
Atmospheric deposition	trace[b]–0.02	0.01	0.2–1.0	0.4
Consumption of petroleum	19–2000	84	130–6000	480
Land-based (river/runoff)	2.6–1900	54	6.8–5000	140
Recreational marine vessels	2.2–9.0	5.6	nd[c]	nd
Spills (nontank vessels)	1.1–1.4	1.2	6.5–8.8	7.1
Operational discharges	0.06–0.60	0.22	90–810	270
Atmospheric deposition	9.1–81	21	23–200	52
Jettisoned aircraft fuel	1.0–4.4	1.5	5.0–22	7.5
Total	110–2300	260	470–8300	1300

Source: National Research Council. (2003). Oil in the Sea III. Inputs, Fate and Effects, National Academy Press, Washington, D.C., p. 265.

[a] Cargo washing is not allowed in U.S. waters but is not restricted in international waters. It was assumed the practice is rare in U.S. waters.

[b] Estimated loads of less than 10 tonnes/year reported as 'trace'.

[c] Worldwide population of recreational vessels was not available.

hydrocarbons. The simple determination of the presence of hydrocarbons may, therefore, be insufficient to determine whether their source is from an isolated oil spill or from biogenic or chronic inputs. To assess the environmental impact of these various sources, their relative importance must be unambiguously identified. While a number of techniques based on molecular and isotopic compositions have been developed to differentiate multiple inputs of hydrocarbons to the environment, a definitive determination of the source of low levels of hydrocarbons in environmental samples is not always possible.

5.2 Physics and Chemistry of Crude and Refined Oils

While the major elements in oil are carbon (80%–87%) and hydrogen (10%–15%), petroleum also contains sulfur (0%–10%), oxygen (0%–5%) and nitrogen (0%–1%) as minor constituents (Hunt, 1979; NRC, 2003), as well as varying concentrations of nickel, vanadium, iron, aluminum, sodium, calcium, copper and uranium (Posthuma, 1977; Hunt, 1979). The hydrocarbons, as well as sulfur (S), oxygen (O) and nitrogen (N) containing compounds found in petroleum, can be subdivided based on their structures into aliphatics (alkanes, branched alkanes, cycloalkanes), aromatics and NSO compounds (Figure 5.1). However, petroleum is not the only source of hydrocarbons in the environment. Various hydrocarbons are biosynthesized by marine and terrigenous organisms including many microbes. Hydrocarbons can also be produced during either natural (i.e., forest and grass fires) or anthropogenic (i.e., combustion of coal, petroleum or wood, or waste incineration) combustion processes.

A number of other parameters can be used to determine the source of a hydrocarbon mixture. These include measurements of bulk parameters such as stable isotope composition, extractable organic matter content, trace element content, the percentages of major hydrocarbon fractions (aliphatic, aromatic, NSO and asphaltene percentages), as well as the molecular composition of fractions of the total extract (e.g., PAH, biomarkers, aliphatic hydrocarbons). In general, a multiparameter approach enhances the ability to determine the source of hydrocarbons. No single parameter or set of parameters is always definitive for all situations, and it is generally easier to determine that a hydrocarbon mixture in a water, sediment, or tissue sample is probably not from a particular source (for example, a specific spill) than to determine definitively that it is.

The areal or vertical distributions of hydrocarbons are also useful in defining sources. For example, a constant trend of hydrocarbons in a several-meter core can indicate that hydrocarbons from natural seepage are a source of hydrocarbons in surficial sediment. Conversely, if hydrocarbon contamination is only present in the top few centimeters of a sediment core, then its source is most likely related to a recent depositional event, such as a spill. Hydrocarbon distribution in sediments around a petroleum production platform can similarly provide indications of the extent of potential contamination from platform operations (Kennicutt et al., 1996).

Comparing hydrocarbons from petroleum with hydrocarbons found in organisms reveals the following differences (which may not apply in all cases) that are useful in detecting the presence of petroleum:

1. Petroleum is a complex mixture of hydrocarbons having a much greater range of molecular structures and molecular weights when compared to the hydrocarbons native to organisms.

2. Petroleum contains many homologous series. Adjacent members of a series are usually present in nearly equal amounts (i.e., even and odd carbon alkanes, the homologous series of C_{12} to C_{20} isoprenoid alkanes). Biogenic inputs are often characterized by a predominance of odd carbon alkanes and a single isoprenoid such as pristane (C_{19}). Since phytane (a similar compound but with 20 carbon atoms) is rarely found in biological material other than some bacteria, most biological hydrocarbons have a pristane/phytane ratio $\gg 1.0$.

3. Petroleum contains a complex mixture of cycloalkanes and aromatic hydrocarbons compared to the small number of these compounds native to marine organisms. Examples are the series of polyalkylated benzenes and naphthalenes.

4. Petroleum normally contains no olefins. Biogenic material can contain polyolefins such as squalene.

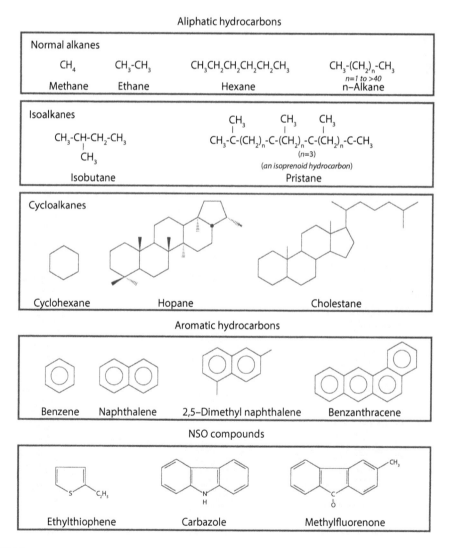

FIGURE 5.1
Typical classes of compounds and structures found in oil.

TABLE 5.2

Selected Hydrocarbon Molecular Markers

Compound Class	Characteristic Marker	Source
n-alkanes	n-C_{23} to n-C_{33}, odd predominance	Plant Waxes
n-alkanes	n-C_{15}, n-C_{17}	Plankton
n-alkanes	n-C_{10} to n-C_{34}^+, no odd predominance	Petroleum
Isoprenoids	Pristane	Plankton/Petroleum
Isoprenoids	Phytane	Petroleum
Cycloalkanes	Unresolved complex mixture	Petroleum
Cyclic Diterpenoids	Retene, Fichtelite Plants	Terrestrial
Triterpenoids, Tricyclic Terpanes, Steranes	Biomarkers	Petroleum
Polynuclear Aromatic	2–5 ring compounds, primarily unsubstituted, no perylene	Combustion
Polynuclear Aromatic	2–3 ring compounds, primarily substituted	Petroleum
Polynuclear Aromatic	Anthracene	Combustion
Thiophenes	Dibenzothiophenes	Petroleum/Combustion

5. Petroleum contains numerous naphthenoaromatic hydrocarbons, together with heterocompounds (S, N, O and metals) and heavy asphaltic material that have not been found in organisms.

6. Petroleum contains biomarker compounds that can be unambiguously linked to biochemicals found in living organisms. However, the biomarkers themselves are not found in organisms and generally represent the analogs of biological precursors (i.e., chlorophyll, sterols) that are generated by hydrogenation or the loss of functional groups from biogenic compounds (such as porphyrins or steranes) during the burial and transformation of organic material into petroleum.

These differences between nonpetroleum and petroleum hydrocarbons have led to the use of molecular marker compounds to differentiate between sources of hydrocarbons in the environment. Some of these are summarized in Table 5.2.

Petroleum introduced into the environment can undergo a series of weathering processes that operate on time scales ranging from days to years. The changes in composition due to these processes make the unambiguous identification of hydrocarbon sources in the environment more difficult. Characteristic effects of weathering on petroleum include the following:

1. Loss of low boiling ($<C_{20}$) aliphatic and aromatic hydrocarbons by evaporation and dissolution.

2. An increase in the relative contribution of the naphthenic and naphthenoaromatic compounds compared to other petroleum constituents.

3. A relative increase in highly branched aliphatic hydrocarbons (i.e., isoprenoids) relative to n-alkanes, since biodegradation selectively degrades n-alkanes before methyl-branched alkanes. A reasonably well-established hierarchy of chemical structure susceptibility to microbial degradation has been established.

Thus, pristane, phytane and farnesane may become the dominant saturated components during certain stages of the weathering process, although further weathering may lead eventually to the loss of these compounds.

4. An increase in alkylated phenanthrene and dibenzothiophene compounds relative to other aromatics.

5. An increase in polycyclic aliphatics (i.e., triterpanes and steranes) relative to other saturated components.

Oil in the environment also undergoes photochemical (if exposed to light) and microbial oxidation. These processes produce a variety of additional reaction products including ketones, aldehydes, carboxylic acids, esters and epoxides (Overton et al., 1979, 1980).

Many investigators have used the n-alkane fraction of aliphatic hydrocarbons extensively to estimate the relative importance of hydrocarbon sources at a given location (e.g., Brassell et al., 1978; Philp, 1985a; Boehm and Requejo, 1986). For example, plankton generally produce a simple mixture of hydrocarbons dominated by $n-C_{15,17,19}$ and pristane, so the presence of these compounds can be used as plankton indicators (Clark and Blumer, 1967; Blumer et al., 1970; Goutx and Saliot, 1980). Petroleum also contains these compounds but usually contains nearly equal amounts of $n-C_{16,18,20}$ and phytane (Farrington and Tripp, 1977; Farrington et al., 1973). Straight-chain biogenic hydrocarbons with 25, 27, 29 and 31 carbons indicate terrestrial or land-derived organic matter (Philp, 1985a,b); these normal alkanes are also found in petroleum accompanied by nearly equal amounts of normal alkanes with 24–33 carbon atoms. Weathered petroleum contains enhanced naphthenic components owing to preferential microbial degradation of normal alkanes. Some commonly used aliphatic source indicators (Boehm et al., 1981) are shown in Table 5.3.

Aromatics account for only a small portion of the total hydrocarbons (4%–9%) in petroleum; however, they are the most toxic components and are widely used as indicators of petroleum contamination in environmental samples. Polynuclear aromatic hydrocarbons (PAHs) are a complex group of compounds composed of substituted and unsubstituted cyclic aromatic rings (Figure 5.1). Several PAHs occur naturally (i.e., retene, picene, perylene and octahydrochrysene), being formed by the aromatization of naturally occurring sedimentary organic matter in near-surface sediments. However, the majority of PAHs in coastal and estuarine ecosystems are derived from petroleum or the incomplete combustion of carbonaceous materials during forest fires and fossil fuel combustion (Youngblood and Blumer, 1973; Lee et al., 1977). Molecular compositions based on parent compound, alkyl homolog, and ring number distributions can be used to differentiate these various sources (Table 5.4). In general, environmental studies only consider PAHs consisting of condensed rings and simple alkylations (e.g., naphthalenes, phenanthrenes, fluoranthenes, chrysenes).

While petroleum PAHs generally are dominated by alkyl-substituted PAHs with two to four rings, PAHs derived from combustion contain homologous series enriched in unsubstituted (parent) compounds, with the abundance of alkylated PAHs generally decreasing with increasing alkylation. Larger numbered ring compounds (i.e., pyrene, fluoranthene, anthracene) are more prevalent in combustion-derived PAHs.

Other indicator compounds, called biomarkers, are used to help identify the specific source of a hydrocarbon sample. These indicators are relatively resistant to alteration by environmental processes, such as microbial degradation, differential solution, or

TABLE 5.3

Aliphatic Hydrocarbon Source Indicators

Parameter	Relevance
ISO/ALK	Measures the relative abundance of branched isoprenoid alkanes to straight chain alkanes in the same boiling range; useful indicators of biodegradation.
L_{ALK}/TOT_{ALK}	Diagnostic alkane compositional ratio used to determine the relative abundance of n-C_{10} to n-C_{20} alkanes (characteristic of light crude and refined oils) to total alkanes, which includes those of biogenic (background) origin.
PRIS/PHY	Source of phytane is mainly petroleum, while pristane is derived from both biological matter and oil. In 'clean' samples, this ratio is very high and decreases as oil is added.
TOT/TOC or n-alkanes/TOC	The ratio of total saturated hydrocarbons (TOT) or n-alkanes (subset of TOT) to total organic carbon (TOC) is used to monitor oil inputs. In sediments receiving 'normal' pollutant inputs within a given region, a specific TOT/TOC ratio is characteristic of the 'geochemical province'. Small (tens of ppm) additions of petroleum to sediment cause the ratio to increase dramatically because n-alkanes increase but TOC does not.
OEPI	Odd-even carbon preference index; describes the relative amounts of odd and even chain alkanes within a specific boiling range; as oil additions increase, the OEPI is lowered.
UCM	Unresolved complex mixture; useful indicator of petroleum.
Abundant Complex Mixture of Steranes and Triterpanes	Typical components of petroleum.
Alkanes/Alkenes	Alkenes are abundant in organisms, so lower ratios are associated with biological hydrocarbons.
Various Compound Ratios	
$C_{17}/C_{31} \gg 1$	Algal alkanes are a major component.
$C_{17}/C_{31} \ll 1$	Higher plant alkanes are major components.
$C_{12}/C_{14} > 1$	Weathered crude oil present.
PRIS + PHY/n-$C_{17} \gg 1$	Degraded crude oil present.

Source: Boehm, P.D., Fiest, D.L. and Elskus, A. (1981). In: *Amoco Cadiz: Consequences d'une Pollution Accidentelle par les Hydrocarbures: Fates and Effects of the Oil Spill.* Centre National Pour L'Exploitation des Oceans. Paris, pp. 159–173; Brassell, S.C. et al. (1978). In: O. Huntzinger, L.H. van Lelyveld and B.C.J. Zoetman (eds) *Aquatic Pollutants, Transformations and Biological Effects,* Pergamon Press, Oxford, pp. 69–86.

evaporation, and can uniquely identify the hydrocarbon source. Because of the wide array of compounds present in petroleum that can only have a biological origin, such as porphyrins, steranes and isoprenoids, we know that petroleum results from the action of temperature and time on the remains of organisms that have been preserved and deposited in sedimentary rocks. These biomarkers form complex mixtures during petroleum generation that provide unique and sensitive fingerprints on the sources of different hydrocarbon mixtures. Studying such fingerprints has shown, for example, that oil from sources other than the *Exxon Valdez* was present on some beaches in Alaska at the time of that spill (Hostettler and Kvenvolden, 1994) and that sediment contamination from the *Arrow* oil spill still remained recognizable 22 years after it occurred (Wang et al., 1994).

TABLE 5.4

PAH Source Indicators

Parameter	Relevance
Alkyl Homologue Distribution (AHD)	Used to assess the relative importance of fossil fuel and combustion PAH sources. The abundance of alkyl-substituted PAHs within a given aromatic series, relative to the parent compound, can be a sensitive indicator of hydrocarbons. Combustion sources are generally characterized by a greater relative abundance of parent compounds while petroleum contains greater relative quantities of alkyl homologues.
FFPI	Fossil Fuel Pollution Index[a]: ratio of fossil fuel-derived PAH to total (fossil + pyrogenic + diagenic) PAH. FFPI for fossil PAH = 100; FFPI for combustion PAH = 0.
Specific PAH ratios	For example, P/D is an excellent indicator of petrogenic input.
PAH/TOC	Analogous to TOT/TOC ratio.

Source: Boehm, P.D., Fiest, D.L. and Elskus, A. (1981). In: *Amoco Cadiz: Consequences d'une Pollution Accidentelle par les Hydrocarbures: Fates and Effects of the Oil Spill.* Centre National Pour L'Exploitation des Oceans. Paris, pp. 159–173.

[a] FFPI = (N + F + P + D)/ΣPAH; ΣPAH = N + F + P + D + FLAN + PYRN + BAA + CHRY + BAP + BEP + BFS + PERY.

5.3 Toxicity of Oil to Marine Life

In 1985 a South African scientist wrote, 'There is possibly no other field of modern scientific research in which so many valueless or almost valueless papers have been published' (Brown, 1985). When marine oil spills occur, the most obvious result is a series of photographs of oil-covered organisms, particularly birds and marine mammals, and articles in the media describing them as major catastrophes to the environment. While such images are very powerful, they give a false impression of the effects of a spill. Oil can affect organisms, including plants as well as animals, in several ways. For example, during major spills, including the recent *Deepwater Horizon (DWH)* spill in 2010, responders went to great lengths to keep oil out of the marshes, as oil will kill ground vegetation, including marsh grasses and mangroves if it gets into their root systems (e.g., Baker, 1982; Baker et al., 1984; Ballou et al., 1987, 1989; Getter and Baca, 1984). Here, we discuss some of the acute and sublethal effects of marine oil spills, drawing on some of the many reports on the topic that have been issued over the years (e.g., Baker, 1983; Brown, 1985; NRC, 2003, 2005 and references therein). Oil can affect organisms at four levels – biochemical and cellular, organismal (integrating physiological, biochemical and behavioral responses within a particular organism), population, and community (NRC, 2003). Of these, the only important ones for the ecosystem are effects at the population and community levels.

Estimated concentration ranges at which oil affects various groups of organisms are shown in Figure 5.2. However, these effects vary, depending on several factors, including bioavailability (how easily the particular toxin can be ingested, adsorbed, or absorbed by an organism) and environmental availability (the particular physical or chemical form of the contaminant). For oil, the most bioavailable forms are, in descending order, the water-soluble fraction, oil in prey and fresh liquid oil droplets. Oil adsorbed on even fine particulates does not seem to be particularly bioavailable. Similarly, oil buried in sediments or liquid oil in thick weathered patches is also generally unavailable, although in the former

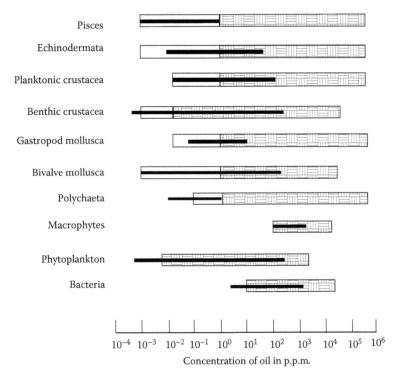

FIGURE 5.2

Concentration of oil causing deleterious effects on different groups of organisms. (Redrawn from Brown, A.C. (1985). *The Effects of Crude Oil Pollution on Marine Organisms: a Literature Review in the South African Context: Conclusions and Recommendations.* Report for Oil Pollution Committee of the South African Committee for Oceanic Research (SANCOR), p. 33.)

case burrowing organisms can ingest oil from the sediment and it can be rereleased to the environment if the sediment is disturbed.

5.3.1 Acute Effects

The most obvious effect of an oil spill is the smothering of organisms in shallow water or onshore. Benthic organisms such as burrowing worms or clams (in sandy or muddy areas) or sessile organisms (snails, barnacles) on rocky coasts are frequently affected. Similarly, birds, turtles and mammals, including seals, sea lions and river and sea otters, can be affected by both floating and beached oil. In the latter case, animals can be affected both by oil that coats fur and feathers thereby affecting waterproofing and thermal regulation, which can lead to death from hypothermia, and by direct ingestion of oil as the animals try to clean themselves. Severe effects can occur with relatively small amounts of oiling on birds such as penguins (less than 10% coverage of plumage), and oil-coated nesting birds can transfer oil to their eggs, thereby affecting embryos (NRC, 2003). Ingestion of oil-contaminated prey can also affect these higher predators. Whales and dolphins are not generally thought to be very susceptible to oil spills, although 122 dead cetaceans were reported during the initial phase of the *DWH* spill, with another 936 deaths reported from the northern Gulf of Mexico between November 2010 and April 2014 (Schleifstein, 2014). The same report lists 1,066 turtle strandings in 2010 and

1,536 between 2011 and 2013 in this region, but it gives no details of any likely oiling. It is known, however, that turtles were affected both by the oil itself and by burning to death during the cleanup operations, and the number of nests of Kemp's ridley sea turtle, a protected species, has decreased considerably since 2010 (P. Plotkin, Texas Sea Grant, pers. comm.), although it is not clear whether this is due to the spill or other causes. Cetacean strandings in the Gulf of Mexico had been increasing prior to the spill, with many corpses showing evidence of brucellosis. However, the only report so far on dolphins (*Tursiops truncatus*) in the region (Schwacke et al., 2013), which suggested that loss of lung function could be caused by exposure to oil, is inconclusive (Jacobs, 2014), similar to a report on orcas after the *Exxon Valdez* spill (Matkin et al., 2008), where it could not be confirmed that the spill was the cause.

Many of the components of crude oil are directly toxic to organisms. These include, in particular, the lighter fractions (up to about eight carbon atoms) and the low-molecular-weight aromatics with one or two rings, which are also the most soluble components of oil. The effects of individual compounds appear to be additive, so that the final effect depends on the total dose absorbed. While multiple ring compounds such as alkylated phenanthrenes and dibenzothiophenes are known to cause cancer in higher organisms, these compounds are present in most oils in considerably lower concentrations than lower-molecular-weight hydrocarbons and are unlikely to cause much damage to marine organisms affected by fresh oil, although their relative concentration will increase as oil weathers and if they are mixed into sediments, where they persist (Capuzzo, 1987). Standard toxicity tests (up to 96 hours in duration) have been carried out on numerous organisms, including shrimp (e.g., Anderson et al., 1980; Franklin and Lloyd, 1982), crabs (e.g., Donahue et al., 1977), fish eggs and larvae (e.g., Benville and Korn, 1977; Linden, 1975), marine snails (e.g., Straughan and Hadley, 1978), and other groups, including microalgae (e.g., Batterton et al., 1978). Such tests generally examine the response of organisms to the water-soluble components of oil, rather than to the larger mass of oil itself. Although these studies, particularly the longer ones, do not represent well what happens in a spill, where dilution and evaporation will reduce the concentrations of the more toxic components within a day or two, we can use them to draw some basic conclusions on the relative toxicity of oil and its components to marine life.

As shown in Figure 5.2, different classes of animals react differently to oil, and there are large variations within the same group of organisms, with different species showing wider variability than different phyla, classes, or orders. In general:

- Younger life stages are more affected than mature organisms, often by one or two orders of magnitude, with effects including increased death rates and abnormalities in embryos
- Shallow-water benthic species are more affected than pelagic species because they are less able to escape from an oil slick
- Crude oils are less toxic than refined oils, which often contain more volatile and water-soluble components
- Weathered oil is less toxic than fresh oil, which again contains more of the volatile, water-soluble components, particularly low-molecular-weight aromatics, which are both toxic and narcotic

As an example, developing killifish (*Fundulus heteroclitus*) embryos are known to be affected similarly by both PAHs and polychlorinated biphenyls (Whitehead et al., 2011), and the same genes are affected in both cases. Induction of these genes leads to developmental abnormalities, decreased hatching success, and decreased embryonic and larval survival. The same effects were found in herring embryos and larvae following the *Exxon Valdez* spill (Hose et al., 1996).

Additional factors that can affect toxicity include temperature, time of year, oil concentration and exposure length, what oil is used in the tests (because of the large variation in chemical makeup of different crudes and refined products), whether organisms accumulate or metabolize oil components, and the effects of these components on biological processes such as nerve transmission or breeding success. For example, a spill during breeding season in an area where large aggregations of a particular organism occur will likely have a much more severe effect than the same spill in the same place at a different time (NRC, 2003). Note, however, that bacteria can be less susceptible than other groups and that bacterial degradation is likely the main route by which oil is broken down in the environment (e.g., Mahmoudi et al., 2013).

5.3.2 Sublethal Effects

From numerous studies on multiple species we can conclude that sublethal oil pollution can not only affect all aspects of behavior and physiology of marine organisms, but also that in many cases, the effects are reversible when the concentration of oil is reduced by dilution, aging, or degradation (Brown, 1985). This is similar to the effects of many other pollutants. Additionally, oil components may be metabolized by an organism, and these metabolites may also affect different processes, with persistent exposure causing permanent changes. Feeding, mobility, and respiration are often the first signs of stress, although mobility and respiration can either increase (Widdows et al., 1982) or decrease (Brown, 1982) depending on concentration. In laboratory studies, symptoms can appear at oil concentrations as low as 10–30 µg/L (Widdows et al., 1982), as shown by changes in heart rate, opercular rates, or increased reflex coughing (for fish), as well as valve closure in bivalves or retraction into the shell (for gastropods). Many early examples of such studies are referenced in Malins (1977), as well as the other review articles listed earlier. Since oil can affect membrane structure because of its lipophilic qualities, it may affect the activity of chemoreceptors or cilia by causing changes in membrane permeability (Sanders et al., 1980). This physical effect likely increases the toxicity of oil components to embryos and larvae because hydrocarbons will accumulate in the yolk sac and can affect how lipids are mobilized and metabolized in young stages of organisms.

While marine organisms may concentrate hydrocarbons during exposure, they can also lose them again by depuration as the concentration in the water decreases. This is known for bivalves in particular, but also for crustaceans, fish, and corals (e.g., Anderson, 1975; Knap et al., 1982; Palmork and Solbakken, 1980; Sericano et al., 1996; Solbakken and Knap, 1986). While depuration rates vary from species to species, complete depuration within about 2 weeks is possible. Many marine animals, especially fish, and possibly some plants, possess mixed-function oxidase enzymes such as cytochrome P450 that can metabolize PAHs, changing them to water-soluble compounds. This helps to increase the depuration rate, but it may not be enough to account for complete depuration (Burns, 1976). After the *Exxon Valdez* spill in 1979, P450 gene expression in fish, sea otters, and ducks was prolonged and suggested persistent exposure to sublethal crude oil concentrations (Peterson et al., 2003). Other changes in immunotoxicity commonly seen following oil pollution incidents include aryl-hydrocarbon receptor binding and calcium mobilization (Reynaud and Deschaux, 2006).

Recent advances in genetics have led to studies of gene expression in a number of organisms. Garcia et al. (2012) and Whitehead et al. (2012) looked at gene expression in the fish *Fundulus grandis*, the local equivalent of the more commonly used *Fundulus heteroclitus*, after the 2010 *DWH* spill, using high-throughput sequencing to determine liver mRNA expression. They found a large set of sequences, of which 1,070 were related to downregulated genes and 1,251 came from upregulated genes. Several genes concerned with choriogenin, consistent with exposure to aromatic hydrocarbons, particularly PAHs, were downregulated, while aryl-hydrocarbon receptor sequences AHR1B and AHR2 were upregulated. Genes concerned with hypoxic stress were also upregulated, in agreement with the idea that oil exposure affects gill membrane activity. Genes concerned with blood vessel morphogenesis and ion transport were also affected, and gene expression was found to take place at oil concentrations of around 20 μg/L oil equivalents. While gene expression of particular enzyme groups *per se* is likely an indicator of pollution by oil or some other pollutant, the question is whether it has any effect on the viability of either the particular organism or the population or is merely a standard biochemical response to an outside stimulus. It may be that continued, chronic exposure to small quantities of oil and its breakdown products is more deleterious to a population than a single large spill, although studies around natural seeps in both the Gulf of Mexico and off the coast of California argue against this hypothesis and suggest that different populations can become resilient to continuous exposure.

After the *DWH* spill, there were numerous anecdotal reports of increased occurrence of diseases in finfish and shrimp. These included eyeless shrimp and major necrosis of fish. A recent study (Murawski et al., 2014) showed that the frequency of lesions in fish, particularly red snapper (*Lutjanus campechanus*), correlated well with PAH concentration in the bile of these fish. The question is whether these problems were caused by the oil itself or whether exposure to oil affected the organisms' underlying metabolism so that the organisms were rendered more susceptible to opportunistic infections by parasites, bacteria, or fungi. Tests on both aquatic invertebrates and fish have shown impaired cellular immunity and lymphocyte proliferation, with immunotoxic effects being correlated with the PAH component of oil (Reynaud and Deschaux, 2006), suggesting that it is infections rather than the oil itself that lead to many of the observed effects.

The effect of oil pollution on populations and ecological communities is generally much harder to judge, because of (1) the need to ensure that other impacts, such as increased temperature or decreased food availability, did not cause the observed changes; (2) normal population variability; and (3) uncertainty in prespill numbers of observed species. This is the case even after large spills, such as the *Exxon Valdez* (Peterson et al., 2003) or *DWH*, where recoveries of different species have taken very different lengths of time. However, individual oil spills can have considerably more impact than chronic oil pollution in certain cases, such as when fish spawn only once a year and their larvae are affected by a spill, thereby reducing the viability of the next year class, or when breeding populations of birds are affected by spills near their breeding area (NRC, 2003). Rapid changes in algal cover on rocky shores can also occur following a spill, mainly because of the death of grazing organisms, such as gastropods, that normally control algal growth, but these can be reversed as the grazing populations recover (Moldan et al., 1979).

One area where oil pollution can be shown to affect populations is that of tainting edible fish and shellfish products. This is of concern to the marine food industry because of the considerable economic damage as people stop buying or consuming these products. This was the case in Scotland following the *Braer* oil spill in the Shetland Islands, when large quantities of farmed salmon had to be discarded (ESGOSS, 1994), as well as in 2010 after the *DWH* spill in the Gulf of Mexico, when not only were fisheries from Louisiana to Florida shut down for several months but the perception of tainting in crabs, shrimp and oysters persisted for considerably longer, well into 2011 (National Commission, 2011; NOAA, 2012). Because the Gulf of Mexico produces about one-third of the U.S. domestic seafood supply, this had a major economic effect on the local human population, even though all tests on edible organisms showed extremely low levels of contamination, none of which were deemed a problem for human health (National Commission, 2011).

5.4 Basic Methodology of Mitigation

Before discussing mitigation efforts, it is necessary to look at what happens during an oil spill. The main fates of an oil spill in the marine environment are shown in Figure 5.3. This figure shows what happens following a surface spill; most spills occur at the surface, so the formation of a deep plume, as was observed following the *DWH* spill (Camilli et al., 2010), is not a factor. While we think of an oil spill as being a coherent mass, once oil is released into the environment, it undergoes several changes, including spreading, evaporation, dispersion, dissolution, weathering, sinking and biological degradation. The most obvious change is that the slick moves, partly because oil spreads out but also because of the influence of wind and currents. However, various oil components also dissolve in water or evaporate at the sea surface, so that for very light crudes with API gravities $>>10°$ up to 50% or more may evaporate within 48–72 hours. This leaves the heavier fractions and increases the density of the remaining oil so that it can eventually form heavy, sticky tar

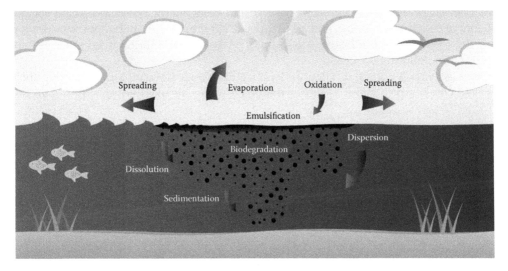

FIGURE 5.3
Factors that cause changes in oil following a spill in the ocean. (ITOPF. (2014). *Handbook, 2014–2015.* International Tanker Owners Pollution Federation, London, p. 52.)

---

* American Petroleum Institute (API) gravity is a measure of the relative densities of different petroleum sources. It is defined by API gravity = 141.5/RD − 131.5, where RD is the relative or specific gravity of the oil, and an API gravity of 10 has the same density as water at 60°F. It defines whether an oil will sink (API < 10) or float (API >10).

balls that can either float or sink, depending on whether they encounter particles that can raise their density to more than that of the water. Other weathering effects include oxidation of certain components by sunlight, which can break them down or make them more soluble, and emulsification, when water mixes with the residual oil to form a relatively stable mixture known as 'chocolate mousse' because of its frothy constituency. Spills generally produce large volumes of mousse, which, after 72–96 hours' exposure to wind and wave action, contains about 60%–70% water, has a higher viscosity than the original oil, and is more difficult to deal with. The eventual fate of mousse is often as floating tar balls that strand on beaches (Smith and Knap, 1985; Wade et al., 1976).

Oil spills became more frequent during the 1960s and 1970s, and more effort was put into planning to lessen their impacts. Two main physical methods frequently used together to contain and remove oil are booming and skimming. Chemically, dispersants are an additional response, while gelling agents have also been used, and burning can be used if conditions are right. These different methods are discussed below. The main problem with almost any oil spill in the marine environment is the sheer scale of the problem; once oil escapes from a ship, well, or pipeline, it spreads rapidly across the sea surface so that containment becomes difficult and collection even more so. Because 1 m³ of oil can cover an area of a hectare (100 × 100 m) with a layer 0.1 mm thick, spills of several thousand cubic meters rapidly become extremely difficult to manage.

5.4.1 Booms and Skimmers

The best way to prevent oil from spreading is to contain it within a boom, which allows responders to collect the oil before it can contaminate the surrounding area, using skimmers, or by burning or adding gelling agents, to prevent it from mixing into the water column. Booms are also used to protect sensitive areas; during the *DWH* spill about 660 miles (770 km) of containment boom and 1,443 miles (2340 km) of sorbent booms were deployed along the coastlines of Louisiana, Mississippi, Alabama and Florida in an attempt to prevent oil from polluting the coast.

Booms essentially consist of a float and a skirt that hangs down into the water (Figure 5.4). The float can be up to about 1–2 m in diameter, while the skirt is typically 1–2 m deep.

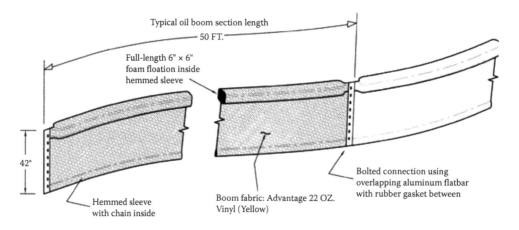

FIGURE 5.4
Standard makeup of a typical oil spill containment boom (downloaded from http://www.bradfordmarinebahamasfabricstructures.com/images/oil-boom-drawing.jpg.)

Containment booms are made from impervious materials so that oil meeting the boom can be directed along it toward collection areas where it is removed by pumping, skimming, or burning. Unfortunately, this only really works well in confined areas where wave, wind and current action are reduced, because waves tend to wash oil over the top of a boom, currents will take it underneath, while strong winds tend to both increase wave height and also move the boom from its original position. Sorbent booms, on the other hand, as their name implies, are made of oleophilic material that traps oil by adsorbing it on the boom's surface. The idea is that these booms can be collected, the oil can be removed, and the booms can be redeployed, although this is not as easy in practice as it sounds, especially in more exposed areas. Many manufacturers have tried to couple booms to boats, either having a boat fitted with a single extension arm holding a boom (the latest models extend and collapse automatically from the side of the boat) or by towing a boom between two boats with a skimmer system in the center. While these can work in confined areas such as rivers and estuaries, in the open sea the main problem is finding a high enough density of oil that can be picked up because of its tendency to spread into very thin films (ITOPF, 2014). Thus, generally less than 10% of a spill in the ocean can be picked up in this way (the official estimate following *DWH* is that only 3% of the total release was picked up) (National Commission, 2011).

Skimmers work on the principle that oil will form a thin layer on a nonporous surface that can be skimmed off. This requires a large contact area, so they can be made with either a drum configuration or a series of belts or multiple rotating discs. As these revolve, so the drum, belts, or discs encounter the oil, which is then wiped off by a blade into a container. For most effective use, all types of skimmers need a relatively thick layer of oil (>1 mm) and a calm surface, preferably in an enclosed area, which essentially limits them to harbors and other situations where the oil can be easily contained. In the open sea, they suffer from the same limitations as booms, although they are often used in these situations.

5.4.2 Dispersants

The *Torrey Canyon* spill in 1967, which at the time was the largest single oil spill in the world (119,000 tons), saw the first large-scale use of dispersants as a control technology. Dispersants operate by breaking oil into small droplets, which are more easily dispersed into the water column and provide a larger surface area for bacteria to colonize, thereby increasing the rate at which oil is degraded. Being liquids, they can be applied from boats, planes and helicopters, and their effectiveness is increased by ensuring that the oil and dispersant mix well, so in most operations at sea either the ship's wake is used to provide turbulence to increase mixing or mixing equipment, such as a series of buoyant bars like a farm gate, is towed behind the ship. In the open ocean, dispersants are only used when wind and wave conditions are sufficient to cause the required mixing, but not so strong as to completely break up a slick. Dispersants are most effective on fresh oils and much less effective once the oil has weathered or formed a mousse. Since they disperse oil into the water column, their use depends on the relative risk to, for example, pelagic organisms or the shoreline (ITOPF, 2014), and this is taken account of in contingency planning, with most countries having strict rules on when and where dispersants can be used.

The first dispersants, used following the *Torrey Canyon* spill, proved to be very efficient at dispersing oil but were also more toxic than the oil itself. Approximately 600,000 imperial gallons (2.7 million L) of BP 1002 dispersant were used on this spill, causing very considerable damage to the intertidal wildlife (Gill et al., 1967). As a result of these effects, there has been an enormous amount of research on dispersant effectiveness and toxicity (e.g., ASTM, 1984; NRC, 1989, 2005), and the more modern dispersants, such as the Corexit 9527 used during the *DWH* spill, are considerably less toxic. They are also somewhat less effective at dispersing oil, the two effects being inversely related. Following the *DWH* spill, some 1.84 million gallons of dispersant were used, both at the surface and at depth, with minimal apparent ill effects on the biota (National Committee, 2011).

Nearly all dispersant usage to date has been on surface spills, where the aim is to remove the oil from the surface to prevent it from impacting shorelines. During the *DWH* spill, however, efforts were made to apply dispersant directly at the wellhead, some 1500 m below the surface. Anecdotal evidence suggests that this had a major effect, reducing considerably the amount of oil that reached the surface during times when dispersant was applied in this manner, although putting more finely dispersed droplets of oil into the water column. An advantage of using dispersant in this manner was that the natural turbulence of the outwelling oil and gas plume caused good mixing at that depth, and it is thought that in any future similar deep spill more concentrated dispersant can be used, reducing the volume that must be transported to the spill site.

5.4.3 Gelling Agents and Sorbents

Oil can be turned into a rubberlike solid by the addition of specialty chemicals, known as gelling agents, which form polymer nets that react with the molecules in the oil. In theory, therefore, one can solidify an oil spill to prevent it from spreading further and use mechanical methods such as nets or suction equipment to remove it from the water surface. While under calm conditions energy is needed to mix the chemicals with the oil, this means that such agents can be used in higher wave states as the wave energy will provide the necessary mixing. In practice, because of the need to add gelling agents in volumes two or three times that of the oil, the logistics of dealing with enough agent to cope with a large spill become impractical.

5.4.4 Burning

Crude oil on the sea surface will burn, despite the high heat capacity of seawater, if it is in a layer several centimetres thick. Following the wreck of the *Torrey Canyon*, the British government experimented to see whether they could use burning to reduce the pollution threat, then bombed the wreck to release the oil, using kerosene and napalm to help ignite the crude and maintain the fire (Gill et al., 1967). While this approach worked, modern response techniques usually only try burning if the oil can be contained within fireproof booms, which is not always an option. During the *DWH* spill, there were 411 controlled burns, which consumed an estimated 5% of the spilled oil (National Commission, 2011). Because the most volatile and flammable components of oil evaporate quickly, this method only works on fresh spills since igniting oil is a problem with older slicks. Burning can, however, be effective for oil spills in icy regions if the oil is either trapped on the ice surface or next to an ice floe. The disadvantages of burning include the air pollution produced by burning oil and the fact that it can produce heavier tarry residues that can sink and contaminate benthic biota.

5.4.5 Bioremediation

Since bacteria are known to break down oil, it is often suggested that adding bacterial cocktails, along with additional nutrients such as nitrogen and phosphorus to aid in their growth, will speed up the process of degradation after a spill. However, despite much interest in using this technique both onshore and at sea, this has not been demonstrated as being either technically feasible or particularly beneficial for large-scale application (ITOPF, 2014). This is mainly because of the time scales needed for bioremediation to be effective; bacteria take months to years to break down crude or refined oils, but most response agencies need a solution that works in hours to days. As stated earlier, in almost all cases natural remediation processes are the final cleanup process for environmental oil spills.

5.4.6 Doing Nothing

Although there is always a demand for 'action' following an oil spill, in many cases the best thing to do is nothing. This is certainly the case for small marine spills of light, refined products, such as diesel oil, because by the time a response can be mounted, almost all biological effects will have occurred and the residual oil will have evaporated. Even for large spills offshore, leaving the oil to disperse naturally can minimize any impacts. This was the case following the *Castillo de Bellver* spill 24 miles off the west coast of South Africa in 1983, when the tanker exploded and released approximately 50–60,000 tons of light crude. Much of the oil burned, and while there was a small amount of dispersant spraying, the residue dispersed naturally with only minimal environmental effects. Although some birds, mainly gannets, were affected initially (1,500 oiled gannets were collected, cleaned, and rereleased within a few weeks of the accident), long-term effects were not observed (Moldan et al., 1985).

5.5 Management Plans for Different Ecosystems

5.5.1 Contingency Planning

In any response to a spill or similar disaster, whether natural or anthropogenically induced, prior planning for such an event allows for an orderly approach to protecting the threatened areas. In the case of oil spills, contingency planning has been available for at least three decades. An oil spill contingency plan is made up of many parts and depends on whether one is concerned about large areas (country to state) or has more local concerns (e.g., town to beach). In either case there are levels of granularity associated with such spills. Because this is a comprehensive subject in its own right, we cannot go into great detail but will list the elements of such a plan. Usually, the plan starts with recognition of the management hierarchy responsible for disaster response. The island of Bermuda, where groundings and spills occur several times a year, is used as an example. As a remote island in the Atlantic it has response issues similar to many small island states.

The oil spill contingency plan usually sets out who is responsible for overall control of response to the spill, the connection with government agencies, media, the management of outside advisors, the ship or other installation (e.g., shore-based holding facilities), the salvors, and the operational teams who deal with booms or other cleanup technology. The plan should also include information on the local ocean currents and knowledge of the environmental sensitivity of the region. Oil spill sensitivity maps are essential in planning which areas of the coast are the most vulnerable environmentally and operationally, as mixed in with environmental concerns are tourist beaches, power plant cooling water intakes, harbors, and other industrial and social infrastructure. Current speeds, direction, and meteorology are important, as these need to be integrated into real-time models to determine where the spill may go and which areas may be the most vulnerable. Tidal regimes are also important for the transport and redistribution of the oil.

Usually the lead for the response is the on-scene commander, who is appointed with a team of people who can provide advisory information. If the spill is from a vessel, the local Coast Guard is often the first to either detect or to be notified of an event. Depending on the thresholds set as to the total risk, there should be a 'call-out' plan by which various representatives are notified at specific times as needed.

One of the key items of the plan is local knowledge of the area and of the tools that may be on hand to deal with the spill, such as booms, skimmers, or dispersants. In the case of the Bermuda plan (Knap et al., 1985), there was an inventory of containment booms, a supply of adsorbent booms and pads, skimmers, and even anchoring information for the key bays and inlets that needed to be protected in case of a spill. High-priority areas to be protected were set through an Environmental Sensitivity Index (ESI). Also, specific areas were designated to be safe for dispersant use. Much of Bermuda is coral reef, and whether or not dispersants can be used depends on an estimate of oil spill thickness. There are only a few mangrove systems in Bermuda, and these are considered to be of more concern than coral reefs because once oil gets into a mangrove swamp, it takes tens of years for the oil to degrade (Jackson et al., 1989). Therefore part of the plan was to set priorities for protection.

The key to all of this is planning for as many scenarios as possible. Oil spill drills (rehearsals) need to be carried out at least every 2 years in full, with desktop exercises once per year (Sleeter et al., 1983). People change jobs, keys get lost, new information on equipment and infrastructure development is always evolving, and this may affect the ESI of a given area, so the plan is quite useless if it is not practiced. Incorporating existing scientific information is essential. In the end, over about a 30-year period, Bermuda's plan was put into motion in a major way only about five times where there were large threats – one was an oil tanker laden with 125,000 tons of crude oil aground on the reefs of Bermuda with a hurricane forecast. Fortunately, thanks to planning – and luck – no oil was spilled (Knap et al., 1985).

5.5.2 Environmental Sensitivity Index

An ESI is an essential part of every plan – basically it describes the most sensitive environments in a coastal area and is concerned with various effects of oil on a specific environment. Linked into the ESI would be the value of a resource to the local community and economy. For example, a sandy tourist beach may change its importance owing to the needs of the local economy even though long-term effects on such a beach are minimal, so although there are specific guidelines based on environmental sensitivity, these can be modified based on environmental and social concerns. The physical energy of the area plays a large role in the decision on priorities of protection. For example, in a mangrove bay with a low current regime, oil can persist for a very long time. Such an area will be far more 'sensitive' than a sea wall or rocky coastline, where strong waves will break up the oil and remove it much more rapidly. A general classification, from most sensitive (those areas of greatest concern from an oil spill) to least sensitive, is given in what follows.

ESIs are generally displayed as environmental sensitivity maps, which can be of three types: strategic, tactical and operational (IPIECA, 2012). Strategic sensitivity maps

■	1	Exposed vertical rocky shores exposed seawalls
▨	2	Exposed rocky platforms
▦	3	Fine-grained sand beaches
□	4	Coarse-grained sand beaches
▱	5	Mixed sand and gravel (shell) beaches/fill
▩	6	Gravel beaches/riprap
▨	7	Exposed tidal flats
□	8	Sheltered rocky shores/seawalls/ vegetated banks, solid man-made structures
▩	9	Sheltered tidal flats
▦	10A	Exposed marshes and/or mangroves
▦	10E	Sheltered marshes and/or mangroves

FIGURE 5.5
Classifications of shorelines used in a typical spill response planning document.

are developed, at a fairly small geographic scale, to provide a broad perspective and to synthesize information, locating and prioritizing the most sensitive sites. Tactical maps provide responders with all required environmental, socioeconomic, logistical and operational information for planning and implementing response and protection operations. Operational maps include information on the general logistical and operational resources. All of these are used for a successful response and can be expanded to provide information on, for example, seabird breeding sites or even migration patterns, fish stocks and other important environmental information. Generally the maps are color coded from Class 1 (purple), the least-sensitive areas such as exposed rocky shores, to Class 10 (red or magenta), low-energy environments such as marshes, swamps and mangroves (IPIECA, 2012).

Figure 5.5 shows a typical ranking of environmental sensitivity. In this case, it is from the state of Florida and is typical of many of the indexes available. As indicated earlier, wave and tidal energy are important for the removal or degradation of oil, so sea walls, rocky shores, and fine- and coarse-grained beaches, as well as shell beaches, are usually by definition high-energy places where there is a great deal of dilution and energy that breaks up oil through wave, current, ultraviolet and microbial degradation. In the case of Figure 5.6, which shows a basic ESI map for part of Normandy, France, there are no mangroves, sea grasses, or coral reefs, which would all be rated as most sensitive (Classes 9 and 10), and the red areas are areas of broken coastal marshes.

5.5.3 Specific Environments

5.5.3.1 Beaches

Beaches tend to be areas that are heavily impacted during an oil spill, but this is generally limited to the area of the tidal regime. In high-energy environments with strong turbulence, the residence time on the beach is limited. However, because these beaches are easily accessed by humans, they are areas of attention during a spill, and while bulldozers and other cleaning activities can be used relatively easily, heavy equipment can affect the beach. The oil is visible, and during spills such as the *Exxon Valdez* or *DWH* blowout, beaches become areas of high activity, especially in cobble beaches where oil penetrates between the pebbles.

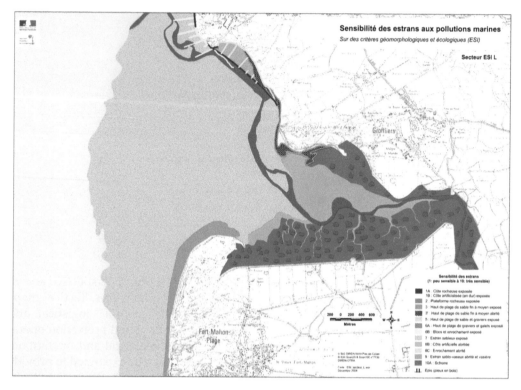

FIGURE 5.6
Basic environmental sensitivity index map for part of Normandy, France. The colors are essentially the same as those listed in Figure 5.5. There are no mangroves or corals in this area, and the red areas are broken marsh. (Taken from IPIECA. (2012). *Sensitivity Mapping for Oil Spill Response*. IPIECA/IMO/OGP Report 477, London, p. 39).

5.5.3.2 Exposed Tidal Flats

Exposed tidal flats are ranked 7 in the ESI code, and oil tends to persist for longer periods because oil is stranded on each tide and can be incorporated into sediments. Because such oiling and dilution occur continuously, they are not as sensitive as other areas, although infauna, wading birds, and other organisms can be affected. Sheltered tidal flats, however, are more vulnerable as the energy to remove the oil is not available as in higher-energy environments. Migratory birds, mollusks and infauna such as shellfish would be at risk for a longer period than on exposed tidal flats. Category 10 regions (exposed or sheltered marshes, together with mangroves and seagrasses) are extremely sensitive, and once oil gets into these areas, it usually takes a very long time to remove it (years to decades).

5.5.3.3 Rocky Shores

Many oil spills over the years have occurred around rocky shores because these areas tend to be the closest to shipping routes and channels. From the *Torrey Canyon, Amoco Cadiz, Atlantic Empress* and *Braer* spills to the *Exxon Valdez* and many others, these areas have been the most affected. However, in many cases, except for perhaps the *Exxon Valdez*, many of these areas are of high energy with breaking waves. In many of these spills, the amount of oil delivered to the area was large, but wave action was also high. Certainly in low-energy rocky shores or on beaches consisting of large cobbles where the oil can penetrate deeply into the substrate, oil can be very persistent (Peterson et al., 2003), and in the case of the *Amoco Cadiz* it took approximately 15 years for some of the populations to recover to prespill conditions (Dauvin, 1998). The case of the *Exxon Valdez*, where rocky shores were badly affected by the oil, is an outlier because there were so many cleanup requirements that much damage was done by the cleanup alone. First the beaches were washed with high-pressure cold water, followed by hot-water high-pressure washing, which killed the organisms and washed the finer sediments out of the area so communities could not recover (Driskell et al., 1996).

5.5.3.4 Marshes

Marshes are ubiquitous and have suffered through many spills. It is worth repeating that toxicity and the effect of oil on organisms depend on the type of oil spilled (e.g., light, heavy, refined), whether the spill is chronic or acute, the amount of oil spilled, and the depth of penetration into the sediments. Weather, season, and cleanup methods all also affect the toxicity and the eventual effects on the environment. Marsh grasses tend to be more affected by light refined oils than crude oil (Hoff, 1996; Webb, 1996), with recovery times from 3 months to 3 years, but can generally withstand being oiled once, although multiple oilings strongly reduce the recovery rate (Baker, 1971). However, oil from the Buzzards Bay spill in Massachusetts in 1969 (Sanders et al., 1980) could be found in sediments many years after the spill (Teal et al., 1992; Culbertson et al., 2007), and thus there can be long-term effects on biota. In colder climates, recovery of ecosystems takes longer, and in the *Amoco Cadiz* spill in northern France, long recovery times were due to (1) a cold environment, (2) high organic content of the soil, (3) a sheltered location in a low-energy environment, (4) heavy oiling, (5) spills of both crude and fuel oil (refined oil) and (6) physical disturbance of the environment during the cleanup (Hoff, 1996).

This last point is a major problem when marshes are oiled. Work carried out in the 1960s and 1970s in the U.K. showed that the best way to clean oiled marshes was simply to cut and remove the oiled plants where possible and leave them to regenerate naturally (Baker, 1971). Recent studies on Louisiana marshes after the *DWH* spill in the Gulf of Mexico reported effects on crabs and arthropods, but that recovery was generally within 1 year, and snails were unaffected (McCall and Pennings, 2012). However, trying to clean oil from a marsh by physical means, particularly when using heavy equipment, is liable to cause considerably more damage by mixing oil into the marsh sediments (Baker, 1971). Two years after the *DWH* spill, Silliman et al. (2012) found that the cleanup attempts following the spill had amplified ongoing coastal erosion, a clear example of how multiple human-induced stressors can work in concert to increase degradation of these environments. This is the theme of many studies in many habitats – 'the straw that broke the camel's back'.

5.5.3.5 Mangroves

Mangroves occur in very low-energy environments and are thus very susceptible to oil spills. The effect of oil on mangroves has been reviewed by Proffitt et al. (1995) and been shown to depend on properties similar to those found with marshes. Oiling effects of mangroves also differ with the age and life history of the trees. The primary effect appears to be the disruption of gas exchange through the stilt roots that, when coated with oil, can no longer supply oxygen to the plant, especially in hypoxic soils. There is also a general disruption of root membranes (Teas et al., 1993; Page et al., 1985). Following a spill at Galeta, on the Atlantic coast of Panama, many adult mangrove trees were affected; adult trees died, seagrass mats disappeared, and erosion increased (Jackson et al., 1989; Keller and Jackson, 1993). Once oil is in these peaty environments, it sticks to the substrate, and at each tidal cycle the whole system of mangroves, seagrasses, and coral reefs is re-oiled. The Galeta spill is very similar to others discussed earlier in that the local environment was already badly affected by river discharge and questionable land-use practices, and the oil spill was the last insult. An unknown aspect of this spill is the effect of dispersants: at least 22,000 gallons of dispersant were purportedly sprayed soon after the spill, although this has never been documented properly (A. Knap, pers. comm.).

One of the longest controlled studies on the effect of oil on tropical systems was the controlled spill on the Atlantic coast of Panama near Bocas del Toro (Knap, 1981). Permission was granted to spill oil experimentally over a coral, seagrass, and mangrove area. At one of those sites, Exxon 9527 (the same dispersant used at the beginning of the *DWH* response) was premixed with Arabian light crude oil and spilled for 24 hours, after which the whole system was monitored for over 20 years. The mangroves in the dispersed oil site fared much better than those that were dosed only with oil. Presumably the dispersant prevented the oil from coating the roots, as discussed by Teas et al. (1993). After many years post-spill, the mangroves in the oil-only area were still affected compared to the controls (Ward et al., 2003).

5.5.3.6 Coral Reefs

Coral reefs are considered to be the most diverse and complex communities in the marine environment. The hermatypic corals (reef builders) are the main structures of the reef, taking calcium carbonate from seawater and converting it to reef mass, and many studies have been focused on them. Corals are complex organisms, as algae (Zooxanthellae) live within the coral tissue and harvest light, converting it to chemical energy and translocating this energy to the corals themselves. So, oil can affect corals through direct contact or by affecting the photosynthesis process and energy transfer within them (Cook and Knap, 1984). The few major studies that have documented the detailed effects of oil and chemically dispersed oil on corals have been reviewed by Knap (1992). Laboratory studies of the very detailed effects of oil and chemically dispersed oil on the brain coral (*Diploria strigosa*) were carried out for 3 years (Knap et al., 1995; Dodge et al., 1984; Wyers et al., 1986). The results suggested that oil contamination of corals affected behavior rapidly but that longer-term effects were absent. Perhaps one of the most detailed studies on a coral reef area happened off Galeta on the Atlantic coast of Panama, as described previously (Jackson et al., 1989; Burns and Knap, 1989). Effects were major and the coral populations, although viable after 2 years, had not recovered after 5 years. However, as discussed earlier in this chapter, these reefs were in very poor health prior to the spill, so this acute insult of oil seriously affected the viability of the system, and the lack of documentation about dispersant use made it difficult to reach definitive conclusions. There have not been many recent studies on the effects of oil and oil dispersants on corals, although reports suggest that deep-water coral communities close to the *DWH* spill site were badly affected (White et al., 2012; Fisher et al., 2014), and Goodbody-Gringley et al. (2013) have investigated the effects of oil and dispersant on coral larvae. Their results suggested that both larval settlement and survival were affected, but a more detailed risk assessment is still required (Knap, 1992).

5.6 Conclusions and Recommendations

The response to oil spills requires trade-offs between protecting different resources and ecosystems, and decisions on the correct response should be developed and documented in response plans before a spill occurs. Despite the general concern over oil spills, in many cases long-term effects on the biota cannot be documented. Effects depend to a large degree on the local ecology affected by the spill, what type of oil is spilled, and how it responds to prevailing weather and ocean conditions. There is a large industry devoted to various aspects of oil spill cleanup, but, as shown by the recent *DWH* incident, the sheer scale of such a disaster means that much of the effort put into trying to collect oil at sea is probably ineffective. Even after more than 40 years of effort, the only effective cleanup method for most open-ocean spills is to apply dispersants to break the oil down into smaller droplets to encourage bacterial decomposition. In the case of the *DWH* spill, dispersant use prevented only about 8% of the oil from reaching sensitive coastal areas, although about twice this amount is thought to have dispersed naturally (National Commission, 2011). These results of spills demonstrate the need for continued research on the effect of dispersed oil on surface and deep ocean ecosystems.

For sheltered areas, such as mangroves and marshes, as well as coral reefs, the best form of response is to prevent the oil from getting into these areas in the first place. If, however, oil does penetrate such areas, then cleanup methodology should be to minimize physical damage to the soil substrate because the use of heavy equipment will cause sediment loss and compaction as well as churning up the surface so that oil is mixed much deeper into the sediments. In all instances, however, having an effective plan of response that is practiced regularly is likely to minimize the damage to the environment.

Acknowledgements

This work was supported in part by the BP/Gulf of Mexico Research Initiative through Grant SA09/GoMRI-006 to Texas A&M University.

References

Anderson, J.W. (1975). Laboratory studies on the effects of oil on marine organisms: an overview. *American Petroleum Institute Publication*, 4249: 1–70.

Anderson, J.W., Kiesser, S.L. and Blaylock, J.W. (1980). The cumulative effect of petroleum hydrocarbons on marine crustaceans during constant exposure. *Rapport et Proces. Verbaux des Reunions Conseil International pour l'Exploration de la Mer*, 179: 62–70.

ASTM. (1984). In: T.E. Allen, (ed.). *Oil Spill Chemical Dispersants: Research, Experience and Recommendations*. American Society for Testing and Materials, Philadelphia, p. 465.

Baker, J.M. (1971). Successive spillages. In: E.B. Cowell (ed.). *The Ecological Effects of Oil on Littoral Communities*. Institute of Petroleum, London, pp. 21–32.

Baker, J.M. (1982). Mangrove swamps and the oil industry. *Oil and Petrochemical Pollution*, 1: 5–22.

Baker, J.M. (1983). Impact of oil pollution on living resources. *The Environmentalist*, 3 (supp. 4): 1–48.

Baker, J.M., Crothers, J.H., Little, D.I., Oldham, J.H. and Wilson, C.M. (1984). Comparison of the fate and ecological effects of dispersed and nondispersed oil in a variety of intertidal habitats. In: T.E. Allen (ed.) *Oil Spill Chemical Dispersants: Research, Experience and Recommendation*. ASTM STP 840, Philadelphia, pp. 239–279.

Ballou, T.G., Dodge, R.E., Hess, S.C., Knap, A.H. and Sleeter, T.D. (1987). *Effects of a Dispersed and Undispersed Crude Oil on Mangroves, Seagrasses and Corals*. American Petroleum Institute Publication 4460, Washington, D.C., p. 198.

Ballou, T.G., Hess, S.C., Dodge, R.E., Knap, A.H. and Sleeter, T.D. (1989). Effects of untreated and chemically dispersed oil on tropical marine communities: a long-term field experiment. *Proceedings of the 1989 Oil Spill Conference, American Petroleum Institute Publication 4479*, Washington D.C., pp. 447–454.

Batterton, J.C., Winters, K. and Van Baalen, C. (1978). Sensitivity of three microalgae to crude oils and fuel oils. *Marine Environmental Research*, 1: 31–41.

Benville, P.E. and Korn, S. (1977). The acute toxicity of six monocyclic aromatic crude oil components to striped bass (*Morone saxatilis*) and bay shrimp (*Crago franciscorum*). *California Fish and Game*, 63: 204–209.

Blumer, M., Souza, G. and Sass, J. (1970). Hydrocarbon pollution of edible shellfish by an oil spill. *Marine Biology*, 5: 195–202.

Boehm, P.D., Fiest, D.L. and Elskus, A. (1981). Comparative weathering patterns of hydrocarbons from the Amoco Cadiz oil spill observed at a variety of coastal environments. In: *Amoco Cadiz: Consequences d'une Pollution Accidentelle par les Hydrocarbures: Fates and Effects of the Oil Spill*. Centre National Pour L'Exploitation des Oceans. Paris, pp. 159–173.

Boehm, P.D. and Requejo, A.G. (1986). Overview of the recent sediment hydrocarbon geochemistry of Atlantic and Gulf Coast outer continental shelf environments. *Estuarine, Coastal and Shelf Science*, 23: 29–58.

Brassell, S.C., Eglinton, G., Maxwell, J.R. and Philp, R.P. (1978). Natural background of alkanes in the aquatic environment. In: O. Huntzinger, L.H. van Lelyveld and B.C.J. Zoetman (eds) *Aquatic Pollutants, Transformations and Biological Effects*, Pergamon Press, Oxford, pp. 69–86.

Brown, A.C. (1982). Pollution and the sandy-beach whelk *Bullia*. *Transactions of the Royal Society of South Africa*, 44: 555–562.

Brown, A.C. (1985). *The Effects of Crude Oil Pollution on Marine Organisms: a Literature Review in the South African Context: Conclusions and Recommendations*. Report for Oil Pollution Committee of the South African Committee for Oceanic Research (SANCOR), p. 33.

Burns, K.A. (1976). Hydrocarbon metabolism in the intertidal fiddler crab, *Uca pugnax*. *Marine Biology*, 36: 5–11.

Burns, K.A. and Knap, A.H. (1989). The Bahia las Minas oil spill: hydrocarbon uptake by reef building corals. *Marine Pollution Bulletin*, 20: 391–398.

Camilli, R., Reddy, C.M., Yoerger, D.R., Van Mooy, B.A., Jakuba, M.V., Kinsey, J.C., McIntyre, C.P. et al. (2010). Tracking hydrocarbon plume transport and biodegradation at Deepwater Horizon. *Science*, 330: 201–204.

Capuzzo, J.M. (1987). Biological effects of petroleum hydrocarbons: Assessments from experimental results. In: D.F. Boesch and N.N. Rabalais (eds) *Long-term Environmental Effects of Offshore Oil and Gas Development*, Elsevier, Applied Science, London, pp. 343–410.

Clark, R.C. Jr and Blumer, M. (1967). Distribution of n-paraffins in marine organisms and sediments. *Limnology and Oceanography*, 12: 79–97.

Cook, C.B. and Knap, A.H. (1984). Effects of crude oil and chemical dispersants on photosynthesis in the brain coral (*Diploria strigosa*). *Marine Biology*, 78: 21–27.

Culbertson, J.B., Valiella, I., Peacock, E.E., Reddy, C.M., Carter, A. and VanderKruik, R. (2007). Long-term biological effects of petroleum residues on fiddler crabs in salt marshes. *Marine Pollution Bulletin*, 54: 955–962.

Dauvin, J.C. (1998). The fine sand *Abra alba* community of the Bay of Morlaix twenty years after the Amoco Cadiz oil spill. *Marine Pollution Bulletin*, 36: 669–672.

Dodge, R.E., Wyers, S.C., Frith, H.R., Knap, A.H., Smith, S.R. and Sleeter, T.D. (1984). The effects of oil and oil dispersants on the skeletal growth of hermatypic coral, *Diploria strigosa. Coral Reefs,* 3: 191–1984.

Donahue, W.H., Welch, M.F., Lee, W.Y. and Nicol, J.A.C. (1977). Toxicity of water soluble fractions of petroleum oils on larvae of crabs. In: C.S. Gian (ed.) *Pollutant Effects on Marine Organisms,* Lexington Books, Lexington, pp. 77–94.

Driskell, W.B., Fukiyama, A.K., Houghton, J.P., Lees, D.C., Mearns, A.J. and Shigenaka, G. (1996). Recovery of Prince William Sound intertidal infauna from Exxon Valdez oiling and shore-line treatments: 1989 through 1992. In: S.D. Rice, R.B. Spies, D.A. Wolfe and B.A. Wright (eds) *Proceedings of the Exxon Valdez Oil Spill Symposium.* American Fisheries Society Symposium 18, Anchorage, AK, pp. 362–378.

ESGOSS. (1994). *The Environmental Impact of the Wreck of the Braer.* Ecological Steering Group on the Oil Spill in Shetland. Scottish Office, Edinburgh, p. 207.

Farrington, J.W., Teal, J.M., Quinn, J.G., Wade, T. and Burns, K. (1973). Intercalibration of analyses of recently biosynthesized hydrocarbons and petroleum hydrocarbons in marine lipids. *Bulletin of Environmental Contamination and Toxicology,* 10: 129–136.

Farrington, J.W. and Tripp, B.W. (1977). Hydrocarbons in western North Atlantic surface sediments. *Geochimica Cosmochimica Acta,* 41: 1627–1641.

Fisher, C.R., Hsing, P.-Y., Kaiser, C.L., Yoerger, D.R., Roberts, H.H., Shedd, W.W., Cordes, E.E. et al., (2014). Footprint of Deepwater Horizon blowout impact to deep-water coral communities. *Proceedings of the National Academy of Sciences,* 111: 11744–11749.

Franklin, F.L. and Lloyd, R. (1982). *The toxicity of twenty-five oils in relation to the MAFF dispersant tests.* Fisheries Research Technical Report 70, Ministry of Agriculture, Fisheries and Food, Lowestoft, p. 13.

Garcia, T.I., Shen, Y., Crawford, D., Oleksiak, M.F. and Whitehead, A. (2012). RNA-Seq reveals complex genetic response to Deepwater Horizon oil release in *Fundulus grandis. BMC Genomics,* 13, 474–482.

Getter, C.D. and Baca, B.J. (1984). A laboratory approach for determining the effect of oils and dispersants on mangroves. In: T.E. Allen (ed.) *Oil Spill Chemical Dispersants: Research, Experience and Recommendation.* ASTM STP 840, Philadelphia, pp. 5–13.

Gill, C., Booker, F. and Soper, T. (1967). *The Wreck of the Torrey Canyon.* David and Charles, Newton Abbot, p. 128.

Goodbody-Gringley, G., Wetzel, D.L., Gillon, D., Pulster, E., Miller, E. and Ritchie, E. (2013). Toxicity of Deepwater Horizon source oil and the chemical dispersant, corexit 9500, to coral larvae. *PLOS.* doi: 10.1371/journal.pone.0045574.

Goutx, M. and Saliot, A. (1980). Relationship between dissolved and particulate fatty acids and hydrocarbons, chlorophyll a and zooplankton biomass in Villefranch Bay, Mediterranean Sea. *Marine Chemistry,* 8: 229–318.

Hoff, R.Z. (1996). Responding to oil spills in marshes: the fine line between help and hindrance. In: C. E. Proffitt and P. F. Roscigno (eds) *Symposium Proceedings: Gulf of Mexico and Caribbean Oil Spills in Coastal Ecosystems: Assessing Effects, Natural Recovery, and Progress in Remediation Research.* OCS Study MMS 95-0063. Department of the Interior, Minerals Management Service, New Orleans, LA, pp. 146–161.

Hose, J.E., McGurk, M.D., Marty, G.D., Hinton, D.E., Brown, E.D. and Baker, T.T. (1996). Sublethal effects of the *Exxon Valdez* oil spill on herring embryos and larvae: morphological, cytogenetic, and histopathological assessments, 1989–1991. *Canadian Journal of Fisheries and Aquatic Sciences,* 53: 2355–2365.

Hostettler, F.D. and Kvenvolden, K.A. (1994). Gechemical changes in crude oil spilled from the *Exxon Valdez* supertanker into Prince William Sound, Alaska. *Organic Geochemistry,* 21: 927–936.

Hunt, J.M. (1979). *Petroleum Geochemistry and Geology.* W.H. Freeman, San Francisco, p. 678.

IPIECA. (2012). *Sensitivity Mapping for Oil Spill Response.* IPIECA/IMO/OGP Report 477, London, p. 39.

ITOPF. (2014). *Handbook, 2014–2015.* International Tanker Owners Pollution Federation, London, p. 52.

Jackson, J., Cubit, J., Batista, V., Burns, K.A., Caffey, H., Caldwell, R., Garrity, S. et al. (1989). Effects of a major oil spill on Panamanian, coastal marine communities. *Science*, 234: 37–44.

Jacobs, L.A. (2014). Comment on: health of common bottlenose dolphins (*Tursiops truncatus*) in Barataria Bay, Louisiana, following the *Deepwater Horizon* oil spill. *Environmental Science and Technology*, 48: 4207–4208.

Keller, B.D. and Jackson, J.B.C. (1993). Long-term assessment of the oil spill at Bahia Las Minas, Panama, Synthesis Report. Volume I: Executive Summary and Volume II, Technical Report. Minerals Management Service, Gulf Of Mexico OCS Region, OCS Study MMS 93-0048, p. 1017.

Kennicutt, M.C. II, Booth, P.N., Wade, T.L., Sweet, S.T., Rezak, R., Kelly, F.J., Brooks, J.M. et al. (1996). Geochemical patterns in sediments near offshore production platforms. *Canadian Journal of Fisheries and Aquatic Sciences*, 53: 2554–2566.

Knap, A.H. (1981). *Bermuda's Delicate Balance. Pollution.* Bermuda National Trust, Hamilton, Bermuda, pp. 239–255.

Knap, A.H. (1987). Effects of chemically dispersed oil on the brain coral *Diploria strigosa*. *Marine Pollution Bulletin*, 18: 119–122.

Knap, A.H. (1992). *Biological Impacts of Oil Pollution: Coral Reefs.* International Petroleum Industry Environmental Conservation Association, London, p. 18.

Knap, A.H., Dodge, R.E., Baca, B.A., Sleeter, T.D. and Snedaker, S. (1995). The effects of oil and oil dispersants on tropical ecosystems: Results of 10 years of research. *Presented at International Maritime Organization Second International Oil Spill Research and Development Forum*, London, May 22–25, 1995.

Knap, A.H., Hughes, I.W. and Sleeter, T.D. (1985). The grounding of the M/T TIFOSO, 1983: a test of Bermuda's contingency plan. *Proceedings 1985 Oil Spill Conference*, Los Angeles, California, (USCG/API/EPA), pp. 289–291.

Knap, A.H., Solbakken, J.E., Dodge, R.E., Sleeter, T.D., Wyers, S.C. and Palmork, K.H. (1982). Accumulation and elimination of (9-14C) phenanthrene in the reef-building coral *Diploria strigosa*. *Bulletin of Environmental Contamination and Toxicology*, 28: 281–284.

Lee, R.F., Furlong, E. and Singer, S. (1977). Metabolism of hydrocarbons in marine invertebrates. Aryl hydrocarbon hydroxylase from the tissues of the blue crab, *Callinectes sapidus* and the polychaete worm, *Nereis* sp. In: C.S. Giam (ed.) *Pollutant Effects on Marine Organisms*, D.C. Heath, Lexington, Massachusetts, pp. 111–124.

Linden, O. (1975). Acute effects of oil and oil/dispersant mixture on larvae of Baltic herring. *Ambio*, 4: 130–133.

Mahmoudi, N., Porter, T.M., Zimmerman, A.R., Fulthorpe, R.R., Kasuzi, G.N., Silliman, B.R. and Slater, G.F. (2013). Rapid degradation of *Deepwater Horizon* spilled oil by indigenous microbial communities in Louisiana saltmarsh sediments. *Environmental Science and Technology*, 47: 13303–13312.

Malins, D.C. (1977). *Effects of Petroleum on Arctic and Subarctic Marine Environments and Organisms.* Vol. II, Biological Effects. Academic Press, New York, p. 500.

Matkin, C.O., Saulifis, E.L., Ellis, G.M., Olesiuk, P. and Rice, S.D. (2008). Ongoing population-level impacts on killer whales *Orcinus orca* following the 'Exxon Valdez' oil spill in Prince William Sound, Alaska. *Marine Ecology Progress Series*, 356: 269–281.

McCall, B.D. and Pennings, S.C. (2012). Disturbance and recovery of salt marsh arthropod communities following *BP Deepwater Horizon* oil spill. *PLoS ONE*, 7: e32735. doi:10.1371/journal.pone.0032735.

McElroy, A.E., Farrington, J.W. and Teal, J.M. (1989). Bioavailability of polycyclic aromatic hydrocarbons in the aquatic environment. In: U. Varanasi (ed.) *Metabolism of Polycyclic Aromatic Hydrocarbons*, CRC Press, Boca Raton, Florida, pp. 1–39.

Moldan, A.G.S., Chapman, P. and Fourie, H.O. (1979). Some ecological effects of the *Venpet – Venoil* collision. *Marine Pollution Bulletin*, 10: 60–63.

Moldan, A.G.S., Jackson, L.F., McGibbon, S., Van der Westhuizen, J. (1985). Some aspects of the *Castillo de Bellver* oil spill. *Marine Pollution Bulletin*, 16: 97–102.

Murawski, S.A., Hogarth, W.T., Peebles, E.B. and Barbieri, L. (2014). Prevalence of external skin lesions and polycyclic aromatic hydrocarbon concentrations in Gulf of Mexico fishes, post *Deepwater Horizon*. *Transactions of the American Fisheries Society*, 143: 1084–1097.

National Commission. (2011). Deep Water: The Gulf Oil Disaster and the Future of Offshore Drilling: Report to the President. National Commission on the BP Deepwater Horizon Oil Spill and Offshore Drilling, p. 380.

National Research Council. (1989). *Using Oil Spill Dispersants on the Sea*. National Academy Press, Washington, D.C., p. 352.

National Research Council. (2003). *Oil in the Sea III. Inputs, Fate and Effects*, National Academy Press, Washington, D.C., p. 265.

National Research Council. (2005). *Oil Spill Dispersants: Efficacy and Effects*. National Academy Press, Washington, D.C., p. 377.

NOAA. (2012). Study: seafood safety after Deepwater Horizon. http://www.nmfs.noaa.gov/stories/2012/02/dwhpaper.html, dated 8 February 2012, accessed 13 June 2014.

Overton, E.B., Laseter, J.L., Mascarella, W., Rashke, C., Noiry, I. and Farrington, J.W. (1980). *Photochemical Oxidation of IXTOC I Oil*. Researcher/Pierce IXTOC I Symposium.

Overton, E.B., Patel, J.R. and Laseter, J.L. (1979). Chemical characterization of mousse and selected environmental samples from the *Amoco Cadiz* oil spill. In: *Proceedings, 1979 Oil Spill Conference*, American Petroleum Institute, Washington, D.C.

Page, D.S., Giliffan, E.S., Foster, J.C., Hotham, J.R. and Gonzalez, L. (1985). Mangrove leaf tissue sodium and potassium ion concentrations as sublethal indicators of oil stress in mangrove trees. *Proceedings of the 1985 Oil Spill Conference*. American Petroleum Institute, Washington, D.C., pp. 391–393.

Palmork, K.H. and Solbakken, J.E. (1980). Accumulation and elimination of radioactivity in the Norway lobster (*Nephrops norvegicus*) following intragastric administration of [9-^{14}C] phenanthrene. *Bulletin of Environmental Contamination and Toxicology*, 25: 668–671.

Peterson, C.H., Rice, S.D., Short, J.W., Esler, D., Bodkin, J.L., Ballachey, B.E. and Irons, D.B. (2003). Long-term ecosystem response to the *Exxon Valdez* oil spill. *Science*, 302: 2082–2086.

Philp, R.P. (1985a). Biological markers in fossil fuel production. *Mass Spectroscopy Reviews*, 4: 1–54.

Philp, R.P. (1985b). *Fossil Fuel Biomarkers: Application and Spectra, Methods in Geochemistry and Geophysics*. Elsevier, NY, 23: 294.

Posthuma, J. (1977). The composition of petroleum. *Rapport et Proces Verbaux. Conseil International pour l'Exploration de la Mer*, 171: 7–16.

Proffitt, C.E., Devlin, D.J. and Lindsey, M.L. (1995). Effects of oil on mangrove seedlings grown under different environmental conditions. *Marine Pollution Bulletin*, 30: 788–793.

Reynaud, S. and Deschaux, P. (2006). The effects of polycyclic aromatic hydrocarbons on the immune system of fish: A review. *Aquatic Toxicology*, 77: 229–238.

Sanders, H.L., Grassle, J.F., Hampson, G.R., Morse, L.S., Garner-Price, S. and Jones, C.C. (1980). Anatomy of an oil spill: long-term effects from the grounding of the barge *Florida* off West Falmouth, Massachusetts. *Journal of Marine Research*, 38: 265–380.

Schleifstein, F. (2014). BP refuses to pay for more research on Deepwater Horizon oil spill effects on dolphins, turtles, oysters. http://www.nola.com/environment/index.ssf/2014/04/bp_refuses_to_pay_for_more_res.html, retrieved 6/11/2014.

Schwacke, L.H., Smith, C.R., Townsend, F.I., Wells, R.S., Hart, L.B., Balmer, B.C., Collier, T.K. et al. (2013). Health of common Bottlenose Dolphins (*Tursiops truncatus*) in Barataria Bay, Louisiana, following the *Deepwater Horizon* oil spill. *Environmental Science and Technology*, 48: 93–103.

Sericano, J.L., Wade, T.L. and Brooks, J.M. (1996). Accumulation and depuration of organic contaminants by the American oyster (*Crassostrea Virginica*). *Science of the Total Environment*, 179: 149–160.

Silliman, B.R., van de Koppel, J., McCoy, M.W., Diller, J., Kasozi, G.N., Earl, K., Adams, P.N. et al. (2012). Degradation and resilience in Louisiana salt marshes after the BP-*Deepwater Horizon* oil spill. *Proceedings of the National Academy of Sciences*, 109: 11234–11239. doi:10.1073/pnas.1204922109

Sleeter, T.D., Knap, A.H. and Hughes, I.W. (1983). Oil spill contingency planning and support coordination in Bermuda: A successful model. In: *Proceedings 1983 API/USCG Oil Spill Conference*, San Antonio, Texas, pp. 149–154.

Smith, S.R. and Knap, A.H. (1985). A significant decrease in the amount of tar stranded on Bermuda. *Marine Pollution Bulletin*, 16: 19–21.

Solbakken, J.E. and Knap, A.H. (1986). The disposition of the xenobiotics phenanthrene and octochlorostyrene in spiny lobsters *Panulirus argus* after intergastric administrations. *Bulletin of Environmental Contamination and Toxicology*, 37: 747–751.

Straughan, D. and Hadley, D. (1978). Experiments with *Littorina* species to determine the relevancy of oil spill data from southern California to the Gulf of Alaska. *Marine Environmental Research*, 1: 135–163.

Teal, J.M., Farrington, J.W., Burns, K.A., Stegeman, J.J., Tripp, B.W., Woodin, B. and Phinney, C. (1992). The West Falmouth oil spill after 20 years: Fate of fuel oil compounds and effects on animals. *Marine Pollution Bulletin*, 24: 607–614.

Teas, J.H., Lessard, R.R., Canevari, G.P., Brown, C.D. and Glenn, R. (1993). Saving oiled mangroves using a new non-dispersing shoreline cleaner. In: *Proceedings, Conference on Assessment of Ecological Impacts of Oil Spills*. American Institute of Biological Sciences, Washington, D.C., pp. 147–151.

Tissot, B.P. and Welte, D.H. (1984). *Petroleum Formation and Occurrence*. Springer-Verlag, New York, p. 699.

Wade, T.L., Quinn, J.G., Lee, W.T. and Brown, C.W. (1976). Source and distribution of hydrocarbons in surface waters of the Sargasso Sea. *Proceedings of the American Institute of Biological Sciences Symposium on Hydrocarbons in the Aquatic Environment*, Washington, D.C., pp. 270–286.

Wang, Z., Fingas, M. and Sergy, G. (1994). Study of 22-year-old *Arrow* oil samples using biomarker compounds by GC/MS. *Environmental Science and Technology*, 28: 1733–1746.

Ward, G.A., Baca, B., Cyriachs, W., Dodge, R.E. and Knap, A.H. (2003). Continuing long-term studies of the Tropics Panama oil and dispersed oil spill site. *Ingternational Oil Spill Conference Proceedings*, Vancouver, April 2003, 259–267.

Webb, J.W. (1996). Effects of oil on salt marshes. In: C.E. Proffitt and P.F. Roscigno (eds.), *Symposium Proceedings: Gulf of Mexico and Caribbean Oil Spills in Coastal Ecosystems: Assessing Effects, Natural Recovery, and Progress in Remediation Research*. OCS Study MMS 95-0063. Department of the Interior, Minerals Management Service, New Orleans, LA, pp. 55–64.

White, H.K., Hsing, P-Y., Cho, W., Shank, T.M., Cordes, E.E., Quattrini, A.M., Nelson, R.K. et al. (2012). Impact of the *Deepwater Horizon* oil spill on a deep-water coral community in the Gulf of Mexico. *Proceedings of the National Academy of Sciences*, 109: 20303–20308.

Whitehead, A., Dubansky, B., Bodinier, C., Garcia, T.I., Miles, S., Pilley, C., Raghunathan, V. et al. (2012). Genomic and physiological footprint of the *Deepwater Horizon* oil spill on resident marsh fishes. *Proceedings of the National Academy of Sciences*, 109: 20298–20302.

Whitehead, A., Pilcher, W., Champlin, D. and Nacci, D. (2011). Common mechanism underlies repeated evolution of extreme pollution tolerance. *Proceedings of the Royal Society, B.* doi:10.1098/rspb.2011.0847.

Widdows, J., Bakke, T., Bayne, B.L., Donkin, P., Livingstone, D.R., Lowe, D.M., Moore, M.N. et al. (1982). Responses of *Mytilus edulis* L. on exposure to the water accommodated fraction of North Sea oil. *Marine Biology*, 67: 15–31.

Wyers, S.C., Frith H.R., Dodge, R.E., Smith, S.R., Knap, A.H. and Sleeter, T.D. (1986). Behavioral effects of chemically dispersed oil and subsequent recovery in *Diploria strigosa* (Dana). *Marine Ecology*, 7: 23–42.

Youngblood, W.W. and Blumer, M. (1973). Alkanes and alkenes in marine benthic algae. *Marine Biology*, 21: 163–172.

6

Modeling for Management of Marine Ecosystem Around an Artificial Structure in a Semiclosed Bay

Daisuke Kitazawa, Shigeru Tabeta and Masataka Fujino

CONTENTS

6.1   Introduction...147
6.2   Basic Concept of Numerical Model..148
    6.2.1   Structure of Numerical Model...148
    6.2.2   Hydrodynamic Submodel 148
    6.2.3   Ecosystem Submodel..149
6.3   Computational Condition...150
    6.3.1   Grid System of Tokyo Bay...150
    6.3.2   Numerical Methods...150
    6.3.3   Simulation Cases...151
    6.3.4   Boundary Conditions...151
6.4   Example of Results...151
    6.4.1   Residual Current 151
    6.4.2   Water Quality 152
    6.4.3   Carbon Cycle 152
6.5   Concluding Remarks..154
References..154

6.1 Introduction

Various forms of marine pollution, such as algal blooming and hypoxic problems, have frequently occurred in semiclosed bays all over the world. These pollution problems are caused by the rapid development of human activities in coastal areas. For example, organic matter and nutrients from anthropogenic sources have been discharged through rivers or channels into coastal seas. The bacterial decomposition of organic matter consumes oxygen and produces nutrients that are utilized by algae to grow. The loading of nutrients directly stimulates algal growth. An algal blooming occurs, and the color of the sea surface changes sometimes accompanied by an unpleasant odor. Then the dead algae sink to the bottom and are decomposed by bacteria, which consume oxygen. Particularly in the case of stratification, oxygen does not diffuse to deeper waters, and the concentration of dissolved oxygen decreases, resulting in hypoxic or anoxic conditions around the sea bottom. Reclamation is also one of the reasons for eutrophication in coastal seas. If shallow waters such as tidal flats and seaweed grounds are reclaimed, the marine organisms living in shallow waters are removed. Some marine organisms serve to purify seawater in coastal seas. Seaweed and seagrass, which compete with algae, absorb nutrients for their growth, resulting in a

TABLE 6.1

List of Numerical Models which are Widely Used in the World

Numerical Model	Source
Estuaries and Coastal Ocean Model (ECOM)	Choi et al. (2014)
Finite-Volume Coastal Ocean Model (FVCOM)	Chen et al. (2003) and Weisberg and Zheng (2006)
HYbrid Coordinate Ocean Model (HYCOM)	
Marine Environmental Committee (MEC) Ocean Model	Mizumukai et al. (2008) and Zhang and Kitazawa (2015)
Modular Ocean Model (MOM)	Giarolla et al. (2005)
Navy Coastal Ocean Model (NCOM)	Barron et al. (2006)
Princeton Ocean Model (POM)	Blumberg and Mellor (1987)
Regional Ocean Modeling System (ROMS)	

decreased concentration of algae. Some animals that are predators of algae, such as bivalves, filter seawater and have particulate organic matter, which also decreases the concentration of algae. But these kinds of purification mechanisms disappear if the shallow waters are reclaimed, and then eutrophication and hypoxic or anoxic problems are induced.

For the preceding case, a control strategy of external loadings of chemical substances and reclamation should be established to maintain sustainable use of coastal seas. Numerical simulation is a useful tool in making quantitative predictions and obtaining information to establish the strategy. Recently, a three-dimensional coupled model of hydrodynamic and ecosystem submodels has been used for the analysis of pelagic and benthic environments with the development of computers. Currently, a lot of hydrodynamic, lower-trophic ecosystem and combined models are available (Table 6.1). The software has been widely used by many researchers. The governing equations are basically common, while the numerical methods such as gridding and discretization are different among the models.

This chapter introduces the basic concept of the numerical model, which combines hydrodynamic and lower-trophic ecosystem submodels. Then the numerical model is applied to the marine environment in Tokyo Bay as an example of the application of the numerical model to a semiclosed bay.

6.2 Basic Concept of Numerical Model

6.2.1 Structure of Numerical Model

The numerical model is composed of hydrodynamic and ecosystem submodels (Figure 6.1). Both submodels are driven by boundary conditions, such as meteorological conditions, river conditions and open boundary conditions. Also, the ecosystem submodel uses the output from the hydrodynamic submodel. Part of the hydrodynamic submodel may also use the information from the ecosystem submodel, such as the extinction coefficient of light depending on the concentration of phytoplankton.

6.2.2 Hydrodynamic Submodel

A description of the hydrodynamic model can be found in articles (e.g., Mellor, 1996; Kundu and Cohen, 2008). For the application of the numerical model to Tokyo Bay, a Cartesian

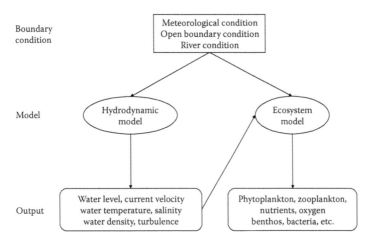

FIGURE 6.1
Structure of numerical model.

coordinate system is adopted. Assuming hydrostatic and Boussinesq approximations, seven governing equations are solved; momentum equations in three directions, an equation of continuity, advection-diffusion equations of water temperature and salinity and a state equation for density as a function of water temperature and salinity.

6.2.3 Ecosystem Submodel

Pelagic and benthic submodels are combined into the ecosystem submodel used for the analysis of the marine environment in Tokyo Bay. A description of the ecosystem submodel can be found, for example, in Kremer and Nixon (1978) and Baretta and Ruardij (1987). In the example here (Figure 6.2), the boxes and arrows represent the state variables and

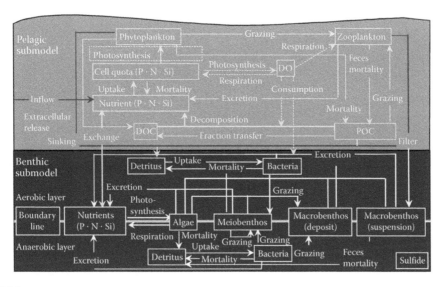

FIGURE 6.2
Ecosystem submodel.

processes (fluxes of materials), respectively. These values are described by the amounts of carbon, phosphorus, nitrogen, silicate and oxygen. They are converted to each other by the ratio of these materials in organic matter. The pelagic submodel contains 13 state variables such as phytoplankton, zooplankton, particulate organic carbon, dissolved organic carbon, phosphate, ammonium, nitrite, nitrate, silicate and dissolved oxygen, and they are converted to each other using the conversion ratios. The benthic submodel contains 15 compartments, such as two types of macrobenthos (deposit feeder and suspension feeder), aerobic and anaerobic bacteria, detritus, phosphate, ammonium, nitrite, nitrate and silicate. The chemical matter and marine organisms in the pelagic environment are transported by the surrounding seawater current, so the time variation in each state variable can be described by the advection-diffusion equation using the output of the hydrodynamic submodel. The chemical matter and marine organisms in the benthic environment are assumed not to move. Detailed information on the complex model can be found in Kitazawa (2002).

6.3 Computational Condition

6.3.1 Grid System of Tokyo Bay

The effects of reclaiming shallow waters are represented as an example of numerical simulation in this chapter. Figure 6.3 shows the grid system of Tokyo Bay in each of the years 1935, 1979 and 1994. In 1935, shallow waters expanded mainly at the head of Tokyo Bay. However, the shallow waters were reclaimed gradually, and most of the shallow waters were reclaimed in 1994. The sea area of Tokyo Bay is latticed by square grids with an edge that is 1000 m long in the horizontal direction. A multilevel model is adopted in the vertical direction, and the number of layers is 10, while the thickness of each layer is different. It is larger in deeper waters.

6.3.2 Numerical Methods

A finite-difference scheme is adopted to solve the governing equations numerically. The explicit time integration method of Euler, QUICK and a second-order central difference method are adopted to time the derivative term, advection term and eddy viscosity and

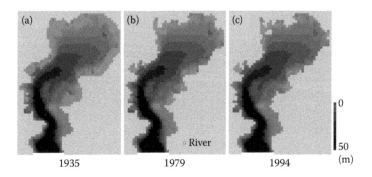

FIGURE 6.3
Grid system of Tokyo Bay.

TABLE 6.2

External Loadings of COD, T-P and T-N for Each Target Year

Case (year)	COD (tons/day)	T-P (tons/day)	T-N (tons/day)
1 (1935)	70	6	70
2 (1979)	405	35	310
3 (1994)	245	20	240

diffusivity terms, respectively. Details on each method can be found in Ferziger and Perić (2002). The time step is limited by the maximum depth and the horizontal grid size and is set at 20 seconds.

6.3.3 Simulation Cases

The conditions of topography and the discharge of chemical matter from rivers are changed for three target years, 1935, 1979 and 1994. Numerical simulations are carried out with the aim of reproducing the marine environment in the summer, when water quality is at its worst during the year.

6.3.4 Boundary Conditions

Boundary conditions averaged in August from 1989 to 1998 are given for the numerical simulation. At the open boundary, variation in the water level induced by the M_2 component tide is given, and the values of the water temperature, salinity and pelagic state variables are estimated from the observations provided by the National Institute for Environmental Studies. Meteorological conditions are determined from the monitoring data of wind, solar radiation and precipitation, and others at Tokyo, Chiba and Yokohama Meteorological Stations, provided by the Japan Meteorological Agency. Six rivers are taken into account (Figure 6.3), and the amount of water inflow from each river is estimated from the measurements at the middle of each river, which are provided by River Bureau of the Ministry of Land, Infrastructure and Transport. The concentrations of chemical matter in each river are estimated based on the external loadings of chemical oxygen demand (COD), total phosphorus (T-P) and total nitrogen (T-N), as summarized in Table 6.2.

6.4 Example of Results

6.4.1 Residual Current

Figure 6.4 shows horizontal distributions of residual current in Cases 1 (1935), 2 (1979) and 3 (1994) at 1 m below the sea surface. The typical residual current of Tokyo Bay in the summer is the inflow of seawater from the outer sea in deeper waters and the outflow of seawater to the outer sea around the sea surface. This is because circulation is induced by stratification and the inflow of freshwater from rivers during the summer season. Horizontal distributions of residual current at 1 m below sea surface are similar in the three cases, while the scale of the clockwise circulation formed at the head of Tokyo Bay in Case 1 is larger than those in Cases 2 and 3. The difference in the residual current field may have a large effect on the transportation of plankton and chemical matter in Tokyo Bay.

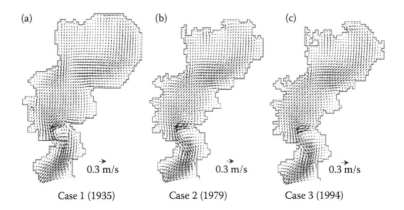

Case 1 (1935)   Case 2 (1979)   Case 3 (1994)

FIGURE 6.4
Horizontal distributions of residual current at 1 m below sea surface.

6.4.2 Water Quality

As an example of water quality, horizontal distributions of chemical oxygen demand (COD) at 1 m below the sea surface are represented in Figure 6.5. COD is the index of the amount of organic matter in seawater and is used widely in Japan to evaluate water quality. It is calculated by the sum of phytoplankton, zooplankton, particulate organic matter and dissolved organic matter in the numerical model. The concentration of COD in Case 2 is quite high, especially at the head of the bay, in comparison with Case 1, owing to the increase in the discharge of chemical matter from the rivers and active photosynthesis. A decrease in seawater-purifying activity due to reclamation of the shallow waters also causes an increase in the concentration of COD. The concentration of COD is slightly lower in Case 3, in comparison with Case 2, owing to the policy of decreasing the external loadings of organic matter and nutrients from the land.

6.4.3 Carbon Cycle

The carbon cycle can be estimated by integrating all the state variables and fluxes in time (Figure 6.6). The boxes and arrows represent total state variables ($\times 10^9$ gC) and fluxes ($\times 10^8$ gC day^{-1}), respectively. The flux of primary production and bacterial uptake of

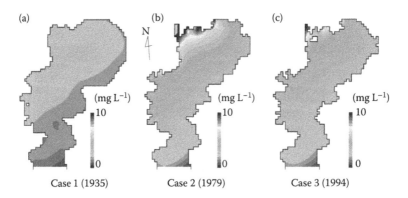

Case 1 (1935)   Case 2 (1979)   Case 3 (1994)

FIGURE 6.5
Horizontal distributions of chemical oxygen demand (COD) at 1 m below sea surface.

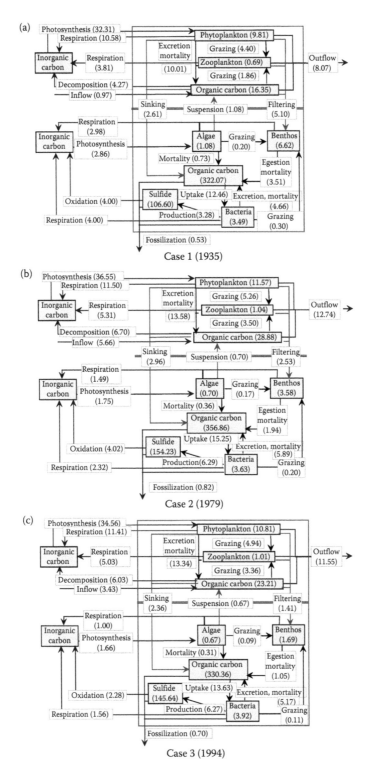

FIGURE 6.6
Carbon cycle in Tokyo Bay for each case.

organic matter are greatest in pelagic and benthic environments, respectively. Comparing the results from Cases 1 and 2, the production of organic matter due to the photosynthesis of phytoplankton and the inflow of chemical matter through rivers increases by 4.24 ($\times 10^8$ gC day^{-1}) and 4.69 ($\times 10^8$ gC day^{-1}), respectively. The primary production of phytoplankton is also active due to the increase in the inflow of nutrients from the land, resulting in an increase in carbon flux by 8.93 ($\times 10^8$ gC day^{-1}). On the other hand, the filtering activities of the organic matter by suspension feeders decrease by 2.57 ($\times 10^8$ gC day^{-1}). This is because the shallow waters were reclaimed from 1935 to 1979, and the biomass of macrobenthos, especially suspension feeders, decreased. The variation in carbon flux caused by the reclamation of shallow waters accounts for about 30% of the variation in carbon flux caused by the increasing loading of chemical matter through the rivers. As a result of reclamation and the increasing loading of chemical matter, the organic matter increased by 14.64 ($\times 10^9$ gC). Organic matter in the benthic environment increased by 34.79 ($\times 10^9$ gC). This is due to the increase in the sinking of organic matter from the pelagic environment. Respiration of aerobic bacteria decreased by 1.68 ($\times 10^8$ gC day^{-1}), and the production of sulfide by anaerobic bacteria increased by 3.01 ($\times 10^8$ gC day^{-1}) owing to the hypoxic conditions caused by bacterial decomposition. Then the result of Case 3 is compared with that of Case 2. Owing to the policy of pollution control, the external loading of chemical matter decreased from 1979 to 1994. As a result, primary production and inflow of carbon through rivers decrease by 1.99 ($\times 10^8$ gC day^{-1}) and 2.23 ($\times 10^8$ gC day^{-1}), respectively. The reduction in the external loading of chemical matter from land led to a modest improvement in water quality in Tokyo Bay. The bacterial decomposition of organic matter was reduced in Case 3, in comparison with Case 2, while it remained much larger in comparison with Case 1.

6.5 Concluding Remarks

This chapter introduces the basic concept of a numerical model that combines hydrodynamic and lower-trophic ecosystem submodels and presents example results of numerical simulations for a polluted and semiclosed bay, Tokyo Bay. The impacts of multiple factors (reclamation and external loadings of organic matter and nutrients in the example case) can be compared using numerical simulation. The numerical model will be developed to combine the models of watershed, atmosphere, higher-trophic ecosystem and so forth. However, the developed model has to be used for the management of coastal seas, such as a semiclosed bay, in reality. A strong combination with political or strategic models may be required in the future.

References

Baretta J. W. and Ruardij P. (1987): *Tidal flat estuaries, Simulation and analysis of the EMS estuary, Ecological Studies 71*. Springer-Verlag, Heidelberg.

Barron C. N., Kara A. B., Martin P. J., Rhodes R. C. and Smedstad L. F. (2006): Formulation, implementation and examination of vertical coordinate choices in the Global Navy Coastal Ocean Model (NCOM). *Ocean Modelling*, 11(3), 347–375.

Blumberg A. F. and Mellor G. L. (1987): A description of a three-dimensional coastal ocean circulation model. *Coastal and Estuarine Science*, 4, 1–16.

Chen C., Liu H. and Beardsley R. C. (2003): An unstructured grid, finite-volume, three-dimensional, primitive equations ocean model: Application to coastal ocean and estuaries. *Journal of Atmospheric and Oceanic Technology*, 20(1), 159–186.

Choi B. J., Hwang C. and Lee S. H. (2014): Meteotsunami-tide interactions and high-frequency sea level oscillations in the eastern Yellow Sea. *Journal of Geophysical Research, Oceans*, 119(10), 6725–6742.

Ferziger J. H. and Perić M. (2002): *Computational methods for fluid dynamics*. Springer-Verlag, Berlin, Heidelberg, New York.

Giarolla E., Nobre P., Malagutti M. and Pezzi L. P. (2005): The Atlantic Equatorial Undercurrent: PIRATA observations and simulations with GFDL Modular Ocean model at CPTEC. *Geophysical Research Letters*, 32, L10607.

Kitazawa D. (2002): Study on the effects of very large floating structure on marine ecosystem by numerical simulation. Doctoral thesis, The University of Tokyo, Tokyo, 377 (in Japanese).

Kremer J. and Nixon S. W. (1978): *A coastal marine ecosystem: simulation and analysis (Ecological studies)*. Springer-Verlag, Heidelberg.

Kundu P. K. and Cohen I. M. (2008): *Fluid mechanics*. Elsevier Inc., New York.

Mellor G. L. (1996): *Introduction to physical oceanography*. Springer, New York.

Mizumukai K., Sato T., Tabeta S. and Kitazawa D. (2008): Numerical studies on ecological effects of artificial mixing of surface and bottom waters in density stratification in semi-enclosed bay and open sea. *Ecological Modelling*, 214(2), 251–270.

Weisberg R. H. and Zheng L. (2006): Circulation of Tampa Bay driven by buoyancy, tides, and winds, as simulated using a finite volume coastal ocean model. *Journal of Geophysical Research, Oceans*, 111, C01005.

Zhang J. and Kitazawa D. (2015): Numerical analysis of particulate organic waste diffusion in an aquaculture area of Gokasho Bay, Japan. *Marine Pollution Bulletin*, 93(1–2), 130–143.

7

The Economics of Ecosystem-Based Fisheries Management

Hans Frost, Lars Ravensbeck, Ayoe Hoff and Peder Andersen

CONTENTS

7.1   A Tale 157
7.2   Challenges and the Role of Economics 159
7.3   Classification of Ecosystem Services 161
7.4   Tools for Economic Analyses 163
7.5   Green Accounting in Theory 164
7.6   Green Accounting in Practice 168
7.7   Methods for Economic Analyses 171
7.8   Inclusion of Ecosystem Services in Green Accounting 172
7.9   Microeconomic Modelling 174
7.10 Prices and Valuation 177
7.11 Bioeconomic Modelling and Management, Brief Overview 180
7.12 Ecosystem-Based Management, Brief Overview 181
7.13 Predator–Prey Relationships and Other Species Interactions 183
7.14 Accounting for Externalities in Ecosystem-Based Fishery Management 185
7.15 Conventional Fisheries Economics Modelling 186
7.16 Conventional Fisheries Management 189
7.17 Dynamic Ecosystem Modelling, Some Results 196
7.18 Conclusion 202
References 203

7.1 A Tale

Imagine a community that has access to a lake with a fish stock. The community exploits the fish stock and consumes the harvest. It is possible for the community to increase the harvest by 3% each year, which is reflected in the growth of the gross national product. Innovative members of the community, who have been taught economics, propose that if some of the harvested fish is not consumed by the fishermen but saved to be consumed by people of the community, who instead of going fishing could build a boat, the fish harvest could then increase even more in the long run. Now with the improved technology the harvest growth rate increases to 4%, which makes it possible to save more fish to invest in boat builders. Everyone is happy because the growth rate of the gross national product is high.

Then one day the growth rate decreases and suddenly the harvest begins to fall. The reaction is to build more boats, that is, save and invest more. But the harvest still decreases. Other members of the community who have studied the ecosystem point out that the fall in harvest may be caused by a declining stock entailing low reproductive capacity. The economists realize that they have failed to include the depreciation of natural capital into the national accounts, which led to the misjudgment about the opportunities for growth. Once this was corrected and the stock recovered, the harvest would be stable and sustainable and form the basis for a long-run growth rate for the entire community's economy at 2%. Now everyone was happy.

Unfortunately, on the other side of the lake, a new community is established. This community exploits another fish stock that serves as prey for the species exploited by the first community. The growth rate of the gross national product of the first community decreases, while it increases for the second community. To meet this challenge, the first community increases its harvests, which leads to a devastating competition between the two communities. The problem is not resolved and sustainable growth rates restored until an agreement between the two communities is made.

This short narrative highlights the issues addressed in this chapter. First we focus on macroeconomic accounting problems, which to some degree include the use of natural capital and to some degree do not. In the second part, we focus on microeconomic issues, which mainly address specific topics as to how natural resources are exploited optimally and what might happen if the ecosystem is left to be controlled by market forces only.

The chapter consists of two main parts. In the first part, Sections 2–8, an array of issues are addressed, including macroeconomic-level focusing on the way the subject is dealt with in national accounting. Furthermore, it discusses how national accounting statistics has been developed and can be further developed to include the consumption of ecosystem services in a comprehensive way. Finally, this section includes a brief discussion of how valuation methods are used, how this can improve national accounting statistics, and how an improved national accounting system can improve the way we use our natural resources. The focus in the first part is therefore mainly on accounting statistics (bookkeeping for nature) on a macroeconomic level.

The second part, Sections 9–17, addresses issues that are associated with the optimal use of natural resources in an economy. In this part, market failures also play an important role because the inclusion and mitigation of these market failures, called externalities, are important for the optimal use of natural resources. The optimal use of non-priced natural resources requires that regulatory measures be imposed. It is shown how the need for regulatory measures arises, but a presentation and discussion of the advantages and disadvantages of the measures are outside the scope of the chapter. In contrast to the first part, the second part applies a bioeconomic model for a fishery comprising three interacting species and two interacting fleets aimed at providing a consistent description and analysis of the problem. The scope of the second main section is to analyze the exploitation of marine resources and ecosystem services and, in particular, the impact of fishing activities on ecosystem services. Fisheries economics theory is well developed, but an extension to include ecosystem services in a broader perspective still needs some work. A more complete understanding is promoted by the use of modelling entailing a deeper understanding of data requirements. This goal has not yet been achieved, but it is shown how the use of a relatively simple but sufficiently complex bioeconomic model that includes realistic parameter values serves the purpose. In this way, a deeper and more consistent understanding of the problems associated with the use and management of marine ecosystem services is attained.

7.2 Challenges and the Role of Economics

This chapter addresses fishing's impact on the ecosystem and whether conventional fisheries' economics theory and management measures are sufficient to analyze and understand the complexities of an ecosystem and how to manage it. Before addressing the subject, it would be expedient to discuss the concepts of *ecology*, *ecosystem*, and *economy*. Ecology may be viewed as the scientific study of interactions among organisms and their environment. An ecosystem is defined as a community of living organisms (plants, animals, and microbes) in interaction with the nonliving components of their environment (air, water, and mineral soil). These biotic and abiotic components are regarded as being linked together through nutrient cycles and energy flows. Hence, ecosystems are composed of dynamically interacting parts, including organisms, the communities they make up, and the nonliving components of their environment. Ecosystem processes, such as primary production, pedogenesis, nutrient cycling, and various niche construction activities, regulate the flux of energy and matter through an environment. These processes are sustained by organisms with specific life history traits, and the variety of organisms is called biodiversity. Biodiversity, which refers to the varieties of species, genes, and ecosystems, enhances certain ecosystem services.

While ecology and ecosystems deal with nutrient cycles and energy flows, economics deals with human well-being (utility), behaviour, and impact; see, for example, Robbins (1932), who defined economics *as a science which studies human behaviour as a relationship between ends and scarce means which have alternative uses.* Hence, ecological economics is related to the study of human activity in terms of acquiring wealth and how this impacts the nutrient cycles and energy flows of our natural resources.

Over the last two decades ecological economics (Daly and Farley, 2004) and ecosystem assessment and management, as represented by the Millennium Ecosystem Assessment (MEA) (2005), have attracted great attention. From an economic point of view, these disciplines challenge parts of the conventional macroeconomics and microeconomics disciplines.

In macroeconomics, economic growth is an important concept because it measures the growth in production represented by the gross national product (GNP). This measure includes the use of natural resources in terms of yield from natural resources, for example, catch of fish and extraction of oil. The extraction of oil lowers the stock, which at some point in time inevitably becomes exhausted. For a renewable natural resource such as fish, a catch lowers the stock if the catch is higher than the natural reproduction of the fish stock. This implies higher fishing costs, and eventually the stock becomes completely exhausted. These issues are not explicitly taken into account in the GNP, and the GNP cannot be used to shed light on what is happening in an ecosystem. Therefore, ecological or ecosystem economics challenges the conventional economic growth concept as an appropriate measure for sustainable development. If economic growth is taking place at the expense of the ecosystem, natural capital is reduced, and this may have severe impacts on the economy. This so-called ecological point of view is challenged by economists arguing that growth and innovation help develop new products and 'resources' that replace old products and 'resources.' Therefore, the net impact is unclear. This discussion about strong and weak sustainability will be addressed later in the chapter.

In microeconomics, optimal use and allocation of scarce resources is central, while the question about exhaustion by itself is of less importance. Daly and Farley (2004) use the image of looking at a ship, and microeconomics deals with how the ship is loaded to assure the optimal balance but does not deal with the problem of overloading the ship with a risk of foundering as a result.

There is a difference between the views of economists and ecologists as to how to define ecosystem services and the importance of each of them (Boyd, 2006; Boyd and Banzhaf, 2007). The general principle in economics is that only the end product is of importance to the consumers and should be counted while all intermediate products should be disregarded because they are embodied in the final product. This means that assets in terms of, for example, fish stocks, forests, land, and oil resources are of no value by themselves if they are not demanded or valued by consumers. The Millennium Ecosystem Assessment includes a list of ecosystem services, but many of the services listed in the MEA (2003) are of indirect importance according to this point of view. Ecologists' view is, in general, that these assets hold a value irrespective of being exploited or not.

The foregoing scene to some extent disregards the fact that economics and sustainability have been the concern of economists, managers, and politicians for a long time, since the birth of classical economists in the eighteenth and nineteenth centuries with the publication of works by, for example, Adam Smith, Jean-Baptiste Say, David Ricardo, Thomas Malthus, and John Stuart Mill. The statement of Hicks (1946) that *'income is the maximum amount an individual can consume during a period and remain as well off at the end of the period as at the beginning...'* aimed at addressing the subject of sustainability, but that subject was, with a few exceptions, not addressed theoretically or practically until the Club of Rome issued its Limits to Growth (Meadows et al., 1972) and the Brundtland Commission launched its report, which included a statement about *'development that meets the needs of the present without compromising the ability of future generations to meet their own needs'* (WCED, 1987). Since then, a substantial amount of effort, within macroeconomics and microeconomics, has been devoted to the area of ecosystem assessment and accounting from a theoretical viewpoint, and many articles on the topic have been published in scientific journals. However, wide gaps remain in the application of the theory, not least because of a lack of practical delineation and lack of data.

Hicks' statement has generally been interpreted as the amount of income that can be spent without depleting the wealth that generates the income. Hence, sustainability requires non-decreasing levels of capital stock over time, or non-decreasing per-capita capital stock. A proper measure of sustainability requires that all assets be included: produced capital, human capital, natural capital, and social capital. For the latter, the problem remains how to define natural capital to avoid overlapping. *Sustainability* can be defined as *strong or weak*.

Weak sustainability requires non-declining well-being over time and further requires only that the combined value of all assets remain constant. It is possible to substitute one form of capital for another, that is, natural capital can be depleted or the environment degraded as long as there are compensating investments in other types of capital: produced capital, human capital, or another type of natural capital. Consequently, the measurement of weak sustainability requires accurate estimation of the monetary value of environmental stocks and functions and all the benefits lost and gained in the transformation of one form of capital into another. The problem is that many services and environmental functions may not have market prices; other resources may have market prices that do not reflect their true value and scarcity. Proponents of weak sustainability recognize that it is extremely difficult to estimate values for many natural assets when (1) resources or environmental functions do not have market prices, (2) resources or environmental functions have current market prices but there are no futures or insurance markets to determine their market prices over time, and (3) there is great uncertainty about the future of ecosystem functioning.

Strong sustainability is based on the concept that the substitutability between produced capital and natural capital is limited. In this case, and when there is no substitution within the natural capital components, and when they are exploited in fixed proportions, production becomes very limited. This concept emphasizes maintaining the environmental functions of natural capital, especially given the uncertainty about the effect of ecosystem degradation on viability and resilience in the future. This approach has led to a set of general guidelines for strong sustainability associated with the precautionary principle: renewable resources, such as fish or forests, should be exploited only at the natural rate of net growth; the use of nonrenewable resources should be minimized and, ideally, used only at the rate for which renewable substitutes are available (e.g., fossil fuel should be replaced by renewable energy over time), emissions of wastes should not exceed the assimilative capacity of the environment, and environmental functions critical to life support should be maintained (FAO 2004: pp. 7–8).

An ecosystem-based perspective also requires that the impacts of other economic activities on the ecosystem be considered. Extending the strong sustainability approach that guides fisheries management requires that neither fishing activities nor other economic activities lead to changes in biological and economic productivity, biological diversity, or ecosystem structure or function from one generation to the next (National Research Council 1999). Strong sustainability is, by itself, very limiting for improvements in economic welfare (FAO 2004: p. 10).

Since World War II, microeconomics in particular has offered a comprehensive framework for analyzing the interaction between human behaviour and ecological components, called services but mainly limited to fish, forests, oil, and land. Over the last two decades, the concept of services has been broadened to include a more detailed classification of services and how they can be addressed (MEA 2005; Holland et al., 2010). Seven recent works require special attention, in particular, in regard to economics: (1) System for Integrated Environmental and Economic Accounting for Fisheries (SEEAF 2004); (2) Ecological Economics (Daly and Farley, 2004); (3) Millennium Ecosystem Assessment (MEA 2005); (4) 'The Economics of Ecosystems And Biodiversity' (TEEB 2010); (5) Report by the Commission on the Measurement of Economic Performance and Social Progress (Stiglitz et al., 2010); (6) UK National Ecosystem Assessment (UK NEA 2011), the latter being a full MEA-type assessment for the UK, carried out between 2009 and 2011; finally, (7) Partnership initiated by the World Bank in 2010 with a broad coalition of UN agencies, governments, international institutes, nongovernmental organizations, and academics called WAVES (Wealth Accounting and the Valuation of Ecosystem Services) to support countries with the move to natural capital accounting (WAVES, 2012). Furthermore, in 2012, governments from more than 90 countries agreed to establish the Intergovernmental Platform on Biodiversity and Ecosystem Services (IPBES), an independent panel of scientists, to assess the latest research on the state of the world's ecosystems.

7.3 Classification of Ecosystem Services

The need to classify and value ecosystem services arises from the fact that resources or their yields, which are important for economic welfare, do not appear explicitly in the utility function of consumers, which leads to a risk of overexploitation if prices do not exactly reflect the opportunity costs (alternative use) of the resources. Economic welfare is based on decision-making, that is, the choice between alternatives that requires that the alternatives be clearly identified and valued.

It is important to distinguish between the lack of recording of value of ecosystems and the sustainability of ecosystems, although recording of information is a precondition for a good assessment of sustainability. Furthermore, ecosystem sustainability is not the same as the optimal (best) use of the resources of the ecosystem. It is possible to have a sustainable ecosystem producing welfare at a low rate of economic growth. Taking the fishery as an example, the use of fish resources, apart from landings, is not recorded. To decide whether a fish stock is exploited on a sustainable basis or not requires recording of changes over time. However, a fish stock can produce fish sustainably across a wide range of fish stock sizes where only one stock size is maximizing welfare.

The classification of ecosystem services of the MEA is as follows (MEA 2003):

1. *Supporting services*, comprising biochemical cycling, primary production, food-web dynamics, diversity, habitat, and resilience;
2. *Provisioning*, comprising food, inedible resources, genetic resources, chemical resources, ornamental resources, energy, space, and waterways;
3. *Regulating*, comprising atmospheric regulation, regulation of local climate, sediment retention, biological regulation, pollution, and eutrophication mitigation control;
4. *Cultural services*, comprising recreation, aesthetic value, science and education, cultural heritage, inspiration by nature's legacy.

Following the MEA, efforts to improve and operationalize the conceptual framework of ecosystem services have been carried out later. Boyd and Banzhaf (2007), Fisher and Turner (2008), and Fisher et al. (2009) advocate a stricter definition with a clear distinction between services and benefits. Bateman et al. (2011) make a clear distinction between the terms *ecosystem services* and *goods*. The term goods includes physical products with a market price, less tangible goods, which may or may not have market prices, as well as nonmarket goods that are valued only for their existence. Moreover, only final services should be counted as ecosystem services, whereas indirect processes and functions – intermediate components – should not be regarded as ecosystem services. The inclusion of intermediate services would imply double counting. The intermediate services will be taken into account properly in the process of considering the final ecosystem services.

More recent studies have utilized modified frameworks for the classification of services in comparison to MEA, among these TEEB and UK NEA. At present, efforts are being made to create and adopt an international standard for ecosystem services, the Common International Classification of Ecosystem Services (CICES, 2013). In CICES, ecosystem services are defined as the contributions that ecosystems make to human well-being. An aim of the framework is to facilitate a standardized ecosystem accounting within the UN System for Integrated Environmental and Economic Accounting (SEEA) environmental accounting system. CICES has a hierarchical structure, but the highest level consists of three categories: (1) provisioning, (2) regulation and maintenance, and (3) cultural ecosystem services. Also, in CICES supporting services are not included as direct services because of the issue of double counting. Still, many of the regulating services may also be treated as supporting services depending on their place in the chain of ecosystem flows (United Nations et al., 2013). In this chapter, MEA is used as the principal methodology because it has been more widely used. In view of these considerations, the application of MEA takes into account potential pitfalls such as double counting.

In this context, under fisheries economics, the provision of food is recorded and analyzed in terms of fish and the (self-)regulating effect in terms of fish stocks being able to reproduce themselves within certain limits. An expansion of pure fisheries economics to an ecological economics approach raises two questions: (1) what is the exact impact on humans of ecosystem services and vice versa and (2) what is the value of ecosystem services that do not fetch a market price, i.e., what are the market failures? Neither of these questions is straightforward to answer. Ecosystem services are generally not valued in a market and have value only if they affect the well-being of humans, and this entails considering the service an end product that is consumed by humans. If services are intermediate products, they are consumed in the end product, and if they are factors of production, then only the changes in the stock of the production factor would count. The way intermediate factors count should be measured by their alternative use (opportunity costs). If there is no alternative use benefiting humans, the opportunity costs are not necessarily zero, with the result that the services are exhausted. It must also be taken into account that instant use of an intermediate factor excludes future use, which also has an opportunity cost.

7.4 Tools for Economic Analyses

To analyze ecological issues from an economic point of view, it is often convenient and sometimes even necessary to use modelling. A number of economic tools are available to integrate economic information and ecosystem services and present the information as management choices and policies. The literature about tools is comprehensive (Holland et al., 2010). In a broad context, the tools comprise:

1. Green accounting (macroeconomics);
2. Pressure, state, response analyses (PSR, DPSIR) (macro-, microeconomics);
3. Bioeconomic modelling (microeconomics)
   a. Types of models: (a) cost-benefit analysis (CBA), (b) cost-effectiveness analysis (CEA), (c) environmental impact assessment (EIA), (d) multi-attribute utility theory (MAUT), and (e) random utility models (RUM)
   b. Valuation methods: (a) revealed preferences, (b) stated preferences, (c) other (hybrid/political) preferences

Green accounting seeks to develop national accounting statistics to take into account the use and degradation of natural resources, aiming at a better measurement of economic indicators, such as gross domestic product. The tool is not convenient for analyses of optimal use but is rather a way to survey the use of resources for the production of commodities.

Useful tools are the Pressure-State-Response (PSR) method and the more comprehensive method Driving force-Pressure-State-Impact-Response (DPSIR), which were developed to identify the cause of the pressure on the state of a system and the impact in order to respond to and to identify the relevant indicators (Organization for Economic Co-operation and Development (OECD), 1994). These methods are particularly useful in data-poor systems because they allow one to establish ordinal and cardinal relationships and to measure the development in relevant indicators, but it is not particularly useful in economic analyses because criteria are needed to evaluate the system. In our context, the DPSIR is used to identify how a fishery (diving force) puts pressure on the impacted ecosystem services and what a possible response might look like. This method's natural successor is an ecosystem model where all indicators (variables) are measured.

Based on DPSIR, it could be argued that supporting and regulating services form the basis for provisioning, while cultural services with respect to fishing relate to tradition and play an independent role. The possible impacts of commercial fishing on recreational fishing and land-based cultural environments related to fishing ports and downstream effects from fishing are so-called secondary effects.

The CBA method measures the benefits and costs of different management options. CBA requires the valuation of all inputs, which requires that valuation methods be applied to non-use goods and services. CEA is a subset of CBA in the sense that targets are specified in advance and then the management approach entailing the lowest costs is chosen. EIA is useful in particular for identifying impacts and causes of particular management approaches.

Secondary effects must be considered carefully to avoid double counting in CBA. Using random utility models (RUM), this problem becomes clear. RUM models include a decision structure in which the agent has to make choices such as whether one wants to go fishing or hunting. Each choice means that something else, at an opportunity cost, is forsaken. The chosen activity is better than the forsaken activity and must be counted as a benefit only by the difference of utility between the two. In a fisheries context, a benefit evaluation of the competition between the commercial fishery and the recreational fishery must take into account that if recreational fishermen want to increase their activity at the expense of commercial activities, recreational fishermen must for go other activities. It is, therefore, often advised that this type of secondary effect be left out of analyses (Holland et al., 2010).

Measurement and valuation pose certain difficulties regardless of the method, be it revealed, stated, or political preferences, that is used. While the theoretical foundation is solid for stated and revealed preferences, it is not for the hybrid/political preferences. However, the latter is easier to work with because a political debate often ends up in a sort of consensus that could substitute stated references. Holland et al. (2010) mention four options regarding measurement: (1) estimate all or the most significant nonmarket values, (2) estimate a targeted set of nonmarket values that are relevant in a political context, (3) use benefit transfer, or (4) do nothing. For many years item 4 has predominated. This has changed, however, in recent decades, but there is still some way to go.

7.5 Green Accounting in Theory

This section briefly outlines the bookkeeping of green accounting and the relation to Standard National Accounts (SNA) in macroeconomics. It underlines the notion that it is the changes in stocks that count, not the value of the stocks. However, often changes are estimated based on the value of stocks. The usual way of measuring economic welfare is the GNP and derivatives thereof. This method has been criticized for many years for not taking into account a number of aspects, among those green accounting and sustainability, in the estimation of economic welfare (Stiglitz et al., 2010). A seminal paper in the green accounting literature is by Nordhaus and Tobin (1973), in which a first example of a correction of the GNP in two steps is presented. The first step derives a measure of economic welfare (MEW) by subtracting from the total private consumption a number of components that do not contribute positively to welfare (such as commuting or legal services) and by adding monetary estimates of activities that contribute positively to welfare (such as leisure or work at home). The second step converts the MEW into a *sustainable measure of economic welfare* (SMEW) that takes into account changes in total wealth. The relation between the MEW and the SMEW is similar to that between GNP and NNP (net national product) in standard national accounts: the SMEW measures the level of MEW that is compatible with preserving the capital stock (FAO, 2004).

The issue was then reopened in the late 1980s by Daly and Cobb (1989) who proposed the *index of sustainable economic welfare* (ISEW), further refined by Cobb and Cobb (1994). The ISEW has much in common with the MEW and the SMEW, but with two important additions, not least with respect to fisheries and ecosystem valuation: (1) an evaluation of natural resource depletion, measured as the investment necessary to generate a perpetual equivalent stream of renewable substitutes, and (2) the distribution of income (Nordhaus and Tobin had themselves acknowledged in their paper that this was one of the dimensions missing from their index).

On the other hand, the ISEW does not include any monetary evaluation of leisure time, for example recreational fishing, because of the difficulty of doing so. The ISEW can be summed up by the simple formula $ISEW = C + P + G + (W - D - E - N)$, where the variables in parentheses are currently left out of conventional GNP measurements. C is consumer spending adjusted for inequality, P is public expenditures excluding defensive expenditures, G is growth in capital and net change in international position, W is non-monetized contributions to welfare, D is defensive (i.e., to prevent events from happening) private expenditures, E is costs of environmental degradation, and N is depreciation of the environmental capital base.

Although the ISEW contributes to sustainable welfare taking into account the cost of environmental degradation and depreciation, it says nothing about whether welfare could be improved by a reorganization of production and use. In the context of ecosystem management, the components W to N play an important role. Compared to ecosystem services, component W relates primarily to *cultural services*, while components D and E relate to *regulating services*. Finally, component N relates to *supporting services* in the MEA framework. In the CICES framework, supporting services are not regarded as ecosystem services, but they are taken into account when estimating ecosystem assets that are important elements of natural capital (United Nations et al., 2013). These issues are subject to theoretical development in regard to natural resources in, for example, Hartwick (1990), Serafy (1997), and Mäler et al. (2008), who are aiming at the delineation of what should be included in the national account statistics.

A handbook on a System for Integrated Environmental and Economic Accounting (SEEA) has been developed by the United Nations (United Nations et al., 2013, 2014), and part of this work with particular reference to fisheries is found in a System for Integrated Environmental and Economic Accounting for Fisheries (SEEAF) FAO (2004). For fisheries and related accounts, the SEEA can be grouped into three main categories:

Monitoring the economic importance of fisheries, which includes the contributions to national income, employment, and foreign exchange earnings of fisheries and their subsectors; the distribution of benefits from fisheries among different groups in society, e.g., commercial, recreational; economic linkages between the fisheries sector and other sectors of the economy; the value of natural assets, in particular commercial fish stocks, and the cost of depletion; value of fishery resources shared with other countries; and monitoring the implementation of international instruments (e.g., UN Law of the Sea, UN Fish Stocks Agreement, Code of Conduct for Responsible Fisheries).

Estimating the full costs and benefits of fisheries and ecosystems, which involves assessing the extent of resource rent recovered by the government, accrued to the private sector, or dissipated on overcapacity and overfishing; assessing the extent of government fisheries management costs and habitat protection costs; assessing environmental externalities caused by fisheries or generated elsewhere in the economy and borne by fisheries (measured in both physical and monetary terms).

Improving fishery management, which involves assessing the economic efficiency of fishing in the subsectors and the potential value of fish under alternative management and policies. Fishery management can then be compared to the management of other resources in the economy; assessing government policies, such as fishery taxes and subsidies, on incentives for the sustainable utilization of fishery resources and on the distribution of access to fisheries and benefits from fisheries. Fishery policies and management can also assess the impact of macroeconomic policies on the fisheries sector, such as economy-wide changes in taxes or interest rates, e.g., are fisheries especially vulnerable to specific policies? (FAO, 2004: p. 3).

The conventional SNA consist of an integrated sequence of accounts that describe the behaviour of the economy from the production of goods and services – generation of income – to how this income is made available to various units in the economy and how it is used by these units. The SNA have identities among each account and between accounts that ensure the consistency and integration of the system. A particularly useful identity for the SEEA involves the total supply and total use of products. In a given economy, a product can be the result of domestic production or production in another territory – imports. Hence: *Total Supply = Domestic Production + Imports*.

On the other hand, the goods and services produced can be used in various ways. They can be used by (1) industries to produce other goods and services (intermediate consumption); (2) households and government to satisfy their needs or wants (final consumption); (3) they can be acquired by industries for future use in the production of other goods and services (capital formation); and finally, (4) they can be used by the economy of another territory (exports). Therefore: *Total Use = Intermediate Consumption + Households/ Government + Final Consumption + Capital Formation + Exports*.

Total supply and total use as defined previously must be equal. In the SNA, this identity is expressed only in monetary terms, but in the SEEA it has to hold also when the accounts are compiled in physical terms. The total supply and use equations only include data information; they do not include any information about sustainability.

An identity of the SNA that is particularly useful in the SEEA involves assets and comes closer to an assessment of sustainability. This identity describes how the stock of some assets at the end of an accounting period is the result of the initial stock level of the asset (opening stock); the value of a producer's acquisitions, less disposals of fixed assets during the accounting period, changes in inventories and acquisition less disposal of valuables (gross capital formation); the consumption of fixed capital; changes in value of assets due to changes in their prices (holding gains/losses on assets); and other changes due neither to transactions between institutional units, as recorded in the capital and financial accounts, nor to holding gains and losses (other changes in the volume of assets) (FAO, 2004: p. 16): *Closing stocks = opening stocks + gross capital formation − consumption of fixed capital + holding gains/losses on assets + other changes in volume of asset.*

The SEEA flow accounts incorporate environmental concerns partly by rearranging items already in the SNA and partly by adding new items: *Expenditures for protection of fish habitats and resource management accounts*. They are already included in the SNA, but the SEEA reorganizes these expenditures in order to make them more explicit and, thus, more useful for policy analysis. In this sense, these accounts are similar to other satellite accounts, such as transportation or tourism accounts, which do not necessarily add new information but reorganize existing information. This set of accounts has three quite distinct components: expenditures for the protection of fish habitat and resource management, by industry and household; activities of industries that provide environmental protection services; and environmental and resource taxes and subsidies (FAO, 2004: p. 19).

An example from fisheries of physical (in contrast to economic) asset accounts shows the fish stocks at the beginning and at the end of the accounting period and, hence, the changes. The accounting period is typically 1 year, corresponding to the accounting period for the national accounts. *Opening stocks* records the volume in tons at the beginning of the accounting period, and *Closing stocks* records the volume at the end of the accounting period. Changes in stocks during the year are divided into changes that result from economic activity (changes in inventory, net growth of stock, or catch) and changes due to other factors (other changes in volume). The role of economic activity is different for cultivated and non-cultivated fisheries, so the treatment of common items, such as catch and natural growth, differs for the two categories of fish.

For cultivated assets, stock raised for harvest over a period of more than 1 year is treated as a work-in-progress, so the changes in stocks are recorded as *Changes in inventories*. For cultivated assets kept as breeding stocks, annual changes are measured as the *Net growth of the breeding stock*, which is equivalent to gross fixed capital formation minus consumption of fixed capital (losses of breeding stock). *Other changes in the volume of assets* include catastrophic losses due to, for example, environmental events or disease, uncompensated seizure, and other factors that are not directly related to economic activity.

For non-cultivated assets, the source of change resulting from economic activity is the annual *Catch*. *Other changes in the volume of assets* includes catastrophic losses but also the net natural growth of the stock (recruitment minus natural mortality). The reason for including net natural growth as part of *other changes in volume* is that this growth is not under the control or management of economic activities, in contrast to cultivated fish stocks (FAO 2004: p. 24). A detailed structure of the asset accounts for fisheries would look as follows:

1. *Opening stocks*

   Changes in stocks

   a. *Total catch*: (a) Commercial, large-scale; (b) Small-scale, commercial, and subsistence; (c) Recreational; (d) Estimated illegal catch

   b. *Net natural growth of stock*: (a) Recruitment minus natural mortality

   c. *Other changes in volume*: (a) Estimated discards, (b) Other

2. *Closing stocks*

   (FAO, 2004, Table III.2)

The use of an accounting framework to develop indicators, such as the SEEAF, would provide credible, compatible, and consistent information for policymaking, regardless of the specific needs for which the information is used.

In the past, environmental-economic accounting was conceived mainly to provide a more accurate measure of sustainable income by revising conventional macroeconomic indicators (usually net domestic product) to reflect the depletion and degradation of natural capital. The approach that is most relevant for fishery resources subtracts the depletion of natural capital from conventional NNP to obtain depletion-adjusted NNP. The adjustment of NNP for asset depletion parallels the treatment of produced assets in the national accounts and is accepted in principle by most economists and statisticians.

It is not easy to measure the depletion of fish stocks. Measuring depletion requires identifying a decline in fish stocks attributable to economic activity, not natural causes, which is not always easy. Even where physical depletion can be quantified, there is not yet a consensus on the correct way to value the depletion of natural capital. Consequently, no country yet includes depletion of fish stocks in its national accounts, or even in experimental accounts.

The SEEA-2003 (United Nations et al., 2003) was developed to provide an information system consistent with the concept, definitions, and classifications of the SNA, from which a consistent and comparable set of indicators could be derived. In recent years, in the case of fisheries, there have been efforts to develop a set of indicators for sustainable fisheries, based on economic, social, ecological, and institutional statistics. FAO published a review of potential indicators for marine capture fisheries and for an ecosystem approach to fishery management under its series Technical Guidelines for Responsible Fisheries (FAO, 1999, 2003).

The SEEAF provides an information system that can be used for the derivation of sustainable development indicators for fisheries. It focuses on the interactions between the economy and the environment, separately identifying transactions related to the fisheries sector and the impacts on the fisheries of other economic activity. Although at present the framework does not specifically address social aspects, these, as well as indicators that cannot be directly derived from the accounts, could be linked to the core accounts through supplementary tables. These supplementary tables could be designed in such a way that the definitions and classifications are consistent with the core framework, thereby generating consistent sets of information (FAO, 2004: p. 63).

Asset accounts for fish stocks contribute to more effective monitoring of both fish resources and national wealth and can be used to improve the management of natural capital from fishery resources.

Finally, SEEAF argues that there are four main areas in which the use of fishery accounts for policy analysis needs to be improved. First, some countries have begun to compile fishery resource accounts, but there are still very few countries with such accounts. Second, most of the hitherto limited work on fishery accounts has focused on commercial fisheries and little has been done on artisanal and subsistence fisheries. Third, the value of fishery resources and their habitats includes many services provided by fish stocks and the associated ecosystem, such as recreation and tourism, as well as existence value. Sustainable fishery management requires addressing all these environmental services, not just the provision of fish for the commercial or subsistence sectors. And fourth, accounts for straddling and highly migratory stocks have not yet been fully established (FAO, 2004: p. 88).

7.6 Green Accounting in Practice

Fisheries are entered into the national accounting statistics with landing, intermediate consumption, and use of production factors in terms of fishing vessels. Fish stocks are currently not entered. To do this, a bioeconomic model and information about fishing costs and prices are required.

However, recording consumption in an accounting system is only one step. Another step is to use this information to decide whether economic welfare could be improved by a reallocation of production factors, goods, and services. Here CBAs that include valuation would help.

It has been shown how stocks could be entered into an accounting system using a shadow price in terms of resource rent as remuneration of the fish stocks parallel with remuneration of labor and vessel capital. What is important, however, is to take into account changes in fish stocks (Danielsson, 2005). If a stock is overexploited compared to maximum economic yield (MEY) or, alternatively, maximum sustainable yield (MSY), the depreciation of the stock is *Depreciation = (Stock at MEY − Current stock) × resource rent/stock unit.*

In green accounting, depreciation is consumption of the fish stock and must be subtracted. On the other hand, if the fish remuneration is not calculated, there is a tendency to overestimate the value of the fleet capital and hence the depreciation. Therefore, using these depreciation costs, some of the fish stock depreciation is already included.

The following example includes the landings (consumption of fish) and the change in the stock biomass, which in this case is viewed as an intermediate consumption of the fish stock production factor (SEEAF, 2004; Danielsson, 2005). Table 7.1 shows the principle of an accounting system for one species (stock). The system could easily be expanded to include many stocks and supporting services to the fish stock as part of the ecosystem assets. However, including supporting services requires a valuation of the service, how it varies with the fish stock, and whether the service is a final or an intermediate product, to avoid double counting.

Assuming there is no discard, landings are equal to catches, and in year 1 the landings are slightly higher than the growth of the stock, and the stock is then reduced at the end of the year. In year 2, the growth is assumed to be lower, but the landings are even higher than in year 1. This could happen if more fishing effort is expended. This means that the harvest value (consumption) increases, which viewed in isolation is beneficial to society. But this happens at the expense of the stock, which is reduced by more than the increase in landings. The net result is that society is worse off in year 2 than in year 1. The upper part of the table shows an example of an indicator system without a criterion as to whether the development is preferable or not. All we know is that year 1 is better than year 2, but we do not know what is best, which requires objectives, modelling, and calculations.

TABLE 7.1
Accounting System for a Fish Stock

Item	Year 1	Year 2	Notes
Opening stock	1000	990	
Forecasted growth	400	396	Say 40% of stock
Landings	410	430	
Stock depletion	−10	−34	
Closing stock	990	956	
Fish price, €/kg	5	5	
Harvest value	2050	2150	
Stock change value	−50	−170	Rated at 100% of fish price
Total value	2000	1980	

In the lower part of Table 7.1, a criterion in terms of value of harvest and net change (depreciation) of the stock is shown. It is assumed that fishing costs are zero in this example. The landing value is rated by the fish price, and the stock change is rated by the same price. In principle, the stock depreciation prices should be estimated as the unit profit at MEY times the MEY stock – the current stock. If stocks are overexploited, this price will be higher than the current catch price. Most often, in practice it is not possible to estimate this price, and therefore market prices may be used. From Table 7.1 it can be seen that if the stock change is rated at a price of 85% of the fish price, the two years perform equally well. But if year 2 is farther away from MEY than year 1, the stock unit price is higher than the catch unit price. Therefore, year 2 is much worse than year 1.

In such a system, the supporting services are intermediate products, i.e., they form the basis for the production of fish stocks. If they are end products, a separate valuation must take place. If they are intermediate products in the production of a fish stock, only the fish stock needs to be assessed. The best possible fish stock is also the best possible indicator for the supporting services. In short-run assessments, a threshold is required and estimates of such a threshold could be included in the price of the fish stock. A high importance of supporting services calls for a high value assigned to the change in the fish stock. The price could be even higher than the sales price of the fish if risk and irreversibility are taken into consideration. These considerations need to be carried out outside the accounting system.

Regulating services are taken into account in the valuation of the fish stocks but could be fully addressed only in cases where cost of fishing is recorded (complete recording of all intermediate products). This makes it possible to calculate the profit from the fishery, which would be the maximum amount that could be spent on sustaining regulating services.

Cultural services related to activities such as recreational fishing, heritage, leisure activities, and others are considered final services that compete with or are supplementary to the fishery. Supplementary means that cultural services vary with the fishing effort, positively or negatively, while competing means that a choice must be made between fishery and cultural services or a trade-off must be established. Cultural services may vary with the stock of certain species. Cultural services must be clearly defined, and it must be established whether they are competing or supplementary, but even if this is done, the economic importance of cultural services may not be high. The contribution of cultural services is already recorded in economic accounts to some extent, such as, for example, the purchase of equipment of recreational fishermen. The next step is to use this information to decide whether the general economic welfare could be improved by reallocating resources. As an example, could the value added by recreational fishermen be increased at the expense of commercial fishermen? This requires economic calculations in which profits (benefits minus costs) of the two activities are compared. While benefits and costs are recorded for fisheries, they are only partly recorded for leisure fishermen. In particular, labor costs are not recorded for leisure fishermen.

Assuming that leisure fishermen are not completely inactive while not fishing, the inclusion of opportunity labor costs would entail that the net benefit from leisure fishing be rather small compared to a well-managed commercial fishery. Therefore, reducing well-managed fisheries in favor of leisure fisheries would very likely lead to an economic loss. Disregarding some cultural services related to leisure fishing in the analyses of the economics of ecosystems would hardly lead to any grievous errors. However, optimization of all values in marine areas could in some cases mean a reduction in fishing activities – provisioning services – in favor of cultural services (e.g., existence, bequest, scientific, and symbolic values) related to the conservation of certain species, often charismatic species like marine mammals. For example, Sanchirico et al. (2013) found that an increase in the population of the endangered Steller sea lion would justify a full or partial closure of fisheries in three important commercial species.

7.7 Methods for Economic Analyses

The macroeconomic accounting systems (SNA and SEEA) suffer from data shortage to be sufficient tools to monitor and, in particular, analyze the consequences for society of the use of ecosystem resources. As a consequence, relatively simple systems are developed. One widely used system is the DPSIR system. DPSIR was initially developed by the OECD (1994). The DPSIR system, as shown in Figure 7.1, helps to define an entire system and to translate it into an ecosystem model.

The *driving force* of the system is the economic gain from the exploitation of the natural resources that the anthropic system may achieve. This entails a *pressure* in terms of effort on the fish stocks and the underlying ecosystem. The *state* is basically the state of the ecosystem described, for example, by the ecosystem services. However, the state has to be detailed further subject to the scope of the study in order to make the system suitable for analysis. For example, the state could be described as biodiversity in terms of sea mammals, fish, plankton, salinity, oxygen contents, and nutrient load, for example. The *impact* could be measured in different ways, but in an economic context, indicators with clear economic relevance need to be identified. While it is easy to establish indicators of the elements that are subject to market price formation, i.e., fish species, it is less easy for indicators of embodied elements that require estimation of importance and valuation. The *response* to remedy undesirable changes would be to remove the driving force, which is normally not possible. Then the pressure could be changed, meaning the fishing effort could be reduced. The state of the system could also be protected, for example, by fishing quotas, closed areas, specific fishing gear provisions, and so forth. Finally, the response could be directed toward the impact in the sense that the deterioration of the state could be alleviated. The output from the DPSIR method is presented in terms of indicators. The development of the indicators over time may be interpreted subject to criteria in, for

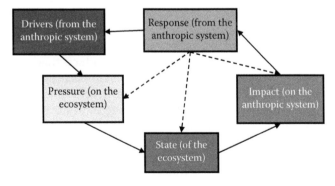

FIGURE 7.1
The DPSIR system. (OECD. 1994. *Environmental Indicators*, OECD, Paris.)

example, traffic-light tables. In many cases, it is preferable to aggregate indicators into one single indicator and even into one single point in time, for example, such as the net present value (NPV) of a stream of payments over time in CBA.

With reference to fisheries, indicators can be derived from the SEEAF methodology. Driving force indicators are related to consumption and production patterns and often reflect resource use intensities and emissions, along with related trends and changes over a period of time. They also include population growth, demographic patterns, and economic growth.

State indicators provide an overview of the situation of the environment and its development over time. Examples of state indicators include, for example, the concentration of pollutants in environmental media, degraded environmental quality and its effect on health, and the status of natural resource stocks and of ecosystems.

Response indicators show the extent to which a society responds to environmental concerns. They refer to individual or collective actions intended to mitigate or prevent the impact of human activities on the environment, halt or revert environmental damage already inflicted, or preserve natural resources. Examples of response indicators include environmental expenditures and environment-related taxes and subsidies.

7.8 Inclusion of Ecosystem Services in Green Accounting

As pointed out earlier, it is not straightforward to combine the services listed in the MEA (2003) with economic accounting systems. Not all of the services listed in the MEA are services in an economic context where services are either end products or intermediate products. While provisioning services are direct contributions to end products together with capital and labor, supporting services are not. They are, rather, functions that contribute to the formation of provisioning services and therefore of intermediate character. However, if they are intermediate factors producing fish stocks, they must not be valued, so as to avoid double counting. If they are valued and entered into the green accounting system, the value must be subtracted from the value of the fish stocks to avoid double counting in the same way as intermediate consumption is subtracted from final consumption.

Table 7.2 represents an attempt to produce an overview of an accounting system, including ecosystem services, inspired by the structure of the asset accounts for fisheries. The first column in Table 7.2 includes products (services), and the first row includes factors that are responsible for the welfare-producing process and hence subject to distributional aspects.

The first five items show the SNA. The total value is expressed by the final consumption, and all five are, in principle, subject to price formation. Products that are consumed but not subject to price formation, which could be the part of leisure fishing beyond the cost (price) of going with a tour boat and paying a license fee, or beyond the cost of one's own boat, could be added to final consumption. In principle, changes in stocks should also be added to final consumption, which is normally not the case. In the next step, final consumption is reduced by intermediate consumption (variable production costs), then the GNP appears. This is of interest if the remuneration of production factors is going to be estimated. The GNP is distributed between labor and (physical) capital. However, if depreciation (or appreciation) of capital is not taken into account, the estimate of the national product is

TABLE 7.2
National Accounting System for an Ecosystem

Product/Process	Large-Scale Fisheries	Small-Scale Fisheries	Recreational Fisheries, Other
1. Output (final consumption)	+/+	+/+	−/−
2. Intermediate consumption	+/+	+/+	−/−
3. Gross National Product (GNP)	+/+	+/+	−/−
3a. Labor remuneration	+/+	+/+	−/−
3b. Capital remuneration	+/+	+/+	−/−
4. Capital change (depreciation)	+/+	+/+	−/−
5. Net National Product (NNP)	+/+	+/+	−/−
6. Fish stock remuneration	−/+	−/+	−/+
7. Fish stock change	−/+	−/+	−/+
8. Supporting service change			
8a. Food-web dynamics	−/−	−/−	
8b. Diversity	−/−	−/−	
8c. Habitat	−/+	−/+	−
8d. Resilience	−/+	−/+	−
9. Regulation			
9a. Biological regulation	+	+	+
10. Cultural			
10a. Recreation	−	−	+
10b. Aesthetic value	−	−	(+)
10c. Cultural heritage	(−)	(−)	(−)
10d. The legacy of nature	−	−	−
11. Ecosystem Net National Product (ENNP)			

Note: +/+: Means recorded/must be recorded; −/−: Means not recorded/do not have to be recorded, etc.

not correct. If depreciation/appreciation is added to GNP, the net national income shows a more complete measure.

The two production factors are not the only ones responsible for the output production, which on the other hand excludes output that is not valued. The next general step is to include fish stocks in the same manner as capital. The resource rent is an estimate of the remuneration of the fish stocks, but the change in the fish stocks also needs to be included. This is included in the SEEAF as the first step.

It is not straightforward to include fish stocks either theoretically or practically. The inclusion of fish stocks affects the valuation of the fishing fleet capital. When fish stocks are not remunerated, the profit allocated to the fishing vessels will be higher than in the case where fish stocks are remunerated. Therefore, the valuing of the fleet capital must be changed when fish stocks are included in the accounts.

This problem becomes even more difficult when ecosystem services are included and there is a need to distinguish between at least three aspects. The first one is whether the ecosystem services are competing end products (complementary) in relation to the fish stocks or intermediate, i.e., supporting the fish stock abundance. The second aspect is that ecosystem services are by definition treated as flow variables. However, ecosystem services may depend on both physical flows and stocks. Fish is both flow (catches) and stock (spawning stock). The third aspect is price setting as the introduction of new items affects the prices of the items already in the accounting system.

If an ecosystem service is considered an end product having a flow value, for example, clean water, the service must be valued and the relationship with the fishery (and the ecosystem) established. If at the same time the stock changes, the net change must be valued too. This is not without complications.

Another question is the importance of biodiversity. A high level of biodiversity may, to some extent, support the productivity of fish stocks, so in this regard it is an intermediate service and is already valued by the valuation of the catches and the fish stocks. If for some reason it is valued, then it must be subtracted from the value of the fish stocks to avoid double counting. However, biodiversity may also provide other services that are of a regulating or cultural nature and not directly linked to catches. The same argument applies to the other supporting services, habitats and resilience. If these are a precondition for productive fish stocks, they are already valued.

With respect to cultural services, another problem is whether recreational fishing should be valued and included as an end product *competing* with commercial fisheries. The immediate answer is no, as an inclusion requires that the cost (intermediated consumption) be valued and subtracted. However, as recreational fishermen refrain from other (leisure) activities, the value of these activities must be subtracted. The end result is, keeping in mind that recreational fishing contributes only a small fraction to the economy, the contribution from recreational fishing is small compared to commercial fishing. To some extent, recreational fishing is embodied in small-scale fishing.

Table 7.2 represents an attempt to list what is already in the SNA, what is possible, and what is not required. Plus (+/) is what is recorded and (/+) is what should/could be recorded. Hence, (−/) shows what is not recorded and (/−) what should not be recorded. There are a number of uncertainties as the services can be either end products or intermediate products or both at the same time.

7.9 Microeconomic Modelling

In this section, the scope is microeconomic modelling ranging from relatively simple CBAs to complex ecosystem models from an economic point of view. Accounting, including green accounting and DPSIR, is, if not necessarily a prerequisite, then at least a good foundation for ecosystem modelling. Often DPSIR is embodied in the development of the ecosystem model in the sense that when the model is constructed, it is considered how causes and effects are structured, what the data demand is, and where data fall short. Usually green accounting systems do not include enough information to run a complete analysis. Estimates must then be used where no data exist on the value of goods and services.

Figure 7.2 shows the structure of a bioeconomic model and how it may be expanded to include ecosystem services. The structure of the model is made with reference to DPSIR. The inner circle (items 1–5) shows the structure of a bioeconomic model, here illustrated by the bioeconomic model Fishrent (Salz et al., 2011; Frost et al., 2013). It is assumed that fishermen want to maximize their profit (item 1). This is done by investing in fishing vessels and engaging in certain fishing behavior. The pressure on the ecosystem is through the harvest of fish, the way fishing is applied, and the technology

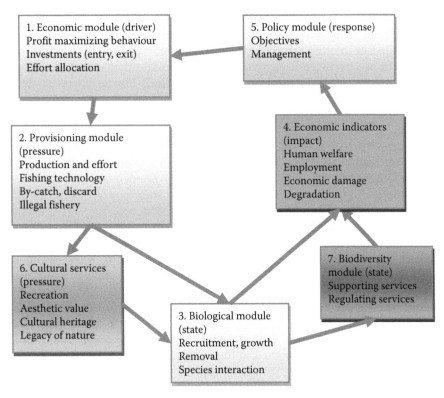

FIGURE 7.2
A bioeconomic model, e.g., Fishrent, based on DPSIR, developed to include ecosystem effects.

used, as well as how this applies pressure in terms of by-catch of mammals and sea birds and by-catch of fish with no commercial value. The state of the stocks is determined by the fishing pressure, recruitment to the stocks, and how the stocks interact with one another. Hence, in a pure fisheries model some ecosystem effects in terms of supporting and regulating services are already included. The result of the model analyses is presented by a selection of indicators, chosen for the particular analyses (item 4). These indicators could be resource rent and employment, but also the degradation of the resource and the environmental damage could be chosen. Finally, the response to unwanted impact is introduced (item 5). The response has the format of policy actions, which need to incorporate public interference because the fishermen do not include externalities in their decisions. If the negative impact can be measured, it could be taken into account in policy measures, and even if the negative impact were not measurable, policy interference would still be possible because it is known how the externalities impact the system.

Cultural supporting and regulating services are included in the outer ring (items 6–7). In particular, some cultural services are already measured in national accounts, but often they are not taken explicitly into account in bioeconomic modelling. Some of the cultural services compete with the provisioning of fish, for example, recreation, while others are produced jointly with provisioning, for example, heritage – if there was no fishery, the heritage would be different. Cultural services are classified under pressure, but the border between pressure and state may be floating.

Supporting services (e.g., food-web dynamics, diversity, habitat, and resilience) and regulating services (e.g., biological recovery, seabed restoration) are to some extent already embodied in the biological module because recruitment, natural mortality, and intrinsic growth are dependent on these services. Some other services are not, but in general it requires considerable effort to determine exactly how and to what extent these services influence the biological module. Furthermore, in some cases, fish stocks are not assessed, and therefore the influence of the supporting and regulating services is difficult to judge.

On top of the structure and the construction of models, models are designed to run several types of solutions, which by themselves require more or less complex models: (1) exogenous simulations, (2) endogenous simulations, (3) stochastic simulations, (4) management simulations, and (5) closed-form management solutions (Arnason, 2000).

Exogenous simulations. A model comprises a causal structure, i.e., what is determined by what. If the model is input driven, the effort determines the landings. The simplest way to solve such a model is by simulations where all the economic decision variables – fishing effort, capital investment and entry/exit – are exogenous. If the causal structure is reversed, the model becomes output driven and it is possible to determine effort as a function of catches/landings, i.e., total allowable catches (TACs)/quotas. Almost all empirical models are either input or output driven. Very few models, one example is Fishrent (Salz et al., 2011; Frost et al., 2013), are constructed in a way that makes it possible to reverse the model. This means that values for either effort or TAC are fixed for each of the simulation years and the model calculates all endogenous variables (indicators).

Endogenous simulations. In endogenous simulations, all of the variables are endogenous and basically determined by the behavior of the fishermen with respect to investments and entry/exit. Only the start values of the model are exogenous. The same causal conditions as mentioned under exogenous simulation apply here, but Fishrent can be reversed, which makes it possible to compare effects of input or output restrictions directly.

Management simulations. When different types of harvest control rules are introduced into either exogenous or endogenous simulations, the effects of management measures could be estimated. The harvest control rules are TAC (output) or effort (input) restrictions or various types of economic management measures.

Stochastic simulations. Fishrent can run with stochastic elements in the equations taking into account that fisheries and, hence, models are subject to uncertainty. This uncertainty, stemming from natural variation, measurement, modelling, and estimation errors, may in many cases be represented by a set of random terms drawn from the appropriate distribution. The solution of each simulation is stochastic, and the final results usually consist of a high number of simulation repetitions in order to obtain estimates of the distribution of outcomes.

Closed-form management solutions (optimal solutions). The final type of simulations is the most ambitious. The aim is to use the model to derive optimal solutions, for example, by maximizing resource rent. The so-called closed-form solutions to management controls provides values for all management controls (e.g., TACs, fishing days, tax rates) as a function only of the state of the system and the vector of exogenous variables in terms of restrictions. Fishrent uses dynamic nonlinear optimization, where the optimization program finds the optimal solution. The advantage of closed-form management solutions is that once they have been derived, there is no need for more simulations, except maybe now and then to check the robustness of the solutions. A closed-form solution provides values for the management controls for any situation that may arise.

7.10 Prices and Valuation

Analyses based on modelling are often carried out without any estimation of the variables and parameters of the model. Such analyses are valuable, and the objective is to determine what is cause and what is effect. Furthermore, the analyses seek to find solution to the problem based on a set of criteria. In economics such criteria are usually applied to, for example, maximize the welfare of humans or maximize employment. It is important to include as many elements as possible, but often the investigated problem is scoped in order to make the analysis more manageable and clear. In ecological economics the choice between what should be included and what should be left out is difficult (MEA, 2003). On one hand, it is tempting to include only what the market wants, i.e., what is subject to price formation on the market. On the other hand, the knowledge that also goods and services that are not marketed are important is essential to take into account in the results of analyses. External effects, i.e., actions by individuals that impact the utility of another individual without this individual's being able to influence these actions, are often considered in economic analyses, but in ecological economics it is imperative to include external effects and even services that are not associated with external effects.

Figure 7.3 shows all the elements that are considered in a complete economic analysis. The right-hand side of the figure shows steps in the valuation procedure. The left-hand side includes a number of elements that are and are not subject to price formation on a

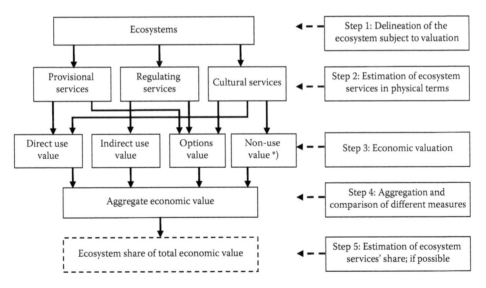

FIGURE 7.3
Valuation of ecosystems and services. (Modified from Hein, L. et al. 2006. *Ecological Economics* 57:209–228.) *Non-use value comprises bequest, altruist, and existence value. Explanation of concepts: direct use value: Results from direct human use of biodiversity (consumptive or non-consumptive); indirect use value: derived from regulation services provided by species and ecosystems; option value: relates to importance given by people to future availability of ecosystem services for personal benefit (option value in a strict sense); bequest value: value attached by individuals to the fact that future generations will also have access to the benefits from species and ecosystems (intergenerational equity concerns); altruist value: value attached by individuals to the fact that other people of the present generation have access to the benefits provided by species and ecosystems (intragenerational equity concerns); existence value: value related to satisfaction individuals derive from mere knowledge that species and ecosystems continue to exist.

market. The second row lists services whose provisioning (e.g., catches of fish) is subject to direct final use and, hence, price formation. Some of the cultural services, for example, recreational fishing and visiting museums, are also examples of final use, but often these services are not subject to complete price formation even if one has to pay for the fishing equipment and the entrance fees to museums.

Regulating and cultural services take various forms in terms of indirect options and non-use value. Supporting services, as explained earlier, do not contribute directly to the different values in Figure 7.3, but they could be values as part of the ecosystem assets. A range of valuation methods has been developed theoretically and applied to make it possible to include these elements in numerical modelling, for example, CBAs, numerical bioeconomic, and ecological modelling. A complete picture of economic value requires valuation of these elements. If that is not possible, these elements should be included based on the way they are assumed to influence the result, for example, based on whether the result has improved or worsened.

Table 7.3 is an attempt to link services as described by the MEA 2003 to the impact of fisheries activities and the most likely evaluation methods. The left-hand-side column

TABLE 7.3

Classification of Ecosystem Services and Fisheries Management Impact on Ecosystem Services

Services	Type	Impact from Fishing	Valuation Methods
Supporting	Biochemical cycling		
	Primary production		
	Food-web dynamics	Fishing down the web	PWTP; DE
	Diversity	By-catch, destructive fishing	PWTP; DE
	Habitat	Bottom-contacting fishing gear, undersized fish	PFM; DE; PWTP
	Resilience	Fish stock abundance	PWTP; RCM
Provisioning	Food	Fish landings	Market
	Inedible resources		
	Genetic resources		
	Chemical resources		
	Ornamental resources		
	Energy	Competition	Market
	Space & waterways	Competition	Market
Regulating	Atmospheric regulation		
	Regulation of local climate		
	Sediment retention		
	Biological regulation	See supporting	All methods
	Pollution control		
	Eutrophication mitigation		
Cultural	Recreation		CVM, TCM, HPM
	Aesthetic value		CVM, PWTP
	Science & education		
	Cultural heritage		PWTP
	Inspiration		
	The legacy of nature		PWTP

Source: Millennium Ecosystem Assessment (MEA). 2003. *Ecosystems and Human Well-being: A Framework for Assessment. World Resources Institute.* Island Press; developed from Barbier, E. B. 2007. *Economic Policy* 22(49):177–229; and Bateman, I. J. et al. 2011. *Environmental Resource Economics* 48(2):177–218.

shows all the services listed by the MEA 2003, but only those in bold face type are impacted by fishing activities. The middle column lists the type of impact, while the right-hand column shows the most likely applicable valuation methods developed by Barbier (2007) and Bateman et al. (2011). WTP: People's willingness to pay is the general driving concept. This is divided into two general methods: revealed preferences (RPs) of consumers (i.e., what they are actually paying) and stated preferences (SPs) (i.e., what they would pay hypothetically if they were asked). These two methods are theoretically well founded but often difficult to apply in practice. A third method, theoretically not well founded, seems more applicable and could, for convenience, be called political willingness to pay (PWTP): what do politicians do and think on behalf of the people who have elected them.

The RP method is further divided into the production function method (PFM), the travel cost method (TCM), the hedonic price method (HPM) (often also called the property value method), and the defensive expenditure (DE) method. The SP method is divided into the contingent valuation method (CVM) and choice experiments (CEs).

Looking at the *supporting services* first from a welfare economics point of view, only if the supporting services (biodiversity, food web, and resilience) are end products should they be valued and included. There is little reason to assume that this will be the case. They are more likely 'intermediate' products with respect to the fish stocks. If the fish stocks are performing optimally, it may be concluded that the intermediate products are being consumed optimally.

As for *regulating services*, their costs (degradation) are largely already included in the national accounting. The costs of management measures, public or private, are included, as are measures such as marine reserves because they change the benefits and costs of the fishery. Inclusion in national accounts is not the same as optimal exploitation of the fish resources, and information about supporting services and their impact on fish stocks and other end products may have to be collected for that specific purpose.

Finally, *cultural services* are, largely, end products that require valuation and inclusion in green accounting. However, apart from recreational fishing and other activities, cultural services are very poorly defined and, hence, difficult to value. Next the value of these services may vary considerably from case to case, and collecting information to value the services may be very costly.

With particular respect to recreational fishing, it could be argued that it should be valued, compared, and managed in relation to commercial fishery. But a recreational fishery is not restricted compared to a commercial fishery. The problem is rather to manage a recreational fishery in order to accrue some of the resource rent.

It is of particular interest to include the value of non-valued goods. A survey shows that few methods are dominant. Table 7.4 shows data that pertain to valuation studies published in the peer-reviewed literature. The total number of valuation studies is 314. Note that specific methods dominate. Of 21 valuation methods, 6–8 are used excessively, and among the 4 services, 2 methods capture more than 50% of all the methods used (TEEB, 2009).

If the 21 valuation methods are aggregated into 5 general methods, as shown in Table 7.4, RPs and SPs are dominant for cultural services, while provisioning is dominated by costs and production methods, reflecting that market prices exist or may be estimated. Regulating services are valued by methods that estimate mitigation or replacement costs, while supporting services are dominated by preference methods. There is no information about the distribution of the method in the 314 articles on the 4 types of services. It is anticipated that most of the studies took place within provisioning and cultural services.

TABLE 7.4
Valuation Approaches Used for Valuing Ecosystem Services (%)

Type of Valuation Approach	Cultural	Provisioning	Regulating	Supporting
Benefits transfer	9	3	4	6
Cost based	5	27	61	17
Production based	1	33	9	0
Revealed preference	38	18	7	28
Stated preference	46	19	19	50
Grand Total	100	100	100	100

Source: TEEB. 2009. *The Economics of Ecosystems and Biodiversity,* Climate Issues Update. 27 p.

Fisheries economics deals with goods that are subject to market prices derived from human utility and the impact on the fish stocks but has left the remaining ecosystem goods and services aside. This has led to the risk of sub-optimization because the reproduction of the fish resources is determined by the direct impact from fishing not only on the fish stocks but also on a number of ecosystem services, which are not subject to market prices or to economic valuation. In our study, the services impacted by fishing are included qualitatively because the valuation of most of these services is extremely difficult, time consuming, or even impossible (EEA report no. 4/2010) (Holland et al., 2010; Spangenberg and Settele, 2010).

7.11 Bioeconomic Modelling and Management, Brief Overview

This section briefly discusses the development of the theory of the exploitation of natural resources with an emphasis on fisheries economics. Bioeconomic modelling has been used extensively to estimate the value of provisioning services in the form of profits or resource rents from fisheries. Also, other ecosystem services can be quantified with the help of bioeconomic models (Barbier, 2007). Bioeconomic modelling is a well-established field within natural resource economics and can be dated back to the more than 150-year-old Faustmann formula, which calculates the present value of the income streams for a forest rotation, and in the fisheries literature to Warming's work on optimal fishing from 1911 (Faustmann, 1849; Warming 1911 in Andersen, 1983; Polasky and Segerson, 2009). The field of bioeconomics, in which the services from ecosystems are converted into the provision of economic goods, has developed extensively since the 1950s, with important contributions from Gordon (1954), Scott (1955), and Schaefer (1957) and later with Clark's (2010) book on mathematical bioeconomics.

The conventional bioeconomic models are concerned with provisioning services alone, such as fisheries or forestry, and typically include a single species. Simple bioeconomic models, such as the Gordon-Schaefer model, in which a biological growth function is coupled with a production and a cost function, have been very useful in providing insight into the dynamics of fishermen's behavior (Gordon, 1954; Bjørndal and Munro, 2012). However, as emphasized by Perrings (2010), there is a need to extend traditional bioeconomic models to multispecies models that take into account the different aspects of biodiversity. Tschirhart (2009) points to the main weakness of the simple bioeconomic models in this respect. They do not assess how harvesting impacts the biodiversity and ecosystem functions, including species interactions. The inclusion of ecosystem services is therefore incomplete. For example, conventional single-species approaches to modelling fish stocks are likely to conclude that marine mammals are detrimental to stocks because they cannot take account of the range of indirect and more complex interactions (Morissette et al., 2012). However, resource economists have applied predator–prey models in which harvesting one or both species requires careful assessment of trade-offs for optimal management depending on the relative prices of the species and the specific interaction (Flaaten, 1998). In brief, management must consider investing the rent from catching an additional fish or leaving it at sea in order to reduce future costs and increase steady-state catches of the same or other species (Hannesson, 1983). Also, spatial models have been designed where habitat patches interact (Sanchirico and Wilen, 1999, 2005). In recent years, both provisioning of ecosystem services, such as fish catch, and regulation and cultural services consisting of habitat maintenance and non-extracting recreational activities are included in the models (Kellner et al., 2011). Similarly, models have been elaborated to incorporate the impact of ecosystem externalities (negative or positive) created by fishing activities (Ryan et al., 2014).

7.12 Ecosystem-Based Management, Brief Overview

Bioeconomic models have traditionally been relatively simple, developed for the analysis of a single species or a single policy measure. Drechsler et al. (2007) examined the differences between the modelling approaches in economics and ecology, assessing 60 randomly selected models that looked into biodiversity conservation issues. The models were classified as mainly ecological, mainly economic, or integrated ecological – economic models. The economic models tend to be relatively simple and typically avoid aspects of space, dynamics, and uncertainty, and they apply analytical methods. They often use simple assumptions, whereas ecological models typically are more complex and often apply simulation. The latter are rather specific and frequently include dynamics, space, and uncertainty. However, they often ignore economic and institutional issues. The integrated ecological – economic models are regarded as having intermediate complexity.

Bioeconomic models integrate a biological fishery model with a model of the economic systems. However, the standard bioeconomic models do not assess how harvesting impacts biodiversity and ecosystem services (Tschirhart, 2009). Over the years, efforts have been made to make models that better reflect complex reality by including additional species or other values than the fishing-related ones in the models. Figure 7.4 shows a continuum of different fishery management types with increasing complexity. As illustrated in the figure, ecosystem-based management can in a simplified way be reflected in a two-species predator–prey model. Generally, though, more complex settings are required for ecosystem-based management (Arkema et al., 2006). The general objective of ecosystem-based fishery management (EBFM) is to avoid the degradation of ecosystems and to consider the requirements of non-target species, protected species, and habitats and take trophic interactions into account (Pikitch et al., 2004). Marine ecosystem-based management furthermore includes the activities of other sectors (Arkema et al., 2006, Curtin and Prellezo, 2010). According to Fogarty (2014), an EBFM should incorporate interrelationships among the different elements of the system, include humans as an integral part of the system and cover the effects of environmental influences. EBFM differs from the ecosystem approach

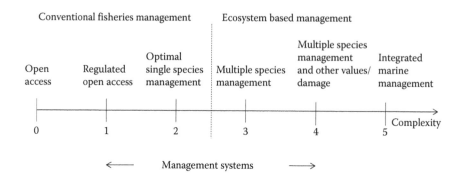

FIGURE 7.4

Continuum of management systems in fisheries. (Modified from Kellner, J. B. et al. 2011. *Conservation Letters* 4:21–30.)

to fishery management (EAFM) which has its focus on individual species or stocks. Ecosystem-based management will require knowledge about the quantitative relationship between stocks of different species. These interconnections influence the ecosystem services and subsequently these services can be assessed economically.

Bioeconomic models for the first steps in Figure 7.4 are well tested, including the predator–prey model (Bjørndal and Munro, 2012). However, for more complex ecosystem-based management there is still limited literature, i.e., for more than two species that interact, the inclusion of damage to marine ecosystems as well as recreational and non-use values. The most complex cases of marine ecosystem-based management would include other sectors' external effects on fisheries and the marine environment, e.g., nitrogen pollution from agriculture, exploitation of raw materials and shipping.

Ecosystem services exhibit important spatial variation, due to issues such as rarity, spatial configuration, size, habitat characteristics and conditions, as well as the demand for ecosystem services generated by the nearby population and its preferences (Polasky and Segerson, 2009). Hence, models have been developed which are spatially explicit. One approach is to model ecosystem services as a function of land use applying ecological production functions that are linked to an economic framework. Another approach is based on mapping of habitat types and subsequent value transfer. The former requires specific knowledge and insight, but should be able to produce rather precise results. The latter approach often ignores the details of spatial variation and assumes a constant value. Mapping of habitats, ecological processes and environmental factors such as soil and climate has advanced more for the terrestrial than for the marine areas. However, in recent years bioeconomic models have been developed to include distinct habitat patches and dispersal of a harvested species between patches (Sanchirico and Wilen, 1999, 2005). These models show important management implications of such variation. An optimal management must set targets in accordance with the optimal resource rent in each patch.

Various approaches can be used for implementation of EBFM (Holland et al., 2010). Conventional economic frameworks such as cost-benefit analysis (CBA) can handle comprehensive assessment of the economic consequences of policy actions. However, the CBA does not contain a specific ecological component and it needs input from natural science to take the effect on the ecosystems into account. Ecosystem models have been developed that aim at modelling the entire ecosystem, such as Ecopath with Ecosim and Atlantis (Fulton, 2010). These ecological models, however, are not constructed for assessing all economic values produced by the ecosystem (Lassen et al., 2013). Bioeconomic models can be expanded by including nonmarket ecosystem services of regulating and cultural character in order to illustrate the management options in coral reef ecosystem with three interacting fish species (Kellner et al., 2011). Hannesson and Herrick (2010) showed how a bioeconomic model can be used to estimate the catch value of sardines in the Pacific Ocean versus value as prey for a number of predators, some of commercial value, such as salmon and tuna, and others with recreational value or existence value, such as seals and whales. Another approach is to link a bioeconomic model with an ecological model to analyse whether the estimated optimal fishing level is within sustainable limits (Lassen et al., 2013).

There are several other ways of integrating economic and ecological aspects. A general equilibrium ecosystem model (GEEM), a parallel to the computable general equilibrium model (CGE), has been applied to a marine ecosystem, comprising eight species, in Alaska (Finnoff and Tschirhart, 2008). Jin et al. (2012) link a CGE model to an ecological model for an ecosystem-based fisheries management of the George Bank ecosystem in New England. An advantage of these models is that they can provide a direct estimate of welfare changes associated with management actions, but they require a large amount of data and specific knowledge. Gascuel et al. (2012) constructed a framework based on stock and fleet data for the operationalization of EBFM. Data on stocks, catch, fishing mortality, recruitment, and mean tropic level were used to construct indices to shed light on the ecological sustainability of fishing activity as well as the dependency on ecosystems by the different segments and their economic performance.

7.13 Predator–Prey Relationships and Other Species Interactions

The most important relationships between species in ecosystems are competition between species and predator–prey relationships, though other forms of interaction exist like mutualism (Flaaten, 1998; Tschirhart, 2009). In fishery models with predator–prey interactions, the predator, the prey, or both are harvested. In these models, the objectives of MSY or MEY are replaced by MSY frontiers and MEY frontiers, respectively. In the predator–prey models with connected ecosystem services, which are of a non-fishing character, harvesting one species affects the ecosystem services provided by another species via a predator–prey relationship (Tschirhart, 2009).

The first analyses of species interactions were published independently of one another by Lotka (1925) and Volterra (1931), later known as the Lotka-Volterra model. The original Lotka-Volterra equations describe the growth rates of two interacting species, where x and y are the biomass densities of prey and predator, respectively. No harvest is included in the model, which describes the change in the prey species over time as a function of recruitment to the prey species and the mortality of the prey species, which is dependent on the size of the biomass of the predator species and the prey species. For the predator species, growth depends on the recruitment to the predator species but also the size of the stock biomass of the prey, whereas mortality is linear in its own biomass. The conventional predator–prey model includes the following basic biological model functions (Flaaten, 1998; Yodzis, 1994): (1) an intrinsic growth rate of the prey population, (2) the amount of prey consumed per unit of time per unit of predator, and (3) the per unit growth rate of the predator population.

Some general economic patterns can be extracted from the predator–prey models. The effect on the predator harvest and resource rent of an increase in prey stock is unambiguously positive, whereas the effect on the prey harvest and resource rent of an increase in predator stock is ambiguous depending on the functional response of the predator, but in most cases it is negative (Flaaten, 1998). To maximize the resource rent from harvesting a prey, the long-run effort should be lower within the two-species framework than for a single-species model. For the predator, the opposite holds. The management objective in the common case of a valuable predator and a cheap prey is to use the prey as forage for the predator, and the level of effort will be close to that under a single-species framework. In the opposite case, with a low-value predator and a high-value prey, the objective is to reduce the predator stock to a low level to leave more prey for the fishermen. Other cases lie in between these two extremes. Following Hannesson (1983), the decision on whether to leave a unit prey in the sea in order to have it converted into an amount of predator will depend on the relative prices of the two species and the ecological efficiency of the predator species (the amount of prey biomass that is converted to predator biomass).

The Lotka-Volterra predator–prey equations are based on the principle of mass action, i.e., the responses are assumed to be proportional to the product of the predator and prey densities (Berryman, 1992). In addition, the intrinsic growth rate of the prey is in theory limitless. Hence, a self-limitation term is normally introduced, often in the form of the logistic growth function (Berryman, 1992). The first to use this modified model was Larkin (1966). Some models include only one carrying capacity of the total system similar to the level at which the prey will stabilize in the absence of the predator (Flaaten, 1989). In such a model, the predator is highly dependent on the particular prey. According to Berryman (1992), this simple model of species interactions may have been one of the main reasons why conventional predator–prey theory failed to explain certain ecological phenomena and resulted in paradoxes (paradox of enrichment and biological control paradox). To obtain more realistic species interaction models, different types of predator functional responses must be considered (Berryman, 1992; Yodzis, 1994). One alternative is to include ratio-dependent predation (Berryman et al., 1995a). It builds on the law of diminishing returns, which manifests itself as an increasing difficulty for a predator to meet its energy demands as its population density increases.

Berryman et al. (1995b) extended the logistic predator–prey model to apply to an arbitrary species in a trophic chain such that each species potentially can have trophic layers both above and below its own level. This model differs from the conventional predator–prey models in that it does not establish a direct relation between the functional response and the numerical response of predator-to-prey density. Instead, it consists of a maximum rate of increase from which is subtracted a number of terms that causes a reduction of the intrinsic growth rate, such as predation and competition for food and other resources. The model contains ratio-dependent species interaction terms. It is described in more detail subsequently in Box 7.1.

The models of competition between species have similarities with the prey component of the predator–prey models in that the growth rates of the involved species are negatively affected by the stock of the competing species (Flaaten, 1991; Clark, 2010). It is important to know the presence and characters of competition beforehand in cases where one of the species is being exploited in such an interspecific competition relationship. The combination of fishing and competition can cause unexpected problems because stock collapse may occur at effort levels that under a single-species framework would have been well within the save limits. For a range of values of species interaction, one species will drive the other to extinction (Tschirhart, 2009; Clark, 2010). However, owing to variation in environmental factors, both species may find living space.

7.14 Accounting for Externalities in Ecosystem-Based Fishery Management

The effects of fishing on marine ecosystems come about in the form of direct damage and as indirect effects via the food web. The indirect effects can be modeled by including species interaction, as discussed in the previous section. With respect to the direct damage, it is clear that a range of habitats and benthic structures are sensitive to fishing activities, such as trawling and dredging. Damage occurs through a number of effects, for example, mortality of non-target fish species, seabirds and mammals, discards, damage to epifauna, smoothing and suspending of sediments, reduction in seabed roughness, removal of species that produce structure, damaging reefs, kelp, and seagrass (Auster and Langton, 1999; Hilborn, 2011). All these effects reduce diversity and nonmarket values and may damage the productivity of stocks and thereby damage future fishing. Large areas are affected by these fishing practices; for example, it has been estimated that some fishing grounds in the southern North Sea probably were swept more than three times a year by beam trawls (Gislason, 1994). However, the extent to which the fishing sector causes externalities upon its own activities by reducing future catches through habitat–fishery interactions is not well known (Armstrong and Falk-Petersen, 2008). Owing to this lack of knowledge about the biological and ecological conditions and their specific impact on the economic value of ecosystem services, it can be challenging to include externalities explicitly in models for ecosystem-based management. Nevertheless, in recent years an increasing interest has been shown for modelling these aspects.

These externalities can be modelled as part of external factors or as an endogenous factor caused by the fishery itself. With respect to the former, factors such as nutrient load and change of habitat areas have been investigated. Huang and Smith (2011) developed a bioeconomic model of optimal harvest in a brown-shrimp fishery subject to environmental disturbance in the form of nutrient pollution from agricultural areas, and they adjust the model to analyze the economic effects of hypoxia on harvest and resource rent. Barbier (2003) developed a model of habitat–fishery interconnection in which habitat is converted to other uses. The model is applied to a case of mangrove loss in Thailand and examines subsequent losses for local fisheries. For a case of positive effects on habitats, Armstrong (2007) showed how a simple expansion of the conventional bioeconomic models could include the potential effects of area closures in the form of marine reserves.

Damage caused to marine ecosystems by fishing activity can be direct owing to the effects of fishing gear on benthic ecosystems, by catch for example, and it can be indirect owing to gear-habitat interactions that make stocks less productive (Armstrong and Falk-Petersen, 2008). The direct effects are the easiest to detect and quantify. However, the indirect damage can be estimated as changes in intrinsic growth rates or carrying capacity. Ryan et al. (2014) examined fishing activities that affect underlying biological productivity and are referred to as ecosystem externalities. They may be positive as well as negative. Fishing may damage habitat in ways that reduce productivity; however, in some cases it may also increase food availability and, hence, increase the productivity of the target species. A simple model incorporates stock and crowding externalities as well as ecosystem externalities, and the effects on system dynamics, open-access equilibria, and optimal fishery regulation are outlined. The effect of ecosystem externalities on harvest is modeled via stock growth. An alternative way of including ecosystem externalities is to use habitat as an additional state variable in a fishery where fishing causes habitat damage, which in turn affects fishery productivity negatively (Janmaat, 2012).

7.15 Conventional Fisheries Economics Modelling

Conventional microeconomics modelling and, hence, fisheries economics modelling is founded on a utility function (demand for commodities and services) and a production function (supply of commodities and services). When ideal prices of commodities and the services and costs of producing them are attached to this system, equilibrium between demand and supply can be estimated (Figure 7.5) by the interaction between the demand and supply curves (Pearce, 2007; Balmford et al., 2011).

In the real world, the market usually does not reflect the ideal. On the demand side, public demand is not reflected by private demand. Similarly, on the supply side, all use of production factors is not reflected in opportunity costs. These market failures can, as discussed in the valuation sections, be included in various ways. However, in ecological economics, it is difficult to account for all market failures.

If market failures are accounted for, i.e., non-priced public demand for certain ecosystem services, the equilibrium formed by the market interaction between demand and supply, [60000; 4.0] in Figure 7.5, is shifted toward a higher equilibrium, [73000; 4.9]. Hence, public demand, for example for bird song in a forest, will drive the supply of forest land upward. On the other hand, if the costs of producing commodities and services do not reflect all costs (external costs), then the inclusion of these costs will shift the supply curve upward, and a now equilibrium will be established as a result of the interaction at [60000; 6.0].

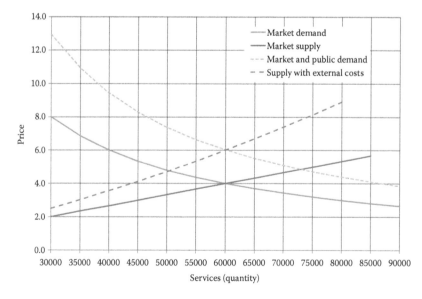

FIGURE 7.5
Demand and supply of ecosystem services.

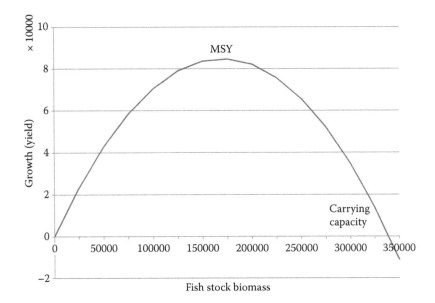

FIGURE 7.6
Yield (catches) and fish stock size as a function of fishing effort.

The problem with the demand supply analyses shown in Figure 7.5 is that nature sets limits on the supply and that these limits are not observed by humans (a market failure). Normally, demand is represented by a demand curve that is sloping downward from left to right, i.e., the lower the price, the higher the demand. On the other hand, supply is represented by a supply curve that is rising from left to right, i.e., the higher the price, the higher the supply. In particular, for natural resources that reproduce themselves in short intervals, for example, fish stocks and animal stocks, the supply is limited, as shown in Figure 7.6.

In a virgin ecosystem, the stocks still end up in a state with balance between the size of the stock biomasses and a net recruitment at zero, i.e., recruitment minus mortality, which is referred to as the carrying capacity of the ecosystem. It is often assumed that the net recruitment is a function of the stock size, as shown in Figure 7.6. The carrying capacity is found where the yield curve intersects the x-axis (around 350000). The problem is that there is no empirical information about virgin ecosystems because ecosystem exploitation has always taken place. Therefore, the carrying capacity stock size can be used only as support in the estimation of the net recruitment, i.e., the yield curve. The relationship between stock size and yield is in balance at any combination of stock size and yield on the curve if the yield is removed each year, for example by fishing or hunting.

Obviously, the system is also in balance if the stock is zero. At this point, an unstable balance is established because the stock will start growing toward carrying capacity if the stock is not totally exhausted.

The system is sustainable if, at no time at any stock size, more fish is removed by harvesting than the yield the stock size provides. To decide whether a system is sustainable, both the stock (production factor) and the yield (provisioning or catch) need to be measured. While the yield could be measured rather easily in terms of catches if there is a market for the service, it is more difficult to measure 'catches' for services where there is no market. It is even more difficult to measure the stock size.

From Figure 7.6 it is clear that one combination of stock size and yield is better than others. This is the peak point on the yield curve and is called MSY. This is the largest supply (provisioning) that can be achieved from a stock. In other words, the supply from a recoverable natural resource cannot increase indefinitely no matter how much effort is expended. This information is used to adjust the supply curves of Figure 7.5.

The supply of a recoverable natural resource is backward bending as seen from consumers' (or fishermen's) point of view, as shown in Figure 7.7. The shape appears as a result of the inverted U-shape of the yield as a function of the size of the natural resource, for example, fish, as shown in Figure 7.6, and it compares to the average market costs of extracting the yield of a natural resource. The underlying explanation of this feature in a fisheries context is that the individual fisherman is not aware that he, by his catch, actually deprives other fishermen of their catches (an externality or market failure). From society's point of view, this externality is taken into account. Therefore, the costs (and supply curve) of society are represented by the marginal costs of exploiting the ecosystem. From Figure 7.7 it is easy to see that if demand for fish follows a path that is equal to or above the backward bending part of the supply curve, there is a strong incentive to exploit the resource, in some cases even to complete exhaustion. The demand curve will move upward when a growing population demands more food. If the demand curve is below the supply curve, as in Figure 7.7, there are two intersections. In an open-access, non-managed case where exploitation of the resource is controlled by market forces, equilibrium appears at [6000; 4]. The optimal solution from society's point of view takes into account that fishermen deprive each other of catches, as represented by the marginal cost curve at the intersection [68000; 3.5].

In open-access, market-driven solution, the price and average costs are equal to one another (cf. Figure 7.7). In the solution that is optimal for society, the price is determined by the intersection of the demand and marginal costs; the price is 3.6 and the average cost is 1.9 at around 67000. Hence, benefit is generated to society in this case. However, interference in the market is required if this result is pursued.

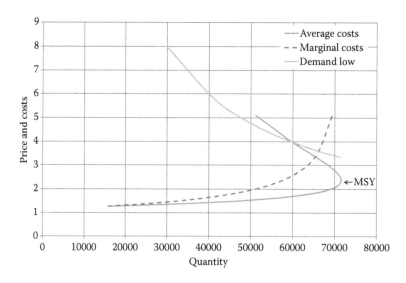

FIGURE 7.7
Backward bending supply of a natural resource.

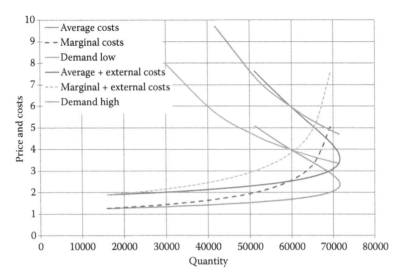

FIGURE 7.8
Backward bending supply of a natural resource with extra demand and externalities.

For recoverable natural resources, market-driven exploitation entails great risk of over-exploitation, which could be alleviated by public interference. However, only some market failures are remedied in this way. Still it is important to include public demand and other costs that are external to the market and to the provisioning of fish or animal products if the optimal solution to society is pursued. If it is, a new set of equilibria arises, as shown in Figure 7.8. First of all, if fishing inflicts damages on an ecosystem, accounting for this activity, for example by a tax on fishing reflecting the damage to the ecosystem, drives the supply curves upward, meaning that the average and marginal costs of fishing will be higher. The equilibrium between low demand and average plus external costs would drive the solution closer to MSY [71000; 3.4] and even below depending on the size of the tax. On the other hand, if some advantages connected with the provisioning of fish are not reflected in the demand for fish, the demand will increase and new equilibria will be set. Still, the optimal solution for society is where the demand for and marginal costs of fishing intersect, in the case of high demand at [65000; 5.3].

Hence, ecological or ecosystem economics and analyses are about constructing models that cover the subject we want to address (cf. Figure 7.4). Qualitative analyses identify functional relationships, including how the solutions of the systems change subject to various assumptions about how the system works. Numerical or empirical models and analyses require that values be assigned to model variables and parameters.

7.16 Conventional Fisheries Management

We now turn to the underlying dynamics of a fishery system to shed light on what is happening and how such a system can be managed. For this purpose a model is constructed to demonstrate a number of features in ecological economics and how they differ from conventional economics. The model includes three species: a predator, prey for the predator,

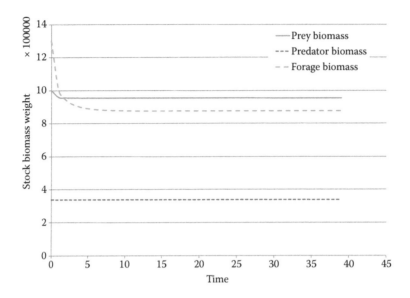

FIGURE 7.9
Stock size with natural species interaction but no external impact such as fishing.

and prey for the second species. The first and second species are subject to exploitation by two fishing fleets, where one fleet is composed of small-scale vessels fishing for the highly valuable predator, perhaps cod, and the other fleet is an industrial fleet fishing for low-value prey, perhaps herring, with the high-value predator as by-catch. While the three species and their interaction reflect supporting and regulating services, the two different fleets reflect aspects within cultural services, i.e., small-scale artisanal fishing and large-scale industrial fishing.

Without any fishery the model in its pure form shows the development of the three species, as shown in Figure 7.9. Assume first that the three species live isolated from one another. In that case, three horizontal straight lines show the development of the stock biomasses over time. Then assume that the borders are removed, allowing the species to interact. Then the predator eats the prey and the prey eats the forage species. This change implies that the prey and the forage species will decline to a new stable level. The model is calibrated in such a way that the system finds a balance between the species. This balance remains steady over time if there is no stochastic recruitment or stochastic natural mortality in the system.

This model forms bases for the graphical presentations in the rest of the paper. An exposition of the model is found in Box 7.1. Such models are applied in different ways, as explained briefly in the next section. Box 7.1 also includes parameter values (Table 7.5). These values are not estimated but fixed in such a way that it reflects an ecosystem exposed to human exploitation.

The links between (1) extended accounting in terms of green accounting, (2) sustainable development, and (3) *optimal* sustainable development are displayed in the subsequent range of figures, which are gradually developed in the chapter. The cases are carried out according to conventional bioeconomic theory (Anderson and Seijo, 2010; Clark, 2010; Bjørndal and Munro, 2012).

Figure 7.10 shows steady-state solutions that are derived from the dynamic system (Figure 7.9). If a virgin system is exposed to fishing by an instant constant fishing effort,

---

BOX 7.1 FISHERY SIMULATION MODEL (GROWTH, INTERACTIONS, FLEETS, ECONOMY)

According to Berryman et al. (1995a,b) and including the fishing mortality rate, F_i, the general food chain equation for stock B_i is given by

$$\frac{1}{B_i}\frac{dB_i}{dt} = a_i - b_i \cdot B_i - \frac{B_i}{c_i \cdot B_{i-1}} - \frac{d_i \cdot B_{i+1}}{B_i} - F_i$$

where B_{i-1} is the stock of a lower trophic level that B_i is preying upon, whereas B_{i+1} is the stock of a species in a higher trophic level that is preying on B_i.

A simplified food chain is assumed with one predator, Y, eating one prey, X, and one prey eating a forage species, Z. The parameters a and b determine recruitment and natural mortality for the individual species, whereas c and d denote the mortality rates inflicted by the species on each other:

$$\frac{1}{x}\frac{dx}{dt} = a_x - b_x \cdot x - \frac{d_x y}{x} - \frac{x}{c_x z} - F_x, \tag{7.1}$$

$$\frac{1}{y}\frac{dy}{dt} = a_y - b_y \cdot y - \frac{y}{c_y x} - F_y, \tag{7.2}$$

$$\frac{1}{z}\frac{dz}{dt} = a_z - b_z \cdot z - \frac{d_z x}{z}. \tag{7.3}$$

In rewritten form:

$$\frac{dx}{dt} = x(a_x - b_x \cdot x) - d_x \cdot y - \frac{x^2}{c_x z} - F_x \cdot x, \tag{7.4}$$

$$\frac{dy}{dt} = y(a_y - b_y \cdot y) - \frac{y^2}{c_y \cdot x} - F_y \cdot y, \tag{7.5}$$

$$\frac{dz}{dt} = z(a_z - b_z \cdot z) - d_z \cdot x. \tag{7.6}$$

In the economic part of the model, revenue and costs are determined by Equations (7.7)–(7.20) in what follows. Here q represents catchability rates and V effort (number of vessels), whereas F is the fishing mortality rate. Harvest is denoted by H, and revenue R is found using a set of prices p. Total costs, U, are assumed to be linear in effort. Profit π is used as a decision variable in optimizations and to determine investments in dynamic simulations. Investment is assumed to be linear in profit (thus positive profits give positive investment, i.e., an increase in effort, whereas negative profit means negative investment, i.e., decrease in effort) and is represented by I:

$$F_x = q_{x,1}V_1 + q_{x,2}V_2, \tag{7.7}$$

(Continued)

BOX 7.1 (*CONTINUED*) FISHERY SIMULATION MODEL (GROWTH, INTERACTIONS, FLEETS, ECONOMY)

$$F_y = q_{y,1}V_1 + q_{y,2}V_2, \tag{7.8}$$

$$H_x = F_x x, \tag{7.9}$$

$$H_y = F_y y, \tag{7.10}$$

$$R_1 = p_{x,1}q_{x,1}V_1 + p_{y,1}q_{y,1}V_1, \tag{7.11}$$

$$R_2 = p_{x,2}q_{x,2}V_2 + p_{y,2}q_{y,2}V_2, \tag{7.12}$$

$$U_1 = u_1 V_1, \tag{7.13}$$

$$U_2 = u_2 V_2, \tag{7.14}$$

$$\pi_1 = R_1 - u_1 V_1, \tag{7.15}$$

$$\pi_2 = R_2 - u_2 V_2, \tag{7.16}$$

$$I_1 = v_1 \pi_1, \tag{7.17}$$

$$I_2 = v_2 \pi_2, \tag{7.18}$$

$$V_{1,t+1} = I_{1,t} + V_{1,t}, \tag{7.19}$$

$$V_{2,t+1} = I_{2,t} + V_{2,t}. \tag{7.20}$$

Note that the model is dynamic, although, for convenience, time is omitted in all equations except for the last two (7.19–7.20).

TABLE 7.5

Parameter values

X carrying capacity prey	1,000,000 (tons)	q_{x1}	0
Y carrying capacity predator	339,000 (tons)	q_{x2}	0.001
Z carrying capacity forage	1,300,000 (tons)	q_{y1}	0.001
a_x	1	q_{y2}	0.00025
a_y	1	p_{x1}	1 (EUR/kg)
a_z	1	p_{x2}	1 (EUR/kg)
b_x	0.000001	p_{y1}	4 (EUR/kg)
b_y	0.00000295	p_{y2}	2 (EUR/kg)
b_z	0.000000769	u_1	400 (1000 EUR/vessel)
c_x	100	u_2	500 (1000 EUR/vessel)
d_x	0.1	v_1	0.00015
c_y	100	v_2	0.00015
d_z	0.05		

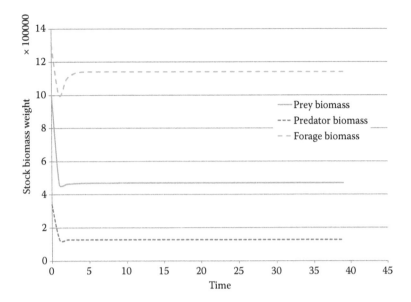

FIGURE 7.10
Development of biotic stocks exposed to fishery; effort constant at 500 vessels of each fleet.

the stock sizes will change. The size of the predator and prey will decrease, whereas the size of the forage species will stay almost constant because lower stocks of the species preying on the forage fish will lower the natural mortality. The link between Figures 7.9 and 7.10 is found keeping in mind that various magnitudes of instant and constant fishing effort over time lower the stock biomasses of the species to varying degrees depending on the effort. If effort is high, then the stocks that are fished will decrease significantly and in some cases become very small if the effort costs are small.

The lessons learned from Figures 7.9 and 7.10, representing an ecosystem, are that an accounting of the provision of a product (valued or not valued) sheds light on what happens, but it is not enough to say anything about sustainability. Sustainability requires a link between the underlying production factors including ecosystem services and the provision of final products. Furthermore, it is necessary to set up management targets and measures to achieve those targets. From an economic point of view, maximization of welfare is usually used as the target. A level below this is demonstrated first in a static context in which time is disregarded and later in a dynamic context over a number of years.

To produce welfare, the yield from the marine resources must be extracted, and this requires the use of production factors in terms of fishing effort, with costs, to bring the yield on shore. Therefore, the yield is transformed into catch as a function of effort in Figure 7.11, which further includes the comparable stock size as a function of effort. No fishing costs are introduced so far, but it is noted that a fish stock at carrying capacity size is not preferable. Of all choices between all the sustainable 'positions' represented by the yield curve, one position is better than all the others. The criterion depicted is the MSY, which is the peak point on the yield curve. Given the assumption used about growth, the coherent stock size is half of the carrying capacity stock size.

A transformation of the ecosystem into an economic system that allows for analyses that take into account the characteristics of both the ecosystem and human behaviour is depicted by the next four figures, Figures 7.11–7.14, which for simplicity depict only one species

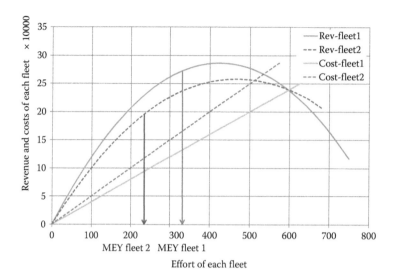

FIGURE 7.11
Maximization of resource rent, the MSY and MEY concepts (see text). Note: the interaction of three curves in [600; 24000] is incidental.

and one fleet, although the underlying system includes three species and two fleets. The methodology is based on neoclassical microeconomics in which maximization of human welfare plays an important role. Maximization requires that human utility be transformed into prices and that the costs of human choices be measured. These costs are measured as foregone opportunities by the efficient allocation of resources. Macroeconomic considerations, in particular growth and employment, are inherent in this approach.

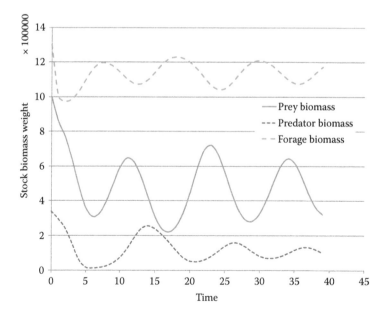

FIGURE 7.12
Development of profit and effort of two fleets with symmetric entry and exit behavior.

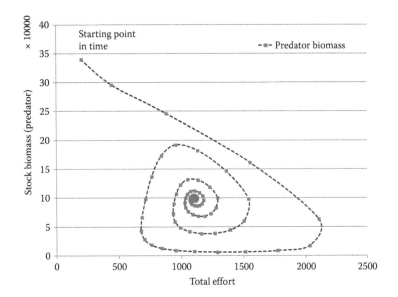

FIGURE 7.13
Phase diagram showing relationship between fishing effort and one stock biomass over 120 years with asymmetric investment behaviour. Note: Dots indicate year.

Central to achieving maximization of welfare is the concept of MEY, which is comparable to the ecological measure MSY. When input factors such as labor, capital, and fish stocks are measured at their opportunity costs and prices reflect the value of services to humans, MEY is compared to the maximization of resource rent, i.e., remuneration of the natural resources. Figures 7.11 and 7.12 resemble Figures 7.6 and 7.7, but now fishing effort

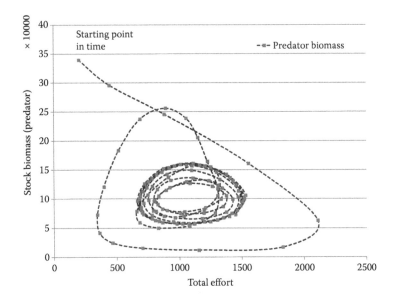

FIGURE 7.14
Phase diagram showing the relationship between fishing effort and one stock biomass over 120 years with symmetric investment behaviour.

is the driving force on the x-axis in Figure 7.11. The concepts of yield from stocks and supply are transformed into revenue as a function of fishing effort.

In a two-species (the third species is not fished), two-fleet case (Figure 7.11), which forms
the basis for conventional bioeconomic thinking, it is possible to identify the MEY and
compare it to MSY. MEY, the maximum economic yield, is a measure of maximum economic welfare that could be derived from the resources. At constant fish prices, MSY in
economic terms is shown by the peak point of the revenue curve, whereas MEY is found
where the distance between the revenue and costs is greatest. While a one-species, one-
fleet case could be represented by the revenue and cost of fleet 1 in Figure 7.11, it is not a
straightforward task to show what happens in a two-species, two-fleet case because both
species and fleets interact. The behaviour of one fleet impacts the results of the other fleet.
In practice, that could lead to devastating competition, which could be observed in high-
seas fisheries where no governments have jurisdiction and no agreements are possible. In
our case, these situations are disregarded, and it is assumed that it is possible to estimate
the effort of each fleet that maximizes economic yield among all the possible combinations of effort of the two fleets. In other words, given the interaction between species and
the technological constraints imposed on the effort, the optimal ratio between the fleets
could be estimated. When this ratio between fleets 1 and 2 is used for all sizes of fleet 1,
the revenue and costs for both fleets can be calculated. This is shown in Figure 7.11, with
an indication of the optimal effort of each fleet. The MEY also indicates the maximum
resource rent, which is the difference between the revenue and the costs, and when the
resource rent by each of the fleets is aggregated, the total welfare derived from the system
is estimated.

However, Figure 7.11 only shows part of the story, in particular with respect to human
behavior. The MSY and MEY solution can only be achieved in a system that is well managed. In a system that is not managed properly, the solutions are different and the risk of
abuse of the ecosystem is obvious.

Before leaving this subject, some behavioral aspects are worth mentioning. In an economic system, the value of marine resources (e.g., fish) can be transformed into other
assets (e.g., a bank account). If the growth rate in a natural resource is high, it pays to
leave the resource in the sea and delay harvest to the future where the growth rate will
decrease to a level comparable to the alternative remuneration of the asset (fish into bank
account). However, if the fisherman is excluded from the catch, if he does not catch the fish
instantly, it will be lost to him and it would pay for him to exploit the resource as much
as possible. The race-for-fish situation occurs. When all fishermen show this behaviour,
the fishery expands to an open-access solution represented by the intersection between
gross revenue and total costs in Figure 7.11. It seems clear that if the exploitation of natural
resources is managed properly and agreements between nations and between fishermen
can be achieved and complied with, the problem may be solved. But things do not always
turn out that way. A further investigation of ecosystem dynamics and management is
explored in the following sections.

7.17 Dynamic Ecosystem Modelling, Some Results

Compared to the conventional static fisheries economics analyses shown in Section 7.15,
ecosystem modelling becomes more complex, at least in principle, although the research

by Drechsler et al. (2007) points out that this is not the case for the models used in past and current analyses. However, modelling by use of Ecopath with Ecosim (Christensen and Walters, 2004, 2011; Christensen et al., 2005) and Atlantis indicates the opposite (Fulton, 2010). The application of economic models, for example Fishrent, has also become more complex and data demanding.

Subsequent figures show development over time of fish stocks and fishing effort with an indication of optimal solutions, the required management, and consequences. The variables are shown either with time on the x-axis or as *phase-diagrams*, where effort is displayed on the x-axis and other variables on the y-axis, with time as the implicit driver of the system.

The model runs over many years, but the results are, in general, shown for a period of 40 years. As the predator eats prey, the growth of the prey is dependent on the stock size of the predator. Furthermore, the growth of the predator stock is dependent on the amount of prey. The prey stock eats forage fish, and its growth is dependent on the size of the forage stock (Berryman, 1992). In addition to biological interactions, the system is impacted by fishing effort, which is controlled by the behaviour of the fishermen. Therefore, even with relatively few variables compared to reality, the model used is rather complex.

In a dynamic context and under open access to exploit biotic resources, the risk of abuse of resources is very high, and a result of a model simulation is shown in Figure 7.12. In this figure, investment behaviour is introduced in the model. When fishermen earn profit, they usually invest all or part of the profit in more fishing because the race for fish under open access incites them to harvest as much as possible. High fishing effort eventually drives the stocks down – in certain cases even to extinction – which means the fishermen will now experience economic deficits. Hence, they leave the industry with more or less speed depending on the other opportunities at their disposal, and the stocks increase again with more or less speed. To highlight ecosystem management problems, incentives to invest and disinvest, i.e., to enter and leave the fishery, in the model runs is assumed to be very strong, which amplifies the variation. In practice, this means that there are no transaction costs and that alternative uses of the vessels are possible. It is noted that, in particular, the predator stock decreases strongly but does not go extinct in our case. The reason for this is the flexibility in the entry and exit of vessels. If there is no alternative use of vessels, staying in the industry in the face of an economic deficit means that the risk of collapse of the predator stock is high.

Further insight into the stability of the ecosystem can be gained by looking at the development in fishing effort and fish stock biomass. The system starts with a high fish stock biomass, resulting in a relatively low yield (Figure 7.6) and low fishing effort. Much depends on the assumptions about the interaction between the species and the economic behaviour of the fishermen. In most cases, the ecosystem approaches equilibrium, as shown in Figure 7.13, which is a phase diagram between total effort and predator stock biomass. It is assumed in this case that few alternative opportunities exist for the use of fishing vessels. This means that exit is less dynamic than entry (asymmetric entry and exit) because fishermen stay in the fishery with small profits and even with deficits for some years. The equilibrium indicated in Figure 7.13 at around [1100, 100000] compares to the open-access solution shown in Figure 7.11 by the intersection between gross revenue and total costs. The path toward equilibrium in Figure 7.13 is one possible path toward that equilibrium.

Although the model predicts stable equilibrium in the future, it takes so many years that it may be difficult to wait so long and accept such a low stock size for many years and even with the risk of exhaustion. With an initial effort given a harvest capacity that exceeds the yield of the stock, it takes only 5 years to deplete the stock significantly. Then it takes almost twice that time for the stock to recover and the effort to be reduced to such a low level that this recovery is sustained. Then it takes more than 50 years for the system to come close to equilibrium.

If, on the other hand, it is assumed that exit reacts stronger to economic deficit than entry reacts to economic profit, the system tends to approach equilibrium with almost the same equilibrium effort and stock biomass as shown in Figure 7.13, but with much smaller variation in the path to equilibrium. It should be noted that investment behaviour and, hence, the magnitude of entry and exit strongly influence the path to equilibrium in an open-access fishery. Under certain circumstances, no equilibrium is obtained before some of the stocks have been driven to exhaustion.

A third case is assuming that investment behaviour is symmetric in regard to entry and exit. Then the fishery may enter into a sort of circular (galactic) path, as shown in Figure 7.14. In this case, equilibrium is not found. Note that under the chosen assumptions, the effort increases rapidly from 100 units of effort in the first year to around 2200, and the stock biomass declines from around 350,000 to around 60,000. Then for the next 6 years the effort decreases while the stock only builds up slowly. In reality, such a situation will be considered critical because no one can be sure what will happen. The situation resembles a situation with asymmetric entry – exit, but the changes are more dramatic.

In Figure 7.15, all three species are included. It is shown what happens in the phase diagram under the assumption about symmetric entry and exit (cf. Figure 7.14). The development in this diagram is shown only for 40 years. It is noted that while the predator species shows right-hand turns all the time, the prey and forage species shift direction depending on the size of the predator stock biomass.

In any case, because it is difficult to see into the future and maintain a strong belief that the ecosystem will eventually reach equilibrium, there is a need to intervene in an open-access system. One argument (economic) is that a higher profit (resource rent) for society is possible by proper management, but another argument (ecological) is that the variability of the ecosystem with respect to the predicted equilibrium is a risky business and may in

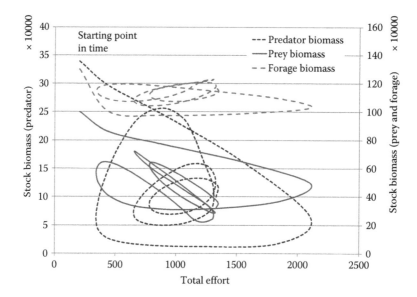

FIGURE 7.15
Phase diagram showing relationship between fishing effort and stock biomasses over 40 years.

some cases even include a risk of exhaustion or at least a reduction of the stock, which may lead to a situation where recovery is not possible.

The foregoing discussion addresses open-access solutions, possible equilibria, and paths toward equilibrium. No interference from governments aimed at managing the system was included. This is changed in the subsequent scenario, in which the impact of management is addressed. From an economic but also from an ecological point of view, profit maximization in terms of the NPV of the exploitation of the biotic resources is considered a reasonable objective. The maximization of NPV resembles the MEY principle but with the exception that the NPV takes time into account explicitly. Furthermore, the adaptation path is highlighted.

If NPV is maximized over a period of 40 years with no restrictions on the entry and exit of vessels, Figure 7.16 shows the development of the stock biomasses. A strong variation appears, often called bang-bang. There is no risk of exhaustion and in particular the forage species is considered safe.

However, this model calculation only shows how the resource rent is maximized. It does not indicate how to manage the stock. An indication of that could be obtained by looking at the development of effort. Bang-bang is also reflected in the development of the fishing fleets, as shown in Figure 7.17. In some years the fishery is simply closed, which allows the stocks to recover with maximum power. The development of the fleet is associated with the profit that could be earned. If the profit is high, the investments will be high and vice versa.

When the stock biomasses are small, the growth potential is very high. At the same time, the profit from the vessels is low. In that case it would pay to close the fishery down to allow the stocks to recover until the growth rate of the stocks falls below the profit rate (Sandal and Steinshamn, 1997).

The management implication is that fisheries are closed down completely in some years and then in other years it pays to fish as much as possible to the extent that the recovery of the stock is not jeopardized. In practice, such a management system is unrealistic because

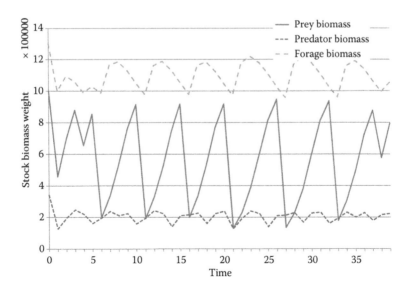

FIGURE 7.16
Optimal stock biomass.

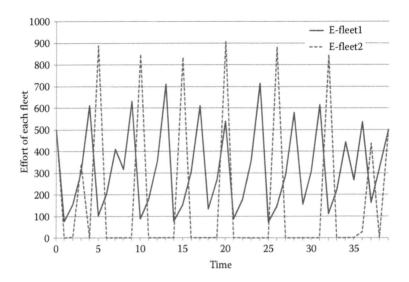

FIGURE 7.17
Optimal fleet.

very often the alternative use of fishing vessels is very limited. Hence, not all costs are saved when the vessels stop fishing, as assumed in the model simulation. Taking that into account, fishermen will continue fishing as long as the variable fishing costs are covered by the revenue from fishing. Furthermore, cessation of fishing will impact the processing industry, which will experience a shortage of raw material. If the raw material for processing cannot be acquired from elsewhere, the employees of both the fishery and the processing plant must be laid off. Furthermore, consumers will be left without a supply of fish.

Therefore, an optimal solution must be found within reasonable restrictions. Theoretically, from an economic point of view, these restrictions may be known but in practice they are often difficult to estimate. One reasonable restriction is that the fleet can only adjust within certain limits keeping in mind that the adjustment of effort will be slow and take place with some delay. Using an upper and lower investment restriction at 5% entry/exit of last year's number of vessels produces the result shown in Figure 7.18 for the biomasses when the NPV is maximized. The bang-bang solution is no longer an option.

However, the proportion of remuneration between the capital assets in terms of biomasses and vessels still prevails, which is indicated by the development in the number of vessels in Figure 7.19. If the system begins with the average values in the optimal solution without restrictions (see also Figure 7.11), the system would gradually build up fleets (within the 5% restriction) until the proportion between the remunerations shifts. Then effort is reduced to the lower limit, and so forth.

The lessons learned from the foregoing discussion sustained by model simulations are that the risk of overexploitation of natural resources is significant in an open-access (market-driven) system. The risk is highest for predator species because intensive exploitation of these species reduces the pressure in terms of natural mortality on the prey species. In our analyses, by-catches of species such as mammals and sea birds are disregarded, but it is straightforward to see that if such species are caught in some fixed proportion with fishing effort, they will become endangered as well.

The foregoing discussion also shows that proper management reduces the risk of overexploitation significantly. However, although some general lessons can be learned from

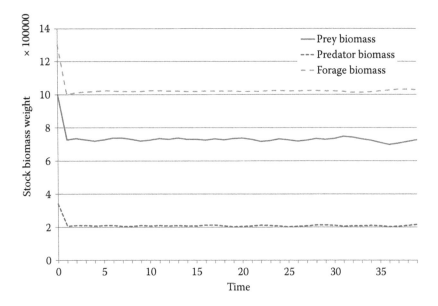

FIGURE 7.18
Optimal stock biomass.

the model simulation, the simulation also shows that changes in parameter values, i.e., assumptions about behaviour in the simulation model, may change the resulting development of the stock biomasses significantly. Therefore, detailed management requires specific information for each ecological system that is subject to investigation. A general outline of management measures and a brief discussion of measures are carried out in the next, concluding section, taking into account the characteristics of specific ecological systems.

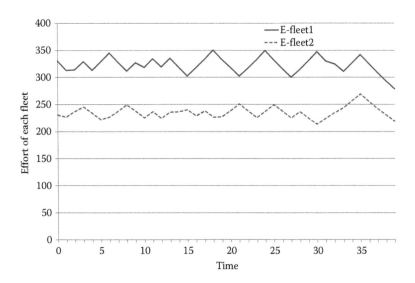

FIGURE 7.19
Optimal fleet.

7.18 Conclusion

Economic accounting has for many years addressed the problem of how to include the use of natural resources in national accounting systems. In particular, in recent years, considerable effort has been devoted to developing methodologies in which the use of natural resources is measured in cases where they are not already serving as intermediate inputs to final products. Correct measurement requires that prices be determined correctly, but this often does not happen because of market failures. Macroeconomic accounting, such as green accounting, reports the use and supply of ecological products. This is a major step forward but does not address the problem of whether the development in the use and supply of products is good or bad. For that purpose, criteria are necessary by which one can judge whether the development is good or bad. Among such criteria is the concept of sustainability, but still more micro-level action by governments to remedy or alleviate market failures is necessary. The literature on the field is growing, and stronger interdisciplinary cooperation between ecological and economic research would advance the field even more.

It is difficult if not impossible to establish and maintain sustainable development in the use of revocable natural resources without interference by society in the pursuit of achieving maximum welfare from the use of natural resources. Microeconomics, in particular, in cooperation with biological and ecological research, is the helpful discipline in this context.

Public interference done to alleviate market failures requires that products that are not subject to price formation on the market be valued and that it be investigated whether they are intermediate or final products in order to avoid double counting. The functioning of an ecosystem must be elucidated to determine cause and effect and whether the interaction between ecosystem services is associated with benefits or costs. If, for example, the food chain is fished down, it will not be possible in advance to judge whether this is good or bad. It will depend on the value to society of the various components of the chain.

Once this is established, CBAs may be carried out aimed at setting up measures that may help to maximize welfare to consumers. A comprehensive literature on the management of fish resources exists, but the field needs to expand beyond fish stocks in order to include ecosystem services that are not yet included in analyses. That would help in judging whether the measures used to manage the exploitation of fish stocks are useful for managing the exploitation of the ecosystem from a broader perspective.

The purpose of this chapter was not to propose and discuss management measures because the literature is comprehensive (Anderson and Seijo, 2010; Clark, 2010; Bjørndal and Munro, 2012), but a few comments are appropriate. In general, three types of management measures are proposed in the literature that aim to put a stop to overexploitation and to maintain fish stocks at a sufficiently high level. Not of least importance is the fact that the literature on achieving maximum economic resource rents is comprehensive. The most widely applied measures are TACs and effort restrictions often referred to as command and control measures. From an economic point of view, these are supplemented by individual transferable and individual nontransferable quotas (ITQ, INTQ) and taxes. A few comments on each of these measures are apropos.

As regards TAC/quota and effort restrictions, fishery management is very often benchmarked against 'open access,' i.e., almost no restrictions. For many years, such a comparison has not been practicable because open access has not been possible in many fisheries. However, in high-seas fisheries as well as with respect to underlying ecosystem services, open access still prevails because no TAC/quotas or effort restrictions have been agreed upon. And even if they had been, enforcement and control of the measures would be difficult and lead to devastating competition and a race for the biggest harvest.

Although TAC/quota and effort restrictions may help to conserve fish stocks, they would not be enough to reduce costs. In some cases, costs could even increase following the introduction of certain measures and put stronger pressure on resources that are not subject to restrictions. Individual or nontransferable and transferable quotas have been used in the northern part of the hemisphere (Europe and North America) effectively since the 1990s, and in other areas of the world these measures have gained in popularity. There is evidence that such regulations have speeded up the reduction of fishing effort and, hence, costs, but there is also the risk that effort is transferred to species and areas where these measures are not in place. Still there is some way to go in practice, although the theory is well established.

The final measure is taxes, either on final products or on the use of production factors. Theoretically, this area is also well established, but the use of the measure is limited. Taxes have not been applied to marine ecosystems so far, at least not in the same way as they have in connection with the extraction of oil and natural gas (and other non-revocable natural resources). Successful use of taxes as well as successful use of ITQ requires that all components that are important in a system be included to avoid tax and ITQ evasion and that effort be moved to other vulnerable ecosystem services.

The purpose of this chapter was to shed light on advances in ecosystem issues with respect to the impact on the ecosystem of fishing activity. In the first part of the chapter, which deals with macroeconomic issues, it was demonstrated that progress has been made in showing how the use of natural capital could be included in national accounting statistics. The work carried out in this field falls short if the aim is to exploit natural resources in a way that maximizes consumer welfare. In the second part, it is demonstrated that extensive management is necessary if overexploitation is to be avoided and a desire for welfare maximization is pursued. Regarding fisheries, the comprehensive literature in this field shows the way to proper management and exploitation, but it is necessary to expand the field to include underlying ecosystem services in analyses. Progress is being made, but further interdisciplinary cooperation between economists and ecologists, in addition to more modelling and data collection and more advice as to how the results could be implemented in practice, is required.

References

Andersen, P. 1983. On rent of fishing grounds: a translation of Jens Warming's 1911 article, with an introduction. *History of Political Economy* 15(3):391–396.

Anderson, L. G. and J. C. Seijo. 2010. *Bioeconomics of Fisheries Management*. Wiley-Blackwell.

Arkema, K. K., S. C. Abramson and B. M. Dewsbury. 2006. Marine ecosystem-based management: from characterization to implementation. *Frontiers in Ecology and the Environment* 4(10):525–532.

Armstrong, C. W. 2007. A note on the ecological–economic modelling of marine reserves in fisheries. *Ecological Economics* 62:242–250.

Armstrong, C. W. and J. Falk-Petersen. 2008. Habitat–fisheries interactions: a missing link? *ICES Journal of Marine Science* 65:817–821.

Arnason, R. 2000. Endogenous optimization fisheries models. *Annals of Operations Research* 94:219–230.

Auster, P. J. and R. W. Langton. 1999. The effects of fishing on fish habitat. In: *Fish Habitat: Essential Fish Habitat and Rehabilitation*, ed. L. Benaka. American Fisheries Society, Bethesda, MD, pp. 150–187.

Balmford, A., B. Fisher, R. E. Green, R. Naidoo, B. Strassburg, R. K. Turner and A. Rodrigues. 2011. Bringing ecosystem services into the real world: an operational framework for assessing the economic consequences of losing wild nature. *Environmental and Resource Economics* 48(2):161–175.

Barbier, E. B. 2003. Habitat–fishery linkages and mangrove loss in Thailand. *Contemporary Economic Policy* 21(1):59–77.

Barbier, E. B. 2007. Valuing ecosystem services as productive inputs. *Economic Policy* 22(49):177–229.

Bateman, I. J., G. M. Mace, C. Fezzi, G. Atkinson and R. K. Turner. 2011. Economic analysis for ecosystem service assessments. *Environmental Resource Economics* 48(2):177–218.

Berryman, A. A. 1992. The origins and evolution of predator-prey theory. *Ecology* 73(5):1530–1535.

Berryman, A. A., A. P. Gutierrez and R. Arditi. 1995a. Credible, parsimonious and useful predator-prey models: a reply to Abrams, Gleeson, and Sarnelle. *Ecology* 76(6):1980–1985.

Berryman, A. A., J. Michalski, A. P. Gutierrez and R. Arditi. 1995b. Logistic theory of food web dynamics. *Ecology* 76(2):336–343.

Bjørndal, T. and G. R. Munro. 2012. *The Economics and Management of World Fisheries.* Oxford University Press, Oxford, UK. 278 p.

Boyd, J. 2006. *The Nonmarket Benefits of Nature. What Should Be Counted in Green GDP?* Resources for the Future, Washington, DC.

Boyd, J. and S. Banzhaf. 2007. *What Are Ecosystem Services?* Resources for the Future, Washington, DC.

Christensen, V. and C. J. Walters. 2004. Ecopath with Ecosim: methods, capabilities and limitations. *Ecological Modelling* 172:109–139.

Christensen, V. and C. J. Walters. 2011. Progress in the use of ecosystem modelling for fisheries management. In *Ecosystem Approaches to Fisheries: A Global Perspective*, ed. V. Christensen and J. Maclean. Cambridge University Press, Cambridge, UK.

Christensen, V., C. J. Walters and D. Pauly. 2005. *Ecopath with Ecosim: A User's Guide.* Fisheries Centre of University of British Columbia, Vancouver, Canada, 154 p.

CICES. 2013. www.cices.eu.

Clark, C. W. 2010. *Mathematical Bioeconomics: The Mathematics of Conservation.* Wiley, New York. Third edition. (First edition 1976, second edition 1990 and reprinted 2005).

Cobb, C. and J. Cobb. 1994. *The Green National Product.* University Press of America.

Curtin, R. and R. Prellezo. 2010. Understanding marine ecosystem based management: a literature review. *Marine Policy* 34:821–830.

Daly, H. and J. B. Cobb 1989. *For the Common Good: Redirecting the Economy toward Community, the Environment and a Sustainable Future.* Beacon Press, Boston.

Daly, H. and J. Farley. 2004. *Ecological Economics, Principles and Applications.* Island Press, Washington, DC.

Danielsson, A. 2005. Methods for environmental and economic accounting for the exploitation of wild fish stocks and their applications to the case of Icelandic fisheries. *Fisheries Environmental and Resource Economics* 31:405–430.

Drechsler, M., V. Grimm, J. Mysiak and F. Wätzold. 2007. Differences and similarities between ecological and economic models for biodiversity conservation. *Ecological Economics* 62(2):232–241.

EEA. 2010. Scaling up ecosystem benefits. A contribution to The Economics of Ecosystems and Biodiversity (TEEB) study. EEA report 4/2010. European Environmental Agency, Copenhagen, Denmark.

FAO. 1999. Indicators for sustainable development of marine capture fisheries. *FAO Technical Guidelines for Responsible Fisheries* No. 8. Rome.

FAO. 2003. Fisheries management. 2. The ecosystem approach to fisheries. *FAO Technical Guidelines for Responsible Fisheries* No. 4 Suppl. 2. Rome.

FAO. 2004. *Integrated Environment and Economic Accounting for Fisheries.* Rome.

Faustmann, M. 1849. Calculation of the value which forest land and immature stands possess for forestry, reprinted in *Journal of Forest Economy* 1995 1(1):7–44.

Finnoff, D. and J. Tschirhart. 2008. Linking dynamic economic and ecological general equilibrium models. *Resource and Energy Economics* 30:91–114.

Fisher, B. R. and R. K. Turner. 2008. Ecosystem services: classification for valuation. *Biological Conservation* 141:1167–1169.

Fisher, B. R., R. K. Turner and P. Morling. 2009. Defining and classifying ecosystem services for decision making. *Ecological Economics* 68:643–653.

Flaaten, O. 1989. The economics of predator-prey harvesting. In *Rights Based Fishing*, ed. P. A. Neher, R. Arnason and N. Mollett. NATO Science Series E: Vol. 169, Springer, Berlin.

Flaaten, O. 1991. Bioeconomics of sustainable harvest of competing species. *Journal of Environmental Economics and Management* 20:163–180.

Flaaten, O. 1998. On the bioeconomics of predator and prey fishing. *Fisheries Research* 37:179–191.

Fogarty, M. J. 2014. The art of ecosystem-based fishery management. *Canadian Journal of Fisheries and Aquatic Science* 71:479–490.

Frost, H. S., P. Andersen and A. Hoff. 2013. Management of complex fisheries: lessons learned from a simulation model. *Canadian Journal of Agricultural Economics* 61:283–307.

Fulton, E. A. 2010. Approaches to end-to-end ecosystem models. *Journal of Marine Systems* 81:171–183.

Gascuel, D., G. Merino, R. Döring, J. N. Druon, L. Goti, S. Guénette, C. Macher, K. Soma, M. Travers-Trolet and S. Mackinson. 2012. Towards the implementation of an integrated ecosystem fleet-based management of European fisheries. *Marine Policy* 36:1022–1032.

Gislason, H. 1994. Ecosystem effects of fishing activities in the North Sea. *Marine Pollution Bulletin* 29(6–12):520–527.

Gordon, H. S. 1954. The economic theory of a common property resource: the fishery. *Journal of Political Economy* 62:124–142.

Hannesson, R. 1983. Optimal harvesting of ecologically interdependent fish species. *Journal of Environmental Economic and Management* 10:329–345.

Hannesson, R. and S. F. Herrick Jr. 2010. The value of Pacific sardine as forage fish. *Marine Policy* 34:935–942.

Hartwick, J. M. 1990. Natural resources, national accounting and economic depreciation. *Journal of Public Economics* 43:291–304.

Hein, L., K. van Koppen, R. S. de Groot and E. C. van Ierland. 2006. Spatial scales, stakeholders and the valuation of ecosystem services. *Ecological Economics* 57:209–228.

Hicks, J. R. 1946. *Value of Capital*. Oxford University Press.

Hilborn, R. 2011. Future directions in ecosystem based fisheries management: a personal perspective. *Fisheries Research* 108:235–239.

Holland, D., J. N. Sanchirico, R. Johnston and D. Joglekar. 2010. *Economic Analysis for Ecosystem Based Management: Applications to Marine and Coastal Environments*. Resources for the Future Press, Washington, DC. 225 p.

Huang, L. and M. D. Smith. 2011. Management of an annual fishery in the presence of ecological stress: the case of shrimp and Hypoxia. *Ecological Economics* 70:688–97.

Janmaat, J. A. 2012. Fishing in a shallow lake: exploring a classic fishery model in a habitat with shallow lake dynamics. *Environmental and Resource Economics* 51:215–39.

Jin, D., P. Hoagland, T. M. Dalton and E. M. Thunberg. 2012. Development of an integrated economic and ecological framework for ecosystem-based fisheries management in New England. *Progress in Oceanography* 102:93–101.

Kellner, J. B, J. N. Sanchirico, A. Hastings and P. J. Mumby. 2011. Optimizing for multiple species and multiple values: trade-offs inherent in ecosystem-based fisheries management. *Conservation Letters* 4:21–30.

Larkin, P. A. 1966. Exploitation in a type of predator-prey relationship. *Journal of the Fisheries Research Board of Canada* 23:349–356.

Lassen, H., S. A. Pedersen, H. Frost and A. Hoff. 2013. Fishery management advice with ecosystem considerations. *ICES Journal of Marine Science* 70(2):471–479.

Lotka, A. J. 1925. *Elements of Physical Biology.* Williams & Wilkins, Baltimore.

Mäler, K. -G., S. Aniyar and Å. Jansson. 2008. Accounting for ecosystem services as a way to understand the requirements for sustainable development. *PNAS* 105(28):9501–9506.

Meadows, D. H., D. L. Meadows, J. Randers and W. W. Behrens III. 1972. *Limits to Growth.* The Club of Rome. Universe Books.

Millennium Ecosystem Assessment (MEA). 2003. *Ecosystems and Human Well-being: A Framework for Assessment. World Resources Institute.* Island Press.

Millennium Ecosystem Assessment (MEA). 2005. *Ecosystems and Human Well-being: Synthesis. World Resources Institute.* Island Press, Washington, DC.

Morissette, L., V. Christensen and D. Pauly. 2012. Marine mammal impacts in exploited ecosystems: would large scale culling benefit fisheries? *PLoS One,* 7(9).

National Research Council. 1999. *Nature's Numbers: Expanding the National Economic Accounts to Include the Environment.* National Academy of Science, Washington, D.C.

Nordhaus, W. and J. Tobin. 1973. Is growth obsolete? In *The Measurement of Economic and Social Performance,* National Bureau of Economic Research.

OECD. 1994. *Environmental indicators.* OECD, Paris.

Pearce, D. 2007. Do we really care about biodiversity? *Environmental and Resource Economics* 37:313–333.

Perrings, C. 2010. The economics of biodiversity: the evolving agenda. *Environment and Development Economics* 15:721–746.

Pikitch, E. K., C. Santora, E. A. Babcock, A. Bakun, R. Bonfil, D. O. Conover, P. Dayton, et al. 2004. Ecosystem-Based Fishery Management. *Science* 305:346–47.

Polasky, S. and K. Segerson. 2009. Integrating ecology and economics in the study of ecosystem services: some lessons learned. *Annual Review of Resource Economics* 1:409–34.

Robbins, L. 1932. *An Essay on the Nature and Significance of Economic Science.* Macmillan, London, 15 p.

Ryan, R. W., D. S. Holland and G. E. Herrera. 2014. Ecosystem externalities in fisheries. *Marine Resource Economics* 29(11):39–53.

Salz, P., E. Buisman, K. Soma, H. Frost, P. Accadia and R. Prellezo. 2011. *Fishrent, Bioeconomic simulation and optimization model for fisheries.* LEI report 2011-024, The Hague.

Sanchirico, J. N. and J. E. Wilen. 1999. Bioeconomics of spatial exploitation in a patchy environment. *Journal of Environmental Economic Management* 37:129–50.

Sanchirico, J. N. and J. E. Wilen. 2005. Optimal spatial management of renewable resources: matching policy scope to ecosystem scale. *Journal of Environmental Economics Management* 50(1):23–46.

Sanchirico, J. N., D. K. Lew, A. C. Haynie, D. M. Kling and D. F. Layton. 2013. Conservation values in marine ecosystem-based management. *Marine Policy* 38:523–530.

Sandal, L. K. and S. I. Steinshamn. 1997. A stochastic feed-back model for optimal management of renewable resources. *Natural Resource Modeling* 10(1):31–52.

Schaefer, M. B. 1957. Some considerations of population dynamics and economics in relation to the management of marine fisheries. *Journal of Fisheries Research Board of Canada* 14:669–681.

Scott, A. D. 1955. The fishery: the objective of sole ownership. *Journal of Political Economy* 63:116–124.

SEEAF. 2004. *Handbook of National Accounting. Integrated Environmental and Economic Accounting for Fisheries.* United Nations Food and Agriculture Organization of the United Nations.

Serafy, S. E. 1997. Green accounting and economic policy. *Ecological Economics* 21:217–229.

Spangenberg, J. H. and J. Settele. 2010. Precisely incorrect? Monetising the value of ecosystem services. *Ecological Complexity* 7:327–337.

Stiglitz, J. E., A. Sen and J -P. Fitoussi. 2010. *Report by the Commission on the Measurement of Economic Performance and Social Progress.* Paris. http://www.stiglitz-sen-fitoussi.fr/documents/rapport_anglais.pdf

TEEB. 2009. The economics of valuing ecosystem services and biodiversity. *The Economics of Ecosystems and Biodiversity,* Climate Issues Update. 27 p.

TEEB. 2010. The Economics of Ecosystem and Biodiversity and Economic Foundation. Edited by Pushpam, Earthscan, London and Washington.

Tschirhart, J. 2009. Integrated ecological-economic models. *Annual Review of Resource Economics* 1:381–407.

UK NEA. 2011. *The UK National Ecosystem Assessment: Synthesis of the Key Findings.* UNEP-WCMC, Cambridge.

United Nations, European Commission, International Monetary Fund, OECD and the World Bank. 2003. Integrated Environmental and Economic Accounting (SEEA), *Handbook of national accounting.*

United Nations, European Commission, International Monetary Fund, OECD and the World Bank 2013. Integrated Environmental and Economic Accounting (SEEA), *Experimental Ecosystem Accounting.*

United Nations, European Commission, International Monetary Fund, OECD and the World Bank. 2014. Integrated Environmental and Economic Accounting (SEEA). *Handbook of national accounting.*

Volterra, V. 1931. Variations and fluctuations of the number of individuals in animal species living together in Animal Ecology. Translated by R. N. Chapman. *Animal Ecology.* Arno, New York, USA.

WAVES. 2012. Moving Beyond GDP. How to factor natural capital into economic decision making. The World Bank, Washington.

WCED. 1987. *Our Common Future.* World Commission on Environment and Development (Brundtland Commission). Oxford University Press.

Yodzis, P. 1994. Predator-prey theory and management of multispecies fisheries. *Ecological Applications* 4(1):51–58.

8

Climate Change Impacts on Marine Ecosystems in Vietnam

Nguyen Quang Hung, Hoang Dinh Chieu, Dong Thi Dung,
Le Tuan Son, Vu Trieu Duc and Do Anh Duy

CONTENTS

8.1 Introduction... 210
    8.1.1 Climate Change in Vietnam.. 210
    8.1.2 Relationship between Climate Change and Marine Ecosystems..................... 211
    8.1.3 Urgency to Assess Impacts of Climate Change on Marine Ecosystems........ 211
8.2 Climate Change Impacts on Coral Reef Ecosystem.. 211
    8.2.1 Types of Coral Reef Related to Climate Change 211
    8.2.2 Impacts of Increasing Water Temperature 212
    8.2.3 Impacts of Sea-Level Rise 214
    8.2.4 Impacts of Storms 214
        8.2.4.1 Research Results on Con Dao Island.. 214
        8.2.4.2 Research Results on Con Co Island... 216
    8.2.5 Impacts of Floods.. 216
    8.2.6 Impact of El Niño and La Niña... 217
8.3 Impacts of Climate Change on Mangrove Ecosystems...................................... 217
    8.3.1 Sensitivity of Mangrove Ecosystems to Climate Change 217
    8.3.2 Impacts of Sea-Level Rise 218
    8.3.3 Impacts of Storms 219
    8.3.4 Impacts of Erosion 219
    8.3.5 Impacts of Salinity Rise 219
    8.3.6 Increase in Estuarine Current.. 220
8.4 Climate Change Impacts on Seaweed and Seagrass Ecosystem........................ 220
    8.4.1 Seaweed/Seagrass Ecosystem Roles Related to Climate Change.............. 220
    8.4.2 Impacts of Sea-Level Rise 220
    8.4.3 Impacts of Temperature Rise 220
    8.4.4 Impacts of Storm 221
    8.4.5 Impacts of Floods and Drought... 221
    8.4.6 Impacts of Increases in CO_2 Concentration 222
    8.4.7 Impacts of Turbidity and Sedimentation Increase................................. 222
8.5 Climate Change Impacts on Estuary Ecosystem 222
    8.5.1 Estuary Types Related to Geo-Climate 222
    8.5.2 Vulnerability of Estuarine Ecosystems to Climate Change 223
    8.5.3 Impacts of Sea-Level Rise 223
    8.5.4 Impacts of Temperature... 223

      8.5.5   Impacts of Typhoons and Floods 223
      8.5.6   Impacts of Erosion 224
8.6   Marine Ecosystems and Climate Change Scenario in Twenty-First Century
     Vietnam..224
8.7   Solutions for Adaptation and Resilience to Climate Change.....................226
      8.7.1   Solutions for Planning and Policy...226
      8.7.2   Solutions for the Protection of Biodiversity 229
      8.7.3   Solutions for Science and Engineering...229
      8.7.4   Solutions for Protection of Coastal Ecosystems against the Rise of Sea
            Level and Erosion 230
      8.7.5   Solutions to Strengthen Resilience of Mangroves to Climate Change 230
      8.7.6   Education, Training and Raising Awareness of Climate Change 232
      8.7.7   Solutions for Strengthening International Cooperation 232
8.8   Conclusions and Recommendations...232
      8.8.1   Conclusions..232
      8.8.2   Recommendations 234
References..234

8.1 Introduction

8.1.1 Climate Change in Vietnam

Vietnam is considered to be a country that will be seriously affected by the impacts of climate change. According to credit rating company Standard and Poor's (2014), Vietnam is one of three countries (Vietnam, Cambodia and Bangladesh) that will suffer the most serious impacts of climate change. In recent decades, the frequency and intensity of storms, floods, rainfall and droughts have increased more dramatically and become harder to forecast. Vietnam is affected by tropical depressions and storms at a frequency of 4.7 times per year. However, these storms are occurring later than in previous years and are moving toward lower latitudes with higher impacts.

In winter, temperatures decrease in the first months of the season and then increase later. In the summer, the average temperature rises significantly, which leads to annual temperature increases. In different periods and regions, there is no clear trend in rainfall. Total monthly and annual rainfall either increases or decreases, but its impact is higher. In many parts of Vietnam, rainfall declines in July and August but rises in September, November and December. For example, heavy rains in November 2008 in Hanoi and nearby areas caused the deaths of 18 people and flooded many houses and streets.

In central Vietnam and the Mekong Delta, floods appear with higher frequency compared to the first half of the twentieth century. In 1999, a historic flood occurred at the end of the rainy season in central Vietnam. In 2007, four enormous floods destroyed all aquaculture facilities in this area during a 3-month period. In contrast, drought occurs annually with higher intensity in southern Vietnam and the Western Highlands during dry season.

Sea-level rise has led to the intrusion of salt water in coastal areas of Vietnam. In the Mekong Delta, 1.77 million ha were inundated by salt water, approximately 45% of the whole area. This area had the highest rate of salt water intrusion in Vietnam. According to climate change scenarios, which predict that sea level will rise +30 cm by 2050 and +100 cm by 2100, this will lead to the loss of land in the Mekong Delta and Red River Delta and jeopardize national food security.

8.1.2 Relationship between Climate Change and Marine Ecosystems

Climate change creates many negative consequences for marine ecosystems, fishery resources and species populations and directly impacts coastal fishing communities. Along with resource depletion, environmental pollution and conflicts among fishing communities, climate change will put more pressure on fishery management so that sustainable development of marine ecosystems can be achieved.

The survival of flora and fauna in many ecosystems is closely linked to climate change because few species can move to a better place to live, or they are at risk as a result of multiple habitat requirements during different stages of life (e.g., egg, juvenile and adult).

8.1.3 Urgency to Assess Impacts of Climate Change on Marine Ecosystems

All levels of authority, management and research need to have information about climate-related impacts on marine ecosystems; their potential effects on marine resources, exploitation and aquaculture in each ecosystem; climate change scenarios (e.g., temperature increase, sea-level rise, high rainfall); and the level of adaptation of marine ecosystems and fishery communities toward climate change. Based on these data, we can develop a better understanding of ecosystem relationships and their threats, such as resource alteration, disease, resource exploitation and mass destruction, as a result of climate change.

Vietnam boasts high biodiversity, with 20 types of ecosystems. Coral reefs, mangroves, seagrass and estuaries are typical ecosystems, but there is no comprehensive research on the impacts of climate change on marine ecosystems. The impacts of climate change are often a part of a small study rather than deep research programmes, whereas the risks and challenges of climate change for marine species might be very high, so immediate solutions are necessary.

To develop a scientific foundation on adaptive solutions to climate change, the Research Institute for Marine Fisheries (RIMF) has been collecting information and analyzing the impact of climate change on marine ecosystems for the twenty-first century. This information becomes scientific knowledge for implementing possible solutions to climate change scenarios. These results were part of the Second National Report in implementing the United Nations Framework Convention on Climate Change (UNFCCC) in 2009.

8.2 Climate Change Impacts on Coral Reef Ecosystem

8.2.1 Types of Coral Reef Related to Climate Change

Tuan et al. (2005) demonstrated that fringing reefs were a typical type of coral reef in coastal areas. Nevertheless, platform reefs and barrier reefs are typical in central Vietnam. The Spratly Islands off the coast of southern Vietnam are famous for their coral reefs, with two distinctive types: atoll and platform reefs (islands that sink under the sea).

Fringing reefs in Vietnam have similar structures to classic fringing reefs. They are composed of five zones: a coastal reef lagoon, back reef, reef crest, buttress zone and deep fore reef zone. The differences in geology, climate and depth in the southern and northern parts of Vietnam have caused particular structures and developments of coral reef, resulting in a diversity of coral reef formations. Typical coral reef structures in the coastal areas of Vietnam are as follows:

- *Coral reef at Western Tonkin Gulf*: The seafloor in this area is narrow and muddy, so the coral reef is short and thin. The reef lagoon and reef crest zone are hard to distinguish. The coral reef in this area is mainly divided into a back reef, buttress and deep fore reef zone. These coral reefs only spread to a depth of 7–10 m. In Ha Long Bay, coral reef only forms in clumps (patch reef) rather than forming a classical type of coral reef. Coral reefs in offshore areas with greater depth, clear water and without a muddy substrate have formed a classical type of coral reef (e.g., in Bach Long Vi and Con Co Island). The reefs in these areas are 500–1000 m wide and 15–20 m deep.

- *Coral reef in central Vietnam*: The marine environment in central Vietnam is more hospitable to coral reefs than the Tonkin Gulf: the seabed is deeper, and water temperature and salinity are higher and more stable. Therefore, coral reefs in this area are bigger, more diverse and more highly developed. However, coral reefs in central Vietnam have an inconsistent form. Fringing reefs can be divided into three types: closed coral reefs, semiclosed coral reefs and open coral reefs. The closed coral reefs exist in wide, flat areas with less diverse features but high coverage. The open coral reefs are distributed in the open ocean, the flat area has a high slope and the coverage of coral reefs is based on seabed topography. The semiclosed coral reefs often have high species diversity and coverage.

- Flatform reefs in the southern part of central Vietnam have never been exposed to the water surface, even in the shallowest tide. This type of coral reef is quite wide with a small coverage percentage. The coastal barrier reefs are small and less common. This type of coral reef is only found in Giang Bo Sea (Khanh Hoa province).

- *Atoll reefs in Spratly Islands*: The Spratly Islands are composed of more than 100 floating islands, shoals and immersed reefs that spread to a width of 180,000 km². The Spratly Islands are the biggest islands in the East China Sea with classical atoll reefs. The coral reefs in the Spratly Islands are divided into two types: closed and open reefs. The open atoll reefs are a group of floating islands, arranged in a chain, covering a lagoon 50 km long, 20 km wide and 50 m deep. The closed reefs are separated islands in a donut shape. There is either a closed lagoon or a connection to outside regions through narrow and shallow creeks, for example, Da Lat, Da Dong, Da Lon, Nui Le, Tien Nu and Da Vanh Khan Islands.

8.2.2 Impacts of Increasing Water Temperature

The average temperature of seawater has increased owing to global warming. Nowadays, almost all coral reefs exist in an environment where the water temperature is higher than optimal for the reefs. Sea temperature rise affects coral reefs by causing coral bleaching (Figure 8.1). Some reefs now live at their upper limit for optimal growth. This phenomenon may cause a reduction or even disappearance of coral reefs because these species cannot survive in such high-temperature conditions. Increasing water temperature will threaten species living in coral reefs when these coral reefs are extinct. The loss of coral reefs will cause a decrease in fishery output and tourism-related revenue.

Coral bleaching was discussed in a report on Con Dao National Park. Data from October 1998 demonstrated that coral bleaching fluctuated between 0% and 90% (average 38%)

FIGURE 8.1
(a) Coral reefs in good conditions. (b) Coral bleaching due to climate change especially water temperature rise. (c) Demolished coral reefs. (Photos from GBRMPA (Great Barrier Reef Marine Park Authority, under Australian Government) www.gbrmpa.gov.au.)

depending on the studied sites (n = 11). In fact, the rate of coral bleaching might be more serious than this number suggests because some species, such as *Acropora* spp., died before the survey began. Soft coral (*Sinularia*), polyp coral (*Millepora*) and hard coral (*Acropora*) are the most sensitive corals and are affected by changes in temperature.

The results from Con Dao National Park were also compared to data collected in Ca Na Bay, where in May 1998 the average water temperature reached as high as 31°C at every studied site, even at a depth of 20 m. The coral was bleached starting in May, and nearly 8% of the coral was dead by September 1998. However, the upwelling water in this area, which occurs annually from June to September, helped to reduce the water temperature and save many coral species. Thus, many coral species in Ca Na Bay were able to recover from the increase in water temperature (Table 8.1).

The positive impacts of the upwelling stream need to be studied to protect coral reefs' biodiversity due to global warming. Thus, an upwelling stream is the best place for coral reefs. Humans can help corals to adapt by creating marine protected areas in these places.

TABLE 8.1

Research Results about Seabed after Coral Bleaching in Con Dao (10/1998) and (4–5/1998)

		After Coral Bleaching at Con Dao (10/1998)			
Study Area	Area	Hard Coral	Dead Coral	Percentage Coral Bleaching	% Recently Deceased Coral
Ben Dam	Shallow	26.3	8.1	26.2	23.8
	Deep	3.19	45.0	37.3	0
Bong Lan	Shallow	7.5	3.1	16.7	8.3
	Deep	19.4	0	74.2	0
Chim Chim	Deep	38.1	6.3	68.8	16.4
Dat Doc	Shallow	1.9	21.3	0	0
	Deep	11.3	8.8	88.9	83.3
Hon Cau	Shallow	0	0	–	0
	Deep	8.8	0	35.7	0
Da Trang	Shallow	41.9	3.7	11.9	9.0
	Deep	41.2	1.9	18.2	3.0

Source: Vo Si Tuan, Nguyen Huy Yet and Nguyen Van Long (2005). Coral reef ecosystem in Vietnam, Science and technology Press, Ho Chi Minh City, Vietnam, p. 212.

8.2.3 Impacts of Sea-Level Rise

Based on projected sea-level rise in the twenty-first century, we need to care more about its impact on the survival of coral reefs. In fact, data show that coral reefs will adapt to sea-level rise. When coral species are grown in conditions that reduce their growth rate and productivity, a small amount of sea-level rise will be a positive factor. However, if the sea level rises suddenly, coral reefs will be seriously and adversely affected.

The relationship between reefs and sea-level rise is another potential issue. For example, coral reefs play an important part in protecting inland areas from wind and tide, so sea-level rise will reduce the effectiveness of this reef function. Moreover, when sea level increases, coral reefs are more submerged, which can accelerate sedimentation and erosion. This problem will directly affect reefs and the form of coastal areas. These changes represent the greatest concern of fishery communities in coastal areas.

Sea-level rise in Con Dao will destroy coastal biodiversity and marine resources including sea turtle breeding grounds and mangrove habitats. Sea turtle breeding grounds in Bay Canh Island are being eroded; other breeding grounds located near the coast are also at risk. Therefore, there is a need to protect young sea turtles. Sea turtle programmes are now focused on moving breeding sites. These actions will affect the natural behaviour of young sea turtles. Instead of moving sea turtle breeding sites, active observation and a better management strategy need to be put in place. Clearly, sea-level rise will reduce beach areas and threaten the existence of important sea turtle breeding grounds on Con Dao Island.

Although sea-level rise is affecting coral reef ecosystems and underwater areas, its level and range are hard to forecast. Suitable technical management needs to be practiced in the operation of Con Dao Marine Protected Area, which can help us cope with sea-level rise and coral bleaching.

8.2.4 Impacts of Storms

Storms and sedimentation from inland represent the most important influences on coral reefs in Vietnam. Storms can destroy corals in shallow water, especially branch corals. Corals are demolished and their coverage reduced following storms. Storms can alter the shallow, muddy seabed where coral reefs are often located. Large storms create strong waves that change the mud that covers reefs. The reduction in salinity as a result of storms with heavy rain will prevent the penetration of sunlight into water, cause negative impacts on the photosynthesis of zooxanthellae and destroy the balance of the corals themselves.

Sedimentation forms inland owing to alluvial currents from rivers and after long, heavy rainfalls. This is due to deforestation, clearance of plantations and the development of infrastructure. Deforestation causes floods to generate more easily in the rainy season. It also makes the freshwater go further to the sea. Freshwater sometimes reach the coral reefs and cause the reduction in salinity and the increase in turbidity. Some coral species that cannot resist these fluctuations might die.

8.2.4.1 Research Results on Con Dao Island

Typhoon Linda in 1997 and a decreasing salinity of the seawater were clear evidence of climate change impacts on the coral reefs on Con Dao, causing destruction to corals (Table 8.2 and Figure 8.2).

TABLE 8.2

Alteration of Hard Coral, Soft Coral Coverage and Dead Coral in Areas Surrounding Con Dao
Island before and after Typhoon Linda, 1997

Survey Areas	1994–1995 (Before Linda)			4/1998 (After Linda)		
	Hard Coral	Soft Coral	Dead Coral	Hard Coral	Soft Coral	Dead Coral
Ben Dam	44.7	0	11.4	31.56	0.31	49.38
Bong Lan	55.4	2.3	29.2	10.73	0.63	50.31
Chim Chim	36.0	23.0	5.2	47.50	19.38	0
Dat Doc	37.5	0	20	19.69	1.25	58.20
Hon Cau	52.0	0	42.0	0.63	0	74.38
Hon Troc	51.5	8.3	6.8	8.76	4.15	65.60
Bai Duong	43.4	8.3	31.6	2.19	0.63	85.31

Source: Vo Si Tuan (2001), Coral reef in Con Dao. Document for WWF, Labour Press.

Note: Typhoon Linda, or Tropical Storm Linda, was the worst typhoon in southern Vietnam in at least 100 years,
killing thousands of people and leaving extensive damage. It formed on 31 October 1997 in the East Sea,
between Indochina and the Philippines. Strengthening as it moved westwards, Linda struck extreme
southern Vietnam on 2 November with winds of 100 km/h (65 mph), dropping heavy rainfall.

Vo Si Tuan (2001) showed that the average coverage of coral reefs on Con Dao was 42.6%
in 1994–1995. About 70% of coral reefs were categorized as high coverage (50%–75%) and
extremely high (>75%). Typhoon Linda badly destroyed coral reefs in this area. Research
results showed that in April 1998 the coverage of coral was 0% to 10% in the places less
affected by the storm. The seawater freshening from Sai Gon River caused coral bleaching;
the bleaching rate averaged 37%.

Research results from the Research Institute for Marine Fisheries in two years (2006
and 2007) showed that many groups of coral in the family Acroporidae were broken into
pieces because of Typhoon Linda around Con Dao Island. However, this report also demonstrated the recovery of hard corals in some studied areas. The recovery rate of other
species was lower in all studied areas, with the young polyp living in the surface of either
branch or block dead corals or even in the seabed. In Hon Tai areas, a high recovery and
growth rate of live polyp and branch was observed.

FIGURE 8.2
Coral reefs were destroyed by Typhoon Linda Storm in 1997 on Con Dao Island. (Vo Si Tuan (2001). Coral reef in
Con Dao. Document for WWF, Labour Press.)

8.2.4.2 Research Results on Con Co Island

The impacts of waves on the coastal areas of Con Co depend on the season and different parts of the island that cause coastal erosion as follows:

- The island's shore (Ben Nghe area) is shrinking from west to east as an erosion consequence of waves. In the Southwest monsoon, the wave directions were mostly southwestward and westward, which made for a strong erosion rate on the western and southwestern coasts, 1000 m long. Over the past 2000 years, the erosion rates in this area have fluctuated from 5 to 7 cm/year. Over a period of 10 years, the erosion rate has been increasing by 5–10 cm/year. The western and southwestern parts of Con Co Island have seen the highest rates of erosion (Van Hai, 2007).

- The floating island is shrinking as a result of sea-level rise: sea-level rise on Con Co Island was predicted to occur at a rate of 2–3 mm/year. This rate was faster compared to that of the geological formation at a rate of 1–1.2 mm/year. High waves with low frequency and shrinking of the island are the crucial factors to create a safe core zone in marine protected areas.

Overall, the above fluctuation trend of Con Co island picked up speed the following years except we have better protection strategies in correlation with shore protection, increasing coastal vegetation and the recovery and development of ecosystems such as coral reefs and intertidal regions. This creates a belt that is a combination of artificial works and natural ecosystems to prevent high waves.

A threat of sediment pollution on Con Co Island was due to the erosion from inland water and two water currents from Cua Tung and Cua Viet River. In 2001, researchers found solid waste that had covered the surface of corals and caused the loss of photosynthetic capability of zooxanthellae. This issue caused the gradual death of coral reefs. Coral reefs in developing areas such as the port, the Ben Nghe area and the northern and northwestern parts of the island often showed a similar phenomenon.

8.2.5 Impacts of Floods

The inconsistent geology in the coastal areas of northern Vietnam creates different seabed types. Therefore, coral species are distributed mainly in the coastal regions of Hai Phong city and Quang Ninh province, where the environment and seabed are suitable for coral growth. However, the impacts of floods and extreme environmental conditions caused a gradual decline of coral reef ecosystems.

Meanwhile, Co To is an island located further offshore, so floods have not affected the coral reef ecosystem. In the inshore areas such as Ha Long, Bai Tu Long and Cat Ba, high turbidity and freshening from floods are extremely detrimental to entire coral reef ecosystems. Because these areas are close to big rivers such as Bach Dang, Cua Cam, Lach Tray, Yen Lap, Cua Luc and Tien Yen in the rainy season, an enormous amount of alluvium goes directly to the coast, causing increased turbidity and decreased salinity. In recent years, increased deforestation and coal mining in Quang Ninh and dredging, expanding infrastructure and heavy rains have caused an increased erosion rate. The high erosion rate has increased turbidity, causing the death of coral reefs in the inshore areas and decline of the coral reefs in the offshore regions.

Because of freshening as a consequence of floods in Hai Phong and Quang Ninh, species numbers and their coverage and distribution changed. In the last few years, coral reefs have spread to a depth of 10 m, but they are now distributed only at a depth of 6–7 m (Nguyen

TABLE 8.3

Alteration of Coverage (%) of Some Typical Coral Reefs in Hai Phong, Quang Ninh
Due to Impacts of Floods

Areas	Coverage (Before 1998)	Coverage (2003)	Rate of Decline (%)
Cong La	29.3	17	42
Ang Tham	55.7	7.4	86.7
Ba Trai Dao	85.7	44.6	48
Van Boi	–	31.1	–
Hang Trai	78.1	65	16.8
Cong Hip	–	75.4	–
Cong Do	28.3	1	96.5
Tung Ngon	64.7	48	25.8
Coc Cheo	68.4	55.9	18.3

Source: Nguyen Dang Ngai (2004). The changes of coral reef ecosystems in Ha Long Bay (Quang
Ninh) and Cat Ba Islands (Hai Phong). In: *Proceedings of Vietnam - Italy workshop on Coastal
Marine Biodiversity Conservation in Vietnam*, National University Press, Ha Noi, Vietnam.

Huy Yet, 1999). Therefore, the distribution areas based on the depth and width of coral reefs
are becoming more narrow, which affects other species living in these coral reefs.

Table 8.3 shows that, overall, the coverage of coral reefs has decreased. Some reefs have
suffered significant rates of decline, such as Ang Tham (86.7%) and Cong Do (96.5%). The
coverage of live corals was low; they were replaced by dead corals and mud. The coverage of Cong La and Ba Tra Dai coral reefs was decreased to one-half of the original values. A significant change occurred in coral populations in these reefs. In the Ba Trai Dao
reefs, before 1998, the *Acropora* spp. were common throughout the area, were distributed
at greater widths and depths, and had a high coverage rate; meanwhile, other species had
low coverage. However, *Acropora* spp. were completely wiped out and replaced by species
of the *Galaxea* genus. Species of this genus became the dominant species and grew rapidly
in the shallow water. Other reefs in this area also changed according to the same pattern.

8.2.6 Impact of El Niño and La Niña

El Niño and La Niña are two phenomena of global warming. They have greatly affected
the occurrence and intensity of floods, droughts and large-scale storms. Vo Si Tuan (2001)
reported on coral bleaching in Con Dao National Park due to El Niño in 1997. This phenomenon had remarkable effects on the Phu Quoc coral reef ecosystem. Local communities faced many challenges when they had to cope with the global issue themselves. A
comprehensive monitoring system would make it possible to support human assessment
of the potential risks of coral bleaching and the rate of recovery from it.

8.3 Impacts of Climate Change on Mangrove Ecosystems

8.3.1 Sensitivity of Mangrove Ecosystems to Climate Change

On a global scale, mangrove forests' distribution is limited by temperature. On a regional
scale, rainfall affects a number of mangrove species and their biodiversity. Hence, global
warming and rainfall alteration have huge impacts on mangrove ecosystems.

Mangrove forests are formed in stable environments, which are the most sensitive ecosystems to climate change and sea-level rise. The most sensitive ecosystem is usually the coastal coral reefs. The second most sensitive ecosystems are lagoons, alluvial areas located behind shorelines. The third most sensitive ecosystem is mangrove forests. However, many factors can increase the sensitivity of mangrove forests in a particular area, including geological factors, for example, the hardness of the seabed and sediment sources.

As for temperature rises, when the average air temperature in the coldest month is higher than 20°C and the variability in temperature in a season is not more than 10°C, mangrove forests can develop properly. Modification of water temperature, salinity, air temperature and fog in connection with drought can cause a reduction in mangrove species. Moreover, temperature influences not only the effectiveness of photosysthesis but also the balance of water in mangrove trees through the opening of stomata and the loss of water, salt absorption and desalinization. Accordingly, the impacts of climate change need to be observed in relation to the alteration of rainfall and direct impact of carbon dioxide in the atmosphere.

8.3.2 Impacts of Sea-Level Rise

Mangrove forests are particularly sensitive to rises in sea level. It is very difficult for mangrove forests to adapt to changes in sea level. However, evidence for the reaction of mangrove forest is unavaiable.

In Vietnam, mangrove forests are mainly distributed in the Red River Delta and Mekong Delta. Pilgrim (2007) showed that these two regions are the most affected areas under a sea-level-rise scenario (Figure 8.3).

Sea-level rise in connection with monsoons, storms and high tides leads to the erosion of coastal regions. In the eastern part of Ca Mau peninsula, a northeast monsoon (northeast

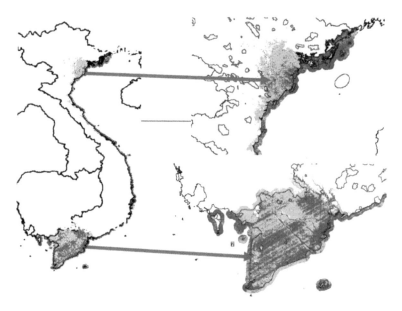

FIGURE 8.3
Sea-level rise scenario on Red River Delta and Mekong Delta ecosystems. Dark blue: underwater regions (<1 m); green: underwater areas (<5 m).

wind) and high tide have caused dozens of kilometers of erosion from Ghenh Hao to Dat Mui. In addition, mangrove trees have lost 20–30 m in width annually, especially gray mangrove species (*Avicennia marina*) in Bo De estuary, Rach Goc and Khai Long. High tides bringing sand to the shore have caused the deaths of many mangrove trees, whose roots become covered by sand (Phan Nguyen Hong, 1991).

In areas where mangrove forests have not developed, the sea-level rise created suitable conditions for mangrove species to become distributed further in agricultural and inland regions, particularly in Quang Binh and southwestern Vietnam. As a result, they affected food productivity and biodiversity and caused the disappearance of freshwater species that were replaced by brackish species.

8.3.3 Impacts of Storms

Storms cause an increase in seawater level. If a storm reaches the shore, the level of sea-level rise can reach its maximum.

Up to now, Vietnam has been subject to a high frequency of storms (5–10 times/year). High storm surges often happen in central Vietnam. For example, the highest storm surge was estimated at more than 3.2 m in Nghe An province; the storm surge fluctuated from 2.5–3 m along the shore of Thua Thien Hue province (central Vietnam); the storm surge of 1–2.5 m was estimated in canals and mangrove forests in southern Vietnam. Even if the sea-level rise was not significant, mangrove ecosystems could suffer tremendous damage (Bui Xuan Thong, 2005).

8.3.4 Impacts of Erosion

The bottoms of mangrove forests are often eroded when the seawater level rises. The layer of land in a mangrove forest can disappear, and the deposition process happens on the outside of the mangrove forest. Erosion normally occurs on the surface of a mangrove bottom. It causes exposure of the tree roots, so the mangrove bottom is lower and the estuaries widen. These alterations cause mangroves to go further inland.

Other types of erosion include forest landslides near the sea, which creates a fragmented forest and then becomes narrow forest strips, and tidal erosion of riverbanks that leads to a loss of mangrove trees. When the sea level is high, erosion that occurs in the bottom of a mangrove forest is a serious problem.

8.3.5 Impacts of Salinity Rise

A rise in salinity puts more pressure on mangrove ecosystems. Some factors in increased salinity include sea-level rise, groundwater-level decrease due to a reduction of freshwater flow, groundwater exploitation and a decline in rainfall. These factors also have negative impacts on mangrove forests. However, two natural features of mangrove trees can help them survive in high-salinity environments:

- Desalinization in mangrove species, which have a special filter system when evaporation and photosynthesis are stopped or decreased by long-term exposure to seawater.
- Salt is removed when mangrove trees photosynthesize using water from evaporation through salt glands in a tree.

8.3.6 Increase in Estuarine Current

The increase in estuarine current affects mangrove ecosystems. This phenomenon is often seen in rainy seasons and during a short period of time. The seawater increase on stormy days and during high tide causes massive damage to coastal communities' property and causes the erosion of coastal areas even if there are defensive mangrove forests in these areas. In the highland, jungle is usually cut, thus, landslide and flash floods occur in heavy rains.

8.4 Climate Change Impacts on Seaweed and Seagrass Ecosystem

8.4.1 Seaweed/Seagrass Ecosystem Roles Related to Climate Change

One of the roles of seaweed/seagrass ecosystems related to climate change is to maintain relative stability for coastal areas as high waves or typhoons happen. In addition, the nutritional relationships are complicated and fluctuate in each region of seagrass ecosystems. Primary production enters the food web through three main channels: marine herbivores, algae and nutrient debris. The habitats that have dominant seaweed/grass are crucial because they contain necessary food and shelter for many species.

8.4.2 Impacts of Sea-Level Rise

Climate change can directly affect seagrass. Because seagrass ecosystems can retain large amounts of sediment, they will be able to withstand sea-level rises in the future. While foliage slows the water flow rate sufficiently for sediment accumulation, seagrass roots can grow strong, which is conducive to bottom depth changes and new upward growth. Seaweeds themselves also generate a large amount of sediment mainly from animal shells and algae on the leaves. Both density and biomass are accumulated by increasing CO_2 in the atmosphere. These species will use light more efficiently in a large range of water depth.

Sea-level rise causes the movement of the water column in coastal areas and estuaries, which affects algae, seagrass and habitat structure. An increase of water flow velocity induces sediment disturbance, turbidity increase and a decrease in the amount of light, all of which have negative impacts on the habitat of seaweed and seagrass. When sea level increases, this will cause saltwater intrusion into land and increase the salinity in estuaries. Each species of seaweed/seagrass has the ability to adapt to different salt concentrations in both tropical and subtropical zones. When the cells of seaweed and seagrass tolerate osmotic pressure in too low or high salt concentration for a long time, they will gradually weaken and be killed.

8.4.3 Impacts of Temperature Rise

Some seaweed and seagrass species of perennial and annual populations have been affected by global warming. Annual populations, which depend on grains to regenerate and grow yearly, are the dominant populations in nonoptimal conditions. Low temperature in winter is a restrictive environmental factor; the seaweed/seagrass grows fast in a few optimal months and very slowly in winter. The perennial population produces grains throughout the year, although this process is not active in the winter, but this does not affect the production of grain very much. Temperature rise will enable perennial species to grow in an expanded distribution to the north.

Temperature rise directly affects metabolic processes, carbon balance, growth and the distribution of seaweed/seagrass. It strongly stimulates and promotes the growth of algae and phytoplankton to the point where it covers the water surface (eutrophication). As a result, turbidity intensifies and the light that penetrates into the substrate decreases, reducing the photosynthesis of seaweed and seagrass. The phenomenon of eutrophication in seawater destroys the balance of epiphytes on seagrass leaves, leading to a decline in the photosynthesis of seagrass. Therefore, eutrophication can destroy seagrass beds in coastal shallow waters. In addition, temperature rise also promotes the growth and development of seaweed, so seaweed will disappear fast. This phenomenon directly affects the metabolic cycle of matter in a water body.

Sea temperature rise has also caused an increased intensity of tropical cyclone winds (Emanuel, 1987) and strong storms (Dyer, 1995). The storms and cyclones have caused disturbances to and impairments of seaweed and seagrass beds in many areas of the world (Short and Wyllie-Echeverria, 1996) and Vietnam (Nguyen Van Tien et al., 2002). Storms induce heavy rains, flash floods, sediment disturbance, strong waves and changes in the quality of seawater. These are factors that cause damage to seaweed and seagrass beds.

8.4.4 Impacts of Storm

The typical influence of storms on seaweed/seagrass ecosystems was seen clearly in Vietnam through Typhoon Linda in 1997. This storm caused a loss and decline of seagrass beds on Con Dao Island. Its recovery so far has happened slowly owing to the impact of human activities such as the rapid development of infrastructure, an increase in sediment taken from urban wastewater and pollution of the marine environment. Then seagrass beds in Con Dao Island receded, causing a lack of food sources for dugong. The degradation of seagrass beds has led to a decline of dugong populations that continues to this day.

Storms, high tides and strong winds cause wave activities in the substrate. The seaweed and seagrass beds surrounding Co To, Cat Ba, Ly Son, Nam Yet, Phu Quy, Con Dao and Phu Quoc Islands have been seriously affected by stormy waves. In 1997, Typhoon Linda destroyed bamboo seagrass beds (*Thalassodendron ciliatum*) that had existed since 1996; Linda covered them in sand. Because of Linda's impact and owing to human activities, approximately 20%–30% of seagrass has been lost on Con Dao Island. Before the storm, *Halophila ovalis* had a density of 2250 buds/m². After the storm, the density decreased to 1551 buds/m² (Nguyen Xuan Hoa and Tran Cong Binh, 2004). Since 2005, the intensity and frequency of storms on the central coast of Vietnam have been increasing and causing considerable damage to the seaweed and seagrass beds surrounding Ly Son, Phu Quy and Con Dao Islands.

8.4.5 Impacts of Floods and Drought

Climate change induces drought in the dry season and floods in the rainy season. Floods in the rainy season increase sedimentation in intertidal areas and coastal estuaries because rivers bring huge alluvial water volumes from the continent. Drought in the dry season decreases flows in rivers and increases turbidity in estuarine waters that previously had seen good growth and development of seaweed/seagrass. In addition, flooding also completely has buried seaweed/seagrass beds under a coarse, thick sediment layer in coastal estuaries. The sudden and prolonged freshet in the rainy season due to flash floods also contributed to the loss of seaweed/seagrass beds in coastal estuarine areas.

However, eutrophication in the water column creates conditions conducive to the fast growth of some seaweed species during the rainy season, especially the growth of hair seaweed species, which prefer high levels of salinity and nutrient content. The development of some seaweed species harms the development of seagrass. They cover the surface of seagrass beds and restrict seagrass growth.

Among climate change factors, precipitation change is one of the biggest factors affecting seaweed/seagrass ecosystems. As rainfall increases, sedimentation flow and turbidity in estuarine areas also rise. Almost all seaweed and seagrass species need sunlight for photosynthesis, so turbidity seriously impacts seaweed/seagrass ecosystems.

8.4.6 Impacts of Increases in CO_2 Concentration

A high concentration of CO_2 in the atmosphere directly impacts plants all over the world (Amthor, 1995), including aquatic plants. Seaweed uses CO_2 during photosynthesis, so high CO_2 concentrations lead to increases in seaweed and seagrass biomass growth. However, the photosynthesis of seaweed and seagrass depends on other environmental factors such as nutrients, temperature and light. Madsen et al. (1996) suggested that the photosynthesis of aquatic plants reduced when high CO_2 concentrations continued over a long period of time, and the photosynthetic rate was in inverse proportion to high CO_2 concentrations. If algae covering seagrass leaves increase during eutrophication, seagrass resources will decrease rapidly.

8.4.7 Impacts of Turbidity and Sedimentation Increase

With huge amounts of alluvial soil travelling from the river to the sea, the turbidity of seawater will increase. This problem restricts the growth of seaweed and seagrass. The increase in seawater turbidity is the major reason for the disappearance of some seaweed/seagrass beds. Turbidity increases during the rainy season, reducing the photosynthetic capacity of seaweed/seagrass due to the mud covering the surface, and can even lead to death.

Van Tien and Yet (2001) conducted experiments to determine the impact of sediment on seagrass on Dau Go Island (Quang Ninh). Two seagrass species are *Zostera japonica* and *Halophila ovalis*. Experimental results showed that both species of seagrass lived normally in conditions without covering by sediments. Simultaneously, both of these grasses were covered by 4 mm of sediments, which they survived. However, with 8–10 mm sediment coverage, *Zoster a japonica* was not well developed. *Halophila ovalis* species showed good response with sediment coverage and could live under a thick mud layer of 2–10 mm in both experiments in 2000 and 2001, but in thick mud of 8–10 mm it existed in a weakened condition. *Zostera japonica* completely died out in sedimentary cover 10–20 mm thick. *Halophila ovalis* died in thick mud coverage of 12–20 mm after 30 days in the experiment.

8.5 Climate Change Impacts on Estuary Ecosystem

8.5.1 Estuary Types Related to Geo-Climate

All rivers empty out into the sea, and each region has different establishment history, geography and climate processes. The interaction between river and sea in each area along the coast of Vietnam is also different. According to Vu Trung Tang et al. (1982) and Vu Trung Tang (1994), Vietnam has four estuarine types as follows:

- Delta estuary: Red River and Mekong Deltas
- Funnel-shaped estuary: Quang Ninh, Hai Phong and Dong Nai
- Central coastal lagoons
- Shallow gulf where freshwater is received from mainland

Each estuarine form has distinct features, their own existing conditions, development trends and unique distribution and biological productivity of particular species or habitats. Thus, the effects of climate change on individual estuaries will vary.

8.5.2 Vulnerability of Estuarine Ecosystems to Climate Change

Coastal ecosystems exist as thin borders around continents and create borders that separate the continents and oceans. Like other border systems, their movements can be calculated using environmental parameters. We use a global climate model system (GCMS) that is too simple to accurately predict climate change and its impacts on coastal systems. As with other coastal systems, identifying the impact of climate warming on estuaries, deltas and coastal saline soils remains a speculative endeavour. However, it is widely believed that the impact of climate will be closely related to human activities in coastal areas, including groundwater extraction, damming and construction.

8.5.3 Impacts of Sea-Level Rise

The productivity of coastal areas is the crucial factor in assessing the local impacts of climate change on estuaries and delta systems. Rising sea levels can lead to an increase in both primary and secondary production levels. For example, marsh grass growth doubles in areas where sea levels rise dramatically. In these cases, the system reaches a threshold value and vascular plant species, mainly composed of phytoplankton species, come to dominate the system. Moreover, advantages can be promoted in areas where there are large trees and embankments, but this development is limited by the distance of the salt marshes that can penetrate into land and upland areas.

8.5.4 Impacts of Temperature

While the direct impact of increases in temperature does not seem to be as important, minor changes in the water cycle can have dramatic impacts. Consequently, a combination of temperature and water can cause multiyear droughts or floods. Moreover, temperature can affect the degree of muddiness of water, and with the warming climate, sediment deposition can occur year round. When the water temperature increases, this leads to an increase in the average rate of annual sediment deposition. The deposition of sediments itself can restrict estuary ecosystems.

8.5.5 Impacts of Typhoons and Floods

When sea surface temperatures rise, a large area of the tropics and subtropics will experience temperatures that exceed 26°C, which is a prerequisite for the formation of tropical storms. Consequently, the warming of the Earth creates favourable conditions for the formation of tropical storms and hurricanes in the Atlantic and the Pacific Oceans. This results in a higher frequency of tropical cyclones, greater storm intensity, a longer rainy season, and an increase in the range of formation and development of tropical cyclones. The resulting storms could affect topography, leading to changes in estuarine currents.

An important characteristic of natural ecosystems is their ability to recover from disasters. However, creating roads, dikes and canals can impact the landscape of coastal estuaries; however, we can limit the effects of such disturbances on wetlands and coastal plains.

8.5.6 Impacts of Erosion

According to the Institute of Marine Environment and Resources, in only 10 years (1990–2000), from Cua Luc (Quang Ninh) to Do Son (Hai Phong) accounting for 43,820 m length of shoreline, there have been 15 erosion locations, where the average erosion speed was 4.4 m/year, and in Cat Hai that number was 13 m/year. Along with erosion, the deposition of sediment in the shipping channel in the port of Hai Phong caused large-scale economic damage. Built in the late nineteenth century, Hai Phong's port served as the largest seaport of Vietnam. But in 1987, the port lost its leading position because of rapid sedimentation. Previously, the system of canals in the funnel-shaped estuary of Bach Dang was able to expel thousands of tons of dredging mass, just over 1 million m³/year. Over the past decade, the deposition of sediment had been increasing at a rate of 3–5 times. Although, the sediment was dredged continuously, shallow streams (from 3.5 to 4 m) still forced huge ships to move cargo to the outermost parts of the port. Thus, opening a new shipping route through Lach Huyen was an inevitable renovation and expansion to the sea port of Hai Phong. Scientists at the Institute for Marine Resources and the Environment Minister also pointed out that the Dinh Vu Dam, an artificial dam, is actively contributing to sedimentation in Nam Trieu. Sedimentation phenomena are also caused by many other factors, but mostly because of human activities, which play a decisive role in the Hai Phong port, as does the plan to build a deep–water, international port in Haiphong in the coming years.

This has been identified as a serious issue that will negatively affect the rise and fall of urban economic centres of the northern coastal region. Sedimentation phenomena also adversely affect marine economic activities, such as river transport, shipbuilding, aquaculture and fisheries and disable irrigation systems. Typically, coastal landing sites at Ngoc Hai (vermilion) formed as a result of decades of intensive sedimentation, causing a stagnation of fishing activities and discouraging tourists from visiting the area because sediment influxes were muddying the beaches.

8.6 Marine Ecosystems and Climate Change Scenario in Twenty-First Century Vietnam

See Figures 8.4–8.6 and Tables 8.4–8.6.

Number	Name of Place Mentioned in Chapter
1	Quang Ninh Province
2	Hai Phong Province
3	Ha Long Bay (Quang Ninh)
4	Co To Island (Quang Ninh)
5	Bai Tu Long Bay (Quang Ninh)
6	Cat Ba Island (Hai Phong)

Continued

7	Bach Long Vi Island (Hai Phong)
8	Tonkin Gulf
9	Quang Binh Province
10	Con Co Island (Quang Tri Province)
11	Ca Na Bay (Ninh Thuan Province)
12	Phu Quy Island (Binh Thuan Province)
13	Southwestern Vietnam
14	Ca Mau province
15	Con Dao Island (Ba Ria – Vung Tau Province)
16	Nam Yet Island (Spratly Archipelago – Khanh Hoa province)
17	Spratly Archipelago
18	Phu Quoc (Kien Giang Province)

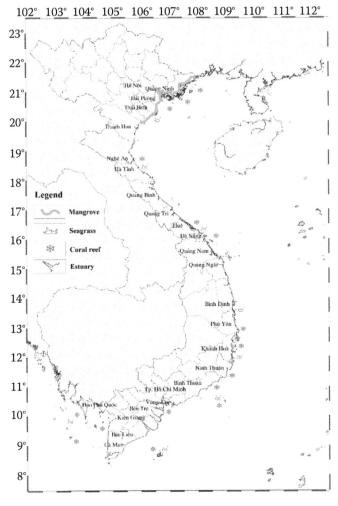

FIGURE 8.4
Distribution of some critical marine ecosystem (mangrove, seaweed/seagrass, coral reef, estuary) in Vietnam. Climate change scenario in twenty-first century in Vietnam. (Adapted from Agency for Meteorology Climate Change Mai Trong Nhuan et al. (2010). *Hazards in the Vietnam coastal zone*, Ha Noi University of Science, Ha Noi, Vietnam, p. 9.)

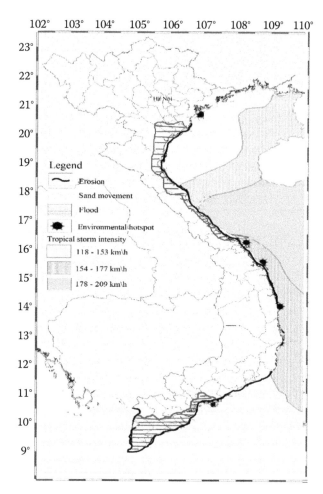

FIGURE 8.5
Some climate change has been observed in Vietnam. (Mai Trong Nhuan et al. (2010). *Hazards in the Vietnam coastal zone*, Ha Noi University of Science, Ha Noi, Vietnam, p. 9.)

8.7 Solutions for Adaptation and Resilience to Climate Change

8.7.1 Solutions for Planning and Policy

- In light of the causes and consequences of climate change mentioned earlier, the general situation is that all countries are struggling to implement the UNFCCC; Vietnam should develop national policies related to climate change issues. The basic objective of such a policy would be to address the causes and negative effects, identify appropriate adaptive measures and fulfill the obligation to work together with the international community on making the most vigorous efforts at mitigating climate change.

- Sea dike strategies should be considered with a view to sustainable economic development in the region, including a combination of marine and mainland economic strategies and ensuring national security. There should be planning for sea dikes with (1) a vision of 20 years, 50 years, 70 years and beyond; (2) wide scope

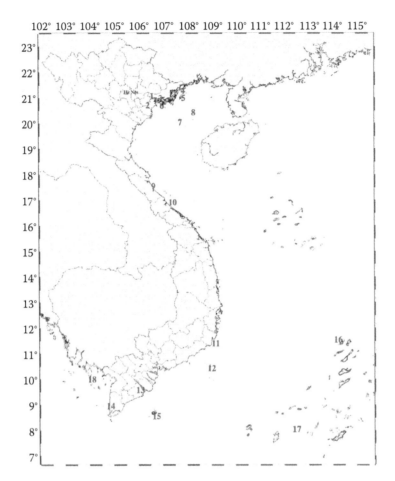

FIGURE 8.6
Place names mentioned in chapter.

and review; (3) consideration of the unstable conditions of sea dikes compared with river dikes (because they are influenced by estuaries, topography and geological formations). Sea dikes are especially affected by climate change.

- Policy should prioritize the tasks of investigating, monitoring and promoting the study of climate change impacts on marine ecosystems. Vietnam already has the

TABLE 8.4

Scenario of Average Temperature Rise (°C) in Vietnam, 1990–2100

Area in Vietnam	1990	2010	2020	2030	2040	2050	2060	2070	2080	2090	2100
Northwest	0	0.275	0.507	0.794	1.134	1.489	1.830	2.178	2.449	2.656	2.779
Northeast	0	0.275	0.499	0.797	1.087	1.478	1.735	2.127	2.349	2.634	2.754
Northern Delta	0	0.275	0.505	0.772	1.097	1.476	1.729	2.103	2.277	2.563	2.703
North Central	0	0.299	0.500	0.742	1.088	1.442	1.716	2.067	2.240	2.501	2.660
South Central	0	0.280	0.484	0.713	1.060	1.363	1.676	1.898	2.158	2.383	2.453
Highlands	0	0.285	0.458	0.706	1.014	1.372	1.621	1.901	2.102	2.343	2.456
South	0	0.273	0.457	0.734	1.015	1.391	1.623	1.945	2.107	2.341	2.455

TABLE 8.5

Scenario of Average Rainfall Rise (%) in Vietnam, 1990–2100

Area in Vietnam	1990	2010	2020	2030	2040	2050	2060	2070	2080	2090	2100
Northwest	0	0.81	1.62	2.54	3.70	4.76	6.01	6.88	8.07	8.24	8.78
Northeast	0	0.87	1.45	2.30	3.21	4.28	5.09	6.26	6.66	7.52	7.98
Northern Delta	0	0.90	1.46	2.35	3.23	4.44	5.12	6.45	6.70	7.84	8.34
North Central	0	0.83	1.59	2.34	3.16	4.45	4.89	5.55	6.59	7.74	8.25
South Central	0	0.67	1.08	1.76	2.45	3.32	3.94	4.93	5.22	5.88	6.27
Highlands	0	0.68	1.13	1.78	2.53	3.39	4.06	4.97	5.37	5.92	6.37
South	0	0.66	0.93	1.33	2.01	2.53	3.19	3.66	4.25	4.39	4.70

TABLE 8.6

Scenario of Sea-Level Rise (cm) in Vietnam, 2000–2100

Area in Vietnam		2000	2010	2020	2030	2040	2050	2060	2070	2080	2090	2100
North	(Hon Dau)	8.75	13.35	17.96	22.56	27.17	31.77	36.38	40.98	45.58	50.19	54.79
Central	(Son Tra)	1.98	4.40	6.81	9.23	11.65	14.07	16.48	18.90	21.32	23.73	26.15
South	(Vung Tau)	5.21	10.71	16.21	21.70	27.20	32.69	38.19	43.68	49.18	54.67	60.17

basis of a system of environmental monitoring surveys using various natural elements (e.g., land, water, forest, sea, climate, hydrology, geology, geophysics), as well as many economic, socio-economic, ethnic and other factors. These systems need to be complete in order to obtain a combination of observations related to climate change. At the same time, a network should be established to investigate the basic characteristics of climate change impacts on marine ecosystems. A laboratory or centre equipped with modern facilities (e.g., modelling, equipment), with enough capacity to conduct specific research in Vietnam and integrating with similar centres in other countries, should be established.

- Policy should aim to raise awareness of organizations and individuals about the impacts of climate change on marine ecosystems (e.g., the causes and consequences of climate change, actions of individuals and organizations related to climate change). Raising awareness of organizations and individuals on climate change is very important because they not only perform actions but also create the favourable conditions for the implementation of solutions related to climate change. Raising awareness of climate change is a very complex process that will take several decades and touch all areas of human lives, at the individual and societal levels. This is a revolutionary movement.

- There must be an appropriate institutional mechanism to ensure the implementation of policies about climate change affecting marine ecosystems. This structure must not only meet organizational requirements of ongoing administrative reforms but also be suitable for current organizational situations.

- Regulations to implement policies addressing the impacts of climate change on marine ecosystems should be issued. There may be separate legislation, but while waiting for a comprehensive regulation, all relevant concerns on content relating to climate change and ecosystems should be mentioned in other regulations. Such legislation must also be compatible with the overall orientation of the United Nations Framework Convention on Climate Change (UNFCCC) and other international agreements.

- To enforce policies on climate change in connection with marine ecosystems, many things should be done: train officers; promote research; raise awareness of individuals; strengthen international cooperation, actively implementing policies aligned with the content of the UNFCCC; guiding and facilitating the national team to fulfill the mandate.

8.7.2 Solutions for the Protection of Biodiversity

- *Forest protection and development of forests*: Strict protection of watershed forests; restoration of destroyed watershed forests; greening of barren land; protection of old-growth forests, nature reserves, river basins and reservoirs.

- *Protection and development of biodiversity*: Conservation and recovery of endangered genetic resources; management of nature reserves and natural landscapes; development and effective management of national parks; protection of specific ecosystems; establishment of a data bank of genetic resources to be protected; building information systems for biodiversity.

- *Preservation and promotion of marine biodiversity*: Restoration of mangroves, estuaries and coastal ecosystems; protecting coral reefs, seaweeds and seagrasses; protection and promotion of marine and island biodiversity; protection of fishing grounds for sustainable exploitation; combination of conservation and development of marine ecotourism.

- *Management of biodiversity*: Investigation and evaluation of biodiversity by nations, regions and provinces depending on type of ecosystem; building a system of nature reserves and national parks; creating a management system of nature reserves using an interdisciplinary and community approach; construction of policy to protect biodiversity; enhancement of efforts to restore biodiversity or prevent destruction of biodiversity if it is threatened; study of biotechnology and its applications; restoration and conservation of national rare genetic resources.

- *Training, education and improving awareness*: Improving knowledge in communities about values and duty to protect biodiversity; training experts and authorities about protection and development of biodiversity and gene bank.

- *International cooperation*: Strengthening commitments to Convention on Biological Diversity and to preserving rare genetic resources; formulation of import and export policy on rare genetic resources; protecting economic interests of biodiversity resources and rare genes; coordinating biodiversity protection with neighbouring countries.

8.7.3 Solutions for Science and Engineering

- Enhancing understanding of sources and substance of climate change impacts on marine ecosystems (e.g., coral reefs, mangroves, seaweed and seagrass, estuaries); carrying out studies of impacts on and damage to ecosystems as a result of climate change in Vietnam.

- Strengthening predictive technologies on the impacts of climate changes in the marine ecosystems to have effective models for the adaptation.

- Building a system of scientific and technological research institutes about climate change and marine ecosystems: establishing research institutes, centres and laboratories related to the environment. Investing in and upgrading research institutes, research facilities and universities.

8.7.4 Solutions for Protection of Coastal Ecosystems against the Rise of Sea Level and Erosion

- Consolidating and building new sea and estuary breakwaters. Vietnam needs to strengthen its existing 2700 km of sea dikes and build 2000 km of new breakwater. The consolidation and building of sea dikes must be achieved in accordance with national standards. Erosion occurs for many reasons, but the main source is the rising sea level. Therefore, strengthening breakwater is an effective measure to prevent erosion.
- Moreover, planting grasses along the slopes of dikes where there is a risk of erosion is needed to protect breakwaters; plans should be made to plant mangroves to protect breakwaters affected by waves; some areas need to build a secondary dike to reduce the risk of flooding when the main dike is broken.
- There is a need to warn coastal residents about (1) the potential areas affected by rising sea levels and (2) the potential damage from rising sea levels in order to prepare measures to cope with and adapt to a scenario of sea-level rise in time.

8.7.5 Solutions to Strengthen Resilience of Mangroves to Climate Change

1. Application of risk-spreading strategies to protect typical mangrove ecosystems: Managers need to identify, protect and breed typical species or ecosystems for standby when a natural disaster happens. The best samples should be kept in nature reserves and classified according to the level of diversity or ecological functions.

2. Identifying and protecting important areas in dealing with climate change: Managers need to protect mangrove ecosystems that have proved highly adaptable to climate change, such as rich sedimentary mangroves with abundant freshwater resources and many mature trees to provide good seed. These mangroves should be included in a system of nature reserves or areas where integrated coastal management is applied. Mainland mangroves should be specially protected because they are easily affected by human activity.

3. Controlling human impacts on mangroves: The distance of more than half the world's mangroves to urban areas (over 100,000 residents) is less than 25 km, and thus, they are gravely threatened by human activities. Human activities that threaten mangroves are expected to increase as sea level rises, so breakwaters or sea walls to prevent soil erosion should be built. These edifices will prevent mangroves from moving toward to mainland when the sea level rises. At the same time, increases in human populations in coastal areas will also increase mangrove destruction for productive land or residence or for timber from trees that protect against erosion. Managers have to identify and solve problems and formulate plans that take into account increases in negative effects, along with the impact of climate change.

4. Establishing green belts and buffer zones may allow mangrove forests to spread when the sea-level rise, mitigating the impact of nearby land-use activities. Green belts must be at least 100 m wide (about 500 m–1 km wide) in coastal areas, 30–50 m wide along estuaries and greater than 10 m for islands, canals. According to research conducted on the Red River Delta, every 5 years, stone breakwater systems must be repaired owing to waves, while a similar dike system that is shielded by 100 m of natural mangroves will survive up to 50 years. Buffer zones should be established around mangroves to reduce the direct impact of humans on protected areas.

5. Restoring areas that have been degraded so they can adapt to climate change: The restoration of degraded areas will help create stable sources of livelihood for local communities, reducing pressure on the surrounding mangroves. Moreover, planting trees to restore mangroves also helps to increase the adaptability of mangroves and brings greater economic efficiency compared to mangrove areas of lower species diversity.

6. Understanding and conserving the linkages between mangroves and freshwater and sediment sources, mangroves and associated ecosystems such as coral reefs and seagrass/seaweed: maintaining freshwater resources in higher areas to provide freshwater and sediment for mangroves. Mangrove areas connected to areas capable of surviving sea-level rise should be protected. If they are well protected, mangroves help to stabilize sediment, hold nutrients and improve water quality for coral reefs, seagrass/seaweed and fish populations.

7. Setting up a database and monitoring responses of mangroves to climate change: Background data on mangroves are essential and must include factors such as vegetation cover, density, richness and diversity of plants and molluscs, primary productivity, hydrological mechanisms, sedimentation rate and relative sea-level rise. The threats from human or current management measures must be listed. This information will be used to evaluate the sensitivity level of mangroves to climate change. Mangroves also need to be monitored to assess their response to natural and artificial effects. Chemical changes (CO_2 and salinity), hydrography (sea level, current, tidal current, wave action) and temperature should be monitored with enough time to determine the trend of climate change.

8. Implementing adaptation strategies to cope with changes in species and environmental conditions: There is a need for a long-term sustainable plan to protect mangroves ahead of global change, including forecasting the locations of future mangrove ecosystems that can be developed, planning to protect these places and forecasting changes in species composition in the ecosystems.

9. Developing alternative livelihoods for communities that rely on mangroves to reduce the destruction of mangroves: It is necessary to encourage local communities to switch to livelihoods that are less harmful to mangroves and protect important aquatic species such as fish or prawns. Other livelihoods could be making charcoal from coconut shells or collecting honey from mangroves. In Vietnam, planting seagrass has become an important alternative source of income to replace income earned from activities that destroyed mangroves. Thanks to alternative livelihoods, communities have more flexibility to adapt to changes in economic, political and social conditions.

10. Developing partnerships with stakeholders to create a source of financial support for dealing with climate change: Coping with climate change always requires collaboration and innovative solutions. There is a need to mobilize resources to support communities at local, regional and global levels. Building partnerships between sectors (e.g., agriculture, tourism, water resource management) combined with conservation and infrastructure development will reduce financial burdens from dealing with these big threats.

Some work and specific implementations of solutions related to climate change are shown in Table 8.7.

8.7.6 Education, Training and Raising Awareness of Climate Change

- Incorporating climate change into education: A climate change education programme should be established for all educational levels integrating the programmes and contents of climate change impacts on fisheries into the effective educational levels.

- Staff training in science, technology and environmental management and forecasting climate change impacts on fisheries are carried out in domestic and foreign nations. Training courses are diversified.

- Raising awareness about the harmful effects of climate change on fisheries to managers, policymakers, social and political institutions, companies and communities: Enhancement of communication tools, expanding the movement for environmental protection and reducing greenhouse gas emissions should be carried out.

8.7.7 Solutions for Strengthening International Cooperation

- Promoting international cooperation for timely information, updated data related to climate change and sea-level rise in Vietnam: There should be cooperation in the training of human resources and cooperation in the investigation and study of scientific issues posed to the region and the world.

- Strengthening international cooperation in the field of scientific research, training of scientists and environmental protection: There should be unified programmes and content about staff training and climate change in the ASEAN region.

8.8 Conclusions and Recommendations

8.8.1 Conclusions

From the results of data collection, analysis and evaluation, some conclusions about the negative impacts of climate change on critical marine ecosystems in Vietnam follow:

- *Coral reef ecosystems*: Storms, flooding and deposition of material from the continent are the most serious impacts. Storms can destroy coral reefs in shallow areas, especially black branch coral (*Antipatharia*). Materials flowing from the continent,

TABLE 8.7

Some Measures for Mangroves with the Ability to Adapt to Climate Change

Measure	Work	Organization of Implementation	Remark
1	Identifying and protecting critical mangrove areas, strategic positioning in dealing with climate change. The mangrove areas near the mainland should receive special attention for protection because they are easily affected by humans. Human activities that affect mangroves should be regulated	Managers	These mangroves have rich sediments, abundant freshwater resources, and many mature trees to provide good seeds. Human activities threatening mangroves are expected to increase as sea level rises, so people can build breakwaters or sea walls to prevent soil erosion.
2	Protection and propagation of species or critical mangrove ecosystems for backup whenever natural disasters occur. The best samples should be kept in nature reserves	Government	Classification according to level of diversity or ecological functions of mangrove areas.
3	Restoration of mangrove areas that have been degraded, creating a stable source of livelihood for local communities, reducing pressure on surrounding mangrove areas	Government Scientists	Planting many kinds of trees to restore mangroves, increasing diversity, adaptation of mangroves.
4	Establishing green belts and buffer zones may allow mangrove forests to spread when the sea level rise, mitigating the impact of nearby land-use activities	Managers	Green belt must be at least 100 m wide (around 500 m–1 km of thickness) for the coastal area, 30–50 m for estuaries.
5	Study vegetation cover, density, richness and diversity of plants and molluscs in mangroves, as well as primary productivity, hydrography, sedimentation rate and sea level rise	Government Scientists	Researching and evaluating sensitivity of mangrove forests to climate change; researching and predicting changes in species composition of mangrove ecosystems.
6	Develop partnerships with stakeholders to create a source of financial support for dealing with climate change	Government Managers Scientists	Encouraging development of alternative livelihoods for communities that rely on mangroves: e.g., collecting honey from mangroves, farming seaweed.

including freshwater, sediment and alluvium from rivers, can cause corals to die out very quickly, causing coral bleaching. Increasing temperature causes coral diseases and the death of fishes. Rising sea levels drown corals owing to the water and energy that are transported through reefs, changing patterns of sediment deposition and erosion and affecting reef and shore structures.

- *Mangrove ecosystems*: These are especially sensitive to the effects of sea-level rise. Mangroves do not seem to adapt to the rate of sea-level rise. Sea-level rise decreases the size of mangrove ecosystems. A number of marine species in coastal forests will migrate to the near-shore, some mangrove plants become invasive species inland, the mangrove floor is eroded when sea level rises, mangrove soils can be lost and deposition processes occur outside of mangrove areas. Factors such as reducing groundwater owing to decreasing freshwater, groundwater extraction and rainfall changes also impact mangroves.

- *Seaweed and seagrass ecosystems*: Changing patterns of rainfall are among the biggest factors affecting seaweed and seagrass ecosystems. As rainfall increases, the turbidity of the surrounding environment of seaweed and seagrass ecosystems also increases. Seaweed and seagrass need light for photosynthesis, so they grow slowly or die from high levels of turbidity. In addition, seaweed and seagrass ecosystems can be destroyed by hurricanes, storm waves and increasing temperature.

- *Estuary ecosystems*: The most sensitive ecosystems, estuaries could be affected by all climate change factors. Rising sea levels can lead to an increase in both primary and secondary production levels. Temperature can affect sedimentation, and with a warming climate, sediment deposition could occur year round. When the water temperature rises, the average rate of annual sediment deposition also increases. The deposition of sediments can shrink estuary ecosystems. Floods, droughts and storm waves also cause changes in topography and population structure in estuary ecosystems.

8.8.2 Recommendations

- *For research institutions*: There is a need for in-depth research on assessment, predicting the impact of climate change on fisheries in Vietnam as well as the causes of climate change as a scientific basis for responsible solutions.

- *For government agencies*: Strengthening inspection and supervision of implementation of environmental laws, the Kyoto Protocol on Climate Change (AWG-KP) and the Climate Convention (AWG-LCA) developing a national policy on climate change, and formulating quick and timely responses to climate change to minimize damage.

- *For education*: Raising awareness about the disaster of global climate change for communities as well as negative impacts of climate change on fisheries among managers, policymakers, political and social institutions, companies and communities. Enhancing communication tools, improving the dissemination of environmental protection and reducing greenhouse gas emissions are crucial measures.

- Climate change is a global problem, not just for one country. Thus, to strengthen international cooperation in scientific research, staff of various institutions should be trained on climate change management and transnational research programmes should be implemented in connection with climate change in order to respond to and mitigate climate change.

References

Amthor, J.S. (1995). Terrestrial higher-plant response to increasing atmospheric [CO2] in relation to the global carbon cycle. *Global Change Biology*, 1, 243–274.

Bui Xuan Thong (2005). Scientific reports about hydrometeorology, in a scientific project under marine science national program 2001–2005.

Dyer, K.R. (1995). Response of estuaries to climate change, *Climate Change: Impact on Coastal Habitation*. In Eisma, D. (ed.) CRC Press: Boca Raton, pp. 85–110.

Emanuel, K.A. (1987). The dependence of hurricane intensity on climate. *Nature*, 326, 483–485.

Ha Van Hai (2007). Some scientific discoveries of new tectonics in Hanoi and sub-areas, *Vietnam geological Journal*, (299), 3–4/2007, pp. 42–49 (Vietnamese).

Madsen, T.V., S.C. Maberly and G. Bowes. (1996). Photosynthetic acclimation of submersed angio-sperms to CO2 and HCO3. *Aquat Bot*, 53, 15–30.

Mai Trong Nhuan, T.Q. Quy, N.T.H. Hue, N.T. Ha, N.T. Ngoc and N.T. Tue. (2010). Hazards in the Vietnam coastal zone, Ha Noi University of Science, Ha Noi, Vietnam, p. 9.

Nguyen Dang Ngai (2004). The changes of coral reef ecosystems in Ha Long Bay (Quang Ninh) and Cat Ba Islands (Hai Phong). In: *Proceedings of Vietnam – Italy workshop on Coastal Marine Biodiversity Conservation in Vietnam*, National University Press, Ha Noi, Vietnam.

Nguyen Huy Yet (1999). Investigation on coral degradation in northern Vietnam in order to pro-pose conservation and restoration solutions, Institute of Marine Environment and Resources (IMER).

Nguyen Van Tien and Nguyen Huy Yet (2001). Recovery of some sea grass communities in coral reef ecosystem in northern Vietnam, Institute of Marine Environment and Resources (IMER). p. 70.

Nguyen Van Tien, Dang Ngoc Thanh and Nguyen Huu Dai (2002). Sea grasses in Vietnam, Science and technology Press, Ha Noi.

Nguyen Xuan Hoa and Tran Cong Binh (2004). Survey on sea grass and dugong population (Dugong dugon) in Con Dao Island from 1998–2002. *Proceeding of South China Sea International Conference. Vietnam natural science and technology Publisher*, pp. 626–637.

Phan Nguyen Hong (1991). Habitats of mangrove plants in Vietnam. Dissertation of marine biology, pp. 115–122.

Pilgrim, J. (2007). Effects of sea level rise on critical natural habitats in Vietnam. In: *Proceedings of the International Symposium on Biodiversity and Climate Change – links with poverty and sustainable development*, Hanoi, 22 –23 May, 2007, German Development Cooperation, p.5.

Standard and Poor's Credit Ratings. (2014). (http://www.21stcentech.com/standard-poor-states-climate-change-impact-credit-ratings-nations-risk/)

Short, F.T. and S. Wyllie-Echeverria (1996). A review of natural and human-induced disturbance of seagrasses. *Environmental Conservation*, 23(1): 17–27.

Vo Si Tuan (2001). Coral reef in Con Dao. Document for WWF, Labour Press.

Vo Si Tuan and Phan Kim Hoang (1996). *Species composition of hard corals (Scleractinia, Hexacorallia, Anthozoa) in Vietnam, Marine Fisheries Proceeding*.

Vo Si Tuan, Nguyen Huy Yet and Nguyen Van Long (2005). Coral reef ecosystem in Vietnam, Science and technology Press, Branch in Ho Chi Minh City, p. 212.

Vu Trung Tang (1994). Estuary ecosystems in Vietnam, Science and technology Press: Ha Noi, Vietnam, p. 271.

9

Management Strategies of St. Martin's Coral Island at Bay of Bengal in Bangladesh

Md. Nazrul Islam and Mamunur Roshid

CONTENTS

9.1 Introduction 238
9.2 Geo-environmental Settings of Saint Martin's Island 238
9.3 Geographical Location 239
    9.3.1 Area 239
    9.3.2 Physiographic Characteristics 239
    9.3.3 Soil Characteristics 239
    9.3.4 Geomorphology 241
    9.3.5 Water Characteristics: Seawater and Freshwater 242
        9.3.5.1 Seawater 242
        9.3.5.2 Freshwater 243
9.4 Results and Discussion 244
    9.4.1 Careless Littering by Tourists Creates Waste on St. Martin's Island 244
    9.4.2 Causes of Environmental Pollution on St. Martin's Island 245
    9.4.3 Dumping Place of Wastes of St. Martin 245
    9.4.4 Environmental Mismanagement in St. Martin's Island 246
    9.4.5 Major Drivers of Ecosystem and Water Quality 246
        9.4.5.1 Overbuilding 247
        9.4.5.2 Population Growth 247
        9.4.5.3 Crowds of Tourists 248
    9.4.6 Resources those Promote to Development in St. Martin's Island 248
9.5 Important Resources and Management Options in St. Martin's Islands 249
    9.5.1 Land 249
    9.5.2 Water 249
    9.5.3 Flora 250
    9.5.4 Fauna 251
        9.5.4.1 Crustaceans (Lobster, Crabs, Shrimp) 251
        9.5.4.2 Seaweeds 251
        9.5.4.3 Molluscs 251
        9.5.4.4 Rocks 252
        9.5.4.5 Coral 252
        9.5.4.6 Fish 252
        9.5.4.7 Birds and Reptile 253

9.6   Integration of Co-management Strategy 253
      9.6.1   Co-Management Framework 253
      9.6.2   Zoning 254
      9.6.3   Managed Resource Zone 255
              9.6.3.1   Sustainable-Use Zone 256
              9.6.3.2   Restricted Access Zone 256
9.7   Recommendations 257
9.8   Conclusion 258
References 258

9.1 Introduction

St. Martin's Island (Narikel Jinjira) is a small island in the northeast of the Bay of Bengal, about 9 km south of Cox's Bazar: Teknaf Peninsula (Ali, 1975; Ahammed et al., 2016). It is a dumbbell-shaped sedimentary continental island. The surrounding coastal environment of the island sustains the existence of coral communities (Tomascik, 1997; Ahammed et al., 2016). The unique geomorphological and biological setting of the island has been under study by several workers (Fattah, 1979; Khan, 1985; Chowdhury et al., 1992; Mahmud and Haider, 1992; Ahmed, 1995), though in some cases the island's remoteness and inaccessibility during the monsoon season (June–September) have hampered regular sampling. Researchers have proposed several management zones for the conservation of the valuable biodiversity of the island, which has national and international significance (Islam and Aziz, 1980; Ali and Parvin, 2010). The island was also under close observation by a couple of protected area specialists in connection with land-use patterns and major biodiversity habitats (Tomascik, 1997; Molony, 2007; Hebara, 2009). St. Martin's Island has unique resources, but recently the island has witnessed severe degradation of its biodiverse habitats owing to intensive pressure from tourism and ownership transfer of unused pristine lands to outsiders (Banglapedia, 2008; Ahammed et al., 2016). Owing to the large crowds and the concomitant increase in the number of people, including tourists, the island has seen overbuilding, which has led to ecosystem degradation (Zakaria and Abdullah, 2002).

9.2 Geo-environmental Settings of Saint Martin's Island

Saint Martin's Island is a small island in the northeast area of the Bay of Bengal, about 9 km south of the tip of Cox's Bazar: Teknaf Peninsula and forms the southernmost part of Bangladesh. It is about 8 km west of the northwest coast of Myanmar at the mouth of the Naf River (Zaman, 2006; Ahammed et al., 2016). The local name of the coconut is *narikel* and the original name of this island is *Narikeljinjira*. The area of the island itself is about 5.9 km², and with rocky platforms extending into the sea, the total area of the island is about 12 km². There are five distinct physiographic areas on the island (Aziz et al., 2002; Banglapedia, 2008).

9.3 Geographical Location

Saint Martin's Island is a union of Teknaf Upazilla and situated between latitude 20°34′, 20°39′N and longitude 92°18′, 92°21′E (Banglapedia, 2008). To the east the Myanmar border (Arakan coastal plain) lies only 4.5 km away, while to the west and southwest is the Bay of Bengal (Ali and Parvin, 2010). The island has a wide variety of land cover types, such as sandy shore, rocky shore, mangrove forest and agricultural land producing rice, watermelon, ground nut, maize, seasonal vegetation, coconut, betel leaf and nuts.

9.3.1 Area

Uttar Para is the northern part of the island, with a maximum length along the north–south axis of 2,134 m and a maximum width (along the east–west axis) of 1,402 m (Figure 9.1). Golachipa is a narrow neck of land connecting Uttar Para with Madhyapara. One of the impressive sunsets on the island can be seen in Golachipa. Madhyapara lies directly south of Golachipa and is 1,524 m long and 518 m wide at its maximum (Chowdhury, 2006; Ahammed et al., 2016). Dakhin Para lies next to the south and is 1,929 m long, with an additional narrow tail of 1,890 m towards the southeast, and at its maximum is 975 m wide (Ali and Parvin, 2010). Cheradia, the southernmost tip of the island and extending south–southeast from Dakhin Para, is a rocky reef approximately 1.8 km long and between 50 and 300 m wide (Banglapedia, 2008). It is separated from Dakhin Para during high tide, and located on this intertidal reef are three small vegetated islands known as Cheradia, of which the middle one is the largest.

9.3.2 Physiographic Characteristics

St. Martin's Island is a dumbbell shaped sedimentary continental island located on the eastern flank of an anticline, which, like Chittagong Naga, is a folded system (Warrick et al., 1993). The surface area of the island is about 8 km^2 depending on the tidal level. The island is almost flat with an average elevation of 2.5 m above mean sea level (MSL), rising to a maximum of 6.5 m high cliffs along the eastern coast of Dakhin Para (Kabir, 2006).

9.3.3 Soil Characteristics

A sequence of marine sedimentary rocks is exposed on the island, ranging in age from Late Miocene (around 11.6 to 5.3 million years before present) to recent. The base rock is grey to bluish-grey Girujan Clay Shale (Pliocene – 5.3 to 2.6 million years before present) interbedded with subordinate sandstone (Aziz et al., 2015). Above this is a layer of St. Martin's Limestone (Pleistocene – 2.6 million to 12,000 years before present), which is coquinoid 1, dirty white, coarse grained, bedded and partly consolidated along with cream-coloured coral clusters, and includes the fossil-bearing Dakhin Para Formation (Molony, 2006; Thompson and Islam, 2010; Ahammed et al., 2016). This is overlain by the Holocene (from 12,000 years before present till today) coquina bed, which is a continuation of the St. Martin's Limestone Formation. The surface deposits (Holocene) of beach sand, which is medium to coarse grained, and light grey to grey with recent shell fragments, lie above the limestone (Banglapedia, 2008). With the gradual relative increase in sea level, dead shell fragments were thrust towards the shore of the island by wave action

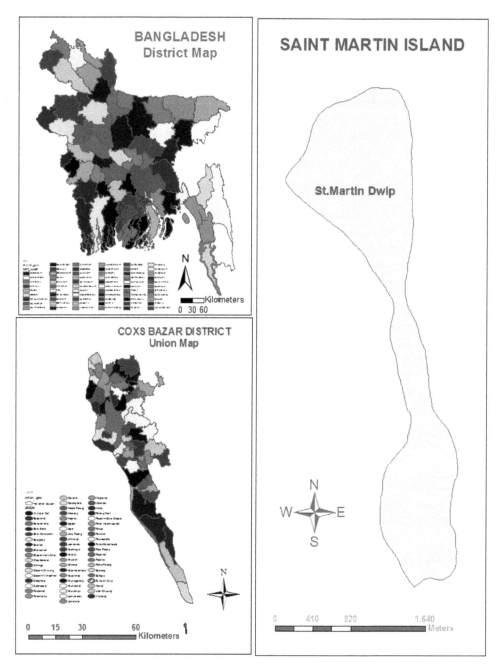

FIGURE 9.1
Map of study area, St. Martin's Island, Bangladesh.

and finally heaped up as a narrow ridge forming the coquina horizon along the present southeast border of the island (Hasan, 2009). Coquina is also known as shelly limestone and is best exposed in a 6.5 m high cliff 166 m in length along the eastern coast of Dakhin Para (Ali and Parvin, 2010). It is composed entirely of broken and crushed shells of molluscs, small crustaceans and corals held together by calcareous cement (Chowdhury, 2006).

In this rock tiny shells are often found unbroken, while the rock overall is brown with a grey weathered surface and is cross-bedded, loose and friable (Khan, 1998). Many of the shells within the rocks are similar to those currently found strewn on the beach. Micro-paleontological investigation of samples of coquina has revealed the presence of foraminifera including *Elphidium crispum* and *Amphistegina radiate*. Radiocarbon dating of a 3 m notch of the coquina limestone cliff located on the central coast of Dakhin Para indicates an age of about 450 years at the base and 292 years at the top (Biodiversity and Eco tourism project of St Martin's, Bangladesh 2010). From this coquina cliff the present mean rate of uplift of the island can be calculated as 19 mm/year.

9.3.4 Geomorphology

The main shoreline features of St. Martin's Island are sandy beaches and dunes, where the main sediments are alluvial sands. The beaches and dunes of the southern part of the island have a higher carbonate content compared to the northern Uttar Para beaches. Most carbonates comprise mollusc shell fragments. The sandy beach in the north and northeast stretches 300–400 m into the sea. The western beach is sandy, but the subtidal area consists of a bed of boulders (Molony, 2006; Thompson and Islam, 2010). Coastal dunes are widespread immediately above the beach and along the shorelines. This dune system is particularly well developed around the middle part of the island. The dunes of St. Martin's are of two types: high and low. High dunes are up to 6 m in height and are mostly found on the western side of Golachipa (Khan, 1985; Chowdhury et al., 1992). Dunes along the northwest and southwest corners of the island are low, undulating and broadly extended. These dune systems act as a natural defence against storms and tidal surges, when they help to save lives and property.

The topsoil of the three main parts of the island (Uttar Para, Madhya Para and Dakhin Para) consists of alluvial sands mixed with marine calcareous (primarily molluscan in origin) deposits. Scattered throughout the area are small clumps of coral colonies, many still in a growth position. A large shallow lagoon is located in the middle of Uttar Para. The lagoon has been largely converted into agricultural fields and is connected to the sea at high tide by a narrow tidal channel on the west coast (Islam et al., 2012). The remaining, flooded, part of the lagoon has an area of about 0.4 km² with a depth of 1 m or less. This bar was formed as a result of deposition and movement of fine- to medium-grained sand materials (consisting of alluvial sand and littoral carbonates) by wave and tidal currents. It connects Dakhin Para with the only smaller islands that form part of St. Martin's Island: three small islands of which the middle one is the largest and which are collectively known as Cheradia (Khan, 1985; Chowdhury et al., 1992). The peaks of these islands are less than 3 m above MSL and become disconnected from Dakhin Para by nearly 1 m deep water at high tide. These small islands are composed of relatively coarser sand particles with frequent shale fragments, broken coral debris and foraminifera and are overlaid on nearly all sides by stony corals and boulders (Ahammed et al., 2016). The middle one of the three small islands has an undulating surface with a subbasin on its top that is slightly submerged during extreme high tides. These three islands can be classified as 'vegetated sand islands' since they developed from the accumulation of both alluvial sands and calcareous littoral deposits.

During low tide, they are connected with Dakhin Para by the spit bar that has accumulated on the top of a rocky intertidal reef. A number of non-vegetated rocky outcrops are found on the northwest coast of Uttar Para. These supratidal outcrops are the seaward continuations of the rocky intertidal zone. Almost the entire coastline of St. Martin's Island is fringed by a rocky intertidal zone that is unique in Bangladesh. The width of the rocky intertidal zone at spring low tides varies from 100 to 400 m. The rocky intertidal zone is formed by small and large boulders, which, according to Alam and Hassan (1998), have a close affinity with the bed rocks of the island. In addition, many of the spherical boulders are calcareous concretions (Islam, 2001). Coral boulders are also present, but these are relatively rare and in no place do they form a coherent feature that can be called a coral reef (IPCC, 2004). The presence of relatively well-preserved dead coral colonies in the upper and middle intertidal zones suggests that the island has been uplifted in relatively recent times.

The recent uplift of St. Martin's is evident from the presence of large Porites micro-atolls, which are found in the lower intertidal zone on the northwest coast of Uttar Para. The morphology of these microatolls suggests that the relative sea level has dropped by about 15 cm during the last 150 years or so (Ali and Parvin, 2010). This rough estimate is based on the size and average growth rates of the microatolls. Clearly, this is an exciting area for new research because it, along with the dating of the cliff sediments, contradicts recent global trends for rising MSLs (Molony, 2006; Thompson and Islam, 2010). The sedimentary boulders, calcareous concretions, sandstone and shale found in the inter-tidal area extend into the subtidal zone. Most of the inshore area around the island is composed of a shelf covered by a layer of sedimentary boulders that vary greatly in size (Khan, 1985; Chowdhury et al., 1992). Though they provide a very suitable substrate for the settlement of coral larvae, as is evident from relatively high recruitment rates of juvenile corals, the boulders are very susceptible to overturning and shifting by the heavy seas that are frequently generated by cyclonic storms and tidal surges (Khan, 2008). The growing corals on the boulders are thus damaged or destroyed when the substrate boulders move. This rocky subtidal zone is much wider along the west coast than along the east coast. A number of offshore rocky reefs along the west coast become exposed during low spring tides.

9.3.5 Water Characteristics: Seawater and Freshwater

9.3.5.1 Seawater

Surface circulation in the Bay of Bengal is determined by the monsoon winds and, to some extent, by the hydrological characteristics of the open part of the Indian Ocean. The prevailing winds reverse twice during the year (Zaman, 2006). They blow from the southwest during May–September and from the northeast during November–January, with the transition taking place during the months in between (Rahman, 1999; Aziz et al., 2010). Forced by these winds, circulation in the Indian Ocean has a general eastward direction during summer and westward during winter (Ali, 1975; Allam, 2003; Islam et al., 2012a). The inflow of freshwater from the Ganges–Brahmaputra Delta into the Bay of Bengal has a significant impact: these reversing currents carry low-salinity water from the Bay of Bengal into more saline Arabian seawater and vice versa, playing a crucial role in maintaining the freshwater–saltwater balance of the North Indian Ocean (Islam, 1999; Vinayachandran and Kurian, 2008). The massive inflow of freshwater and sediment from the Ganges and Brahmaputra Rivers, and locally from the Naf River, is also an important factor influencing the flora and fauna of the island. Thus, coral reef development is inhibited by the low water salinity, high turbidity and soft substrates present.

The island experiences normal semidiurnal tides, i.e., two high and two low tides over a period of 24 hours and 52 minutes (Banglapedia, 2008). The mean tidal range at Shahpuri Island (about 9 km northeast of St. Martin's Island) in the Naf estuary is 1.87 m. It is expected that somewhat similar, probably lower, tidal ranges occur at St. Martin's Island. The surface salinity in the coastal parts of the Bay of Bengal oscillates from 10 to 25 parts per thousand (ppt), i.e., grams per kilogram of seawater. Coastal seawater is significantly diluted with freshwater throughout the year, although the inflow of river water is greatly reduced during winter. The coastal water salinity of St. Martin's Island, as measured during the dry season (Tomascik, 1997), fluctuates between 26 and 35 ppt. It is believed that the salinity drops below this level due to increased freshwater discharge from the Naf River during the rainy season (July–October). Water transparency measured in December fluctuated from 0.62 m near St. Martin's Bazar, where the water is heavily affected by human activity, to 3.9 m at Galachipa (Hossain, 2006).

This low light penetration is the consequence of many factors. In addition to silt discharged by the Naf, the combined action of wind-generated waves, ocean swell and high-velocity tidal currents causes resuspension of bottom sediments (fine sand, silts and mud). A sea beach depth of over 7 m is required for optimal growth of reef-building corals (Khan, 1985; Chowdhury et al., 1992; Islam et al., 2012a). Since corals are light-sensitive organisms, the turbid coastal waters of St. Martin's Island are a key environmental factor limiting the development of coral reefs. The effects of aquatic pollution on coral communities are not well understood; however, there is evidence that pollution and other human activities have degraded the quality of water surrounding St. Martin's Island and have adverse impacts on the health, development and survival of corals and associated biodiversity (Hossain, 2006). Dissolved oxygen (DO) concentration in the surface waters around St. Martin's Island ranges from 4.56 to 6.24 mg/L in December. The highest value of 6.24 mg/L of DO was found at Badam Bunia, whereas the lowest value was recorded at St. Martin's Bazar (Hossain, 2006).

9.3.5.2 Freshwater

Being very porous and permeable, the shelly limestones of the island provide an excellent aquifer wherever they occur beneath the alluvium. The shelly limestone and recent marine sand are the chief sources of freshwater (Islam et al., 2012a). The rocks underlying these two formations are mostly impervious Tertiary shale and calcareous sandstone. Because rain water cannot flow downwards through these rocks, it accumulates either in the shelly limestone or in the marine sand. The shelly limestone that underlies the village of Jinjira averages 1.2 m (4 ft) thick and is overlain by 0.6–1.2 m (2–4 ft) of soil. Drinking and irrigation water is obtained by sinking shallow wells 1.5–3 m (5–10 ft) to the level of the Tertiary rocks. However, there is now a scarcity of drinking water on the island (Alam and Hassan, 1998; Kabir, 2006; IUCN, 2008). Only a few ponds and several tube wells supply water for drinking as well as for cultivation. Deforestation and large-scale expansion of agriculture have adversely impacted the groundwater lens of the island (Ali and Parvin, 2010). The shallow wells used for crop irrigation may reduce the availability of potable water. Increasing salinity in some tube wells has been reported. Throughout the wet season until January, the water table on St. Martin's Island is within a range of 0.3–2.1 m (1–7 ft) below the surface.

However, in March and April most of the wells go dry and groundwater is confined to areas where the shelly limestone is more than 1.8 m (6 ft) thick (Islam et al., 2012a). The needs of the local population and the large annual influx of tourists during the dry season have created a great demand for freshwater, leading to a drop in the water table. This demand will only increase in the near future. Already there is a shortage of water for 2 or 3 months before the wells are replenished by summer rains in May and June.

9.4 Results and Discussion

9.4.1 Careless Littering by Tourists Creates Waste on St. Martin's Island

We all know that waste is the ultimate driver of one of the largest pollution sources throughout the world. In particular on St. Martin's Island this issue is of great concern because the significant amount of natural resources on the island may be subject to degradation. According to a Department of Environment (DoE) report, unplanned tourism, marine pollution and illegal artificial structures pose a serious threat to the biodiversity of Bangladesh's only coral island, Saint Martin's.

A survey was taken among 35 experts in the study area about the level of littering by tourists who dump their waste on the island. The results of the survey show that 34.29% of tourists are careless when it comes to disposing of trash and 42.86% are moderately careless, which are alarming figures for ecotourism and development (Table 9.1). On the other hand, Table 9.1 also shows that 8.57% of tourists are concerned about waste dumping and believe the biodiversity and ecosystem health of St. Martin's Island are important.

TABLE 9.1

Tourist Littering Waste Level

Level	Respondents	Percentage (%)
Careful	3	8.57
Careless	12	34.28
Highly Careful	2	5.714
Highly Careless	3	8.57
Moderately Careless	15	42.85
Total	35	100

TABLE 9.2

Different Causes of Environmental Pollution

Cause of Pollution	Respondents	Percentage (%)
Accidental Oil	3	8.57
Beachgoers	3	8.57
Cigarette Butts	2	5.71
Plastic Bottles	2	5.71
Plastic Packets	4	11.42
Polluted Runoff	5	14.28
Polythene Bags	3	8.57
Sanitary Sewerage Overflow	4	11.42
Storm Water Runoff	3	8.57
Vessel Discharges	4	11.42
Wildlife and Pet Waste	2	5.71
Total	35	100

9.4.2 Causes of Environmental Pollution on St. Martin's Island

Knowledge of pollution is very important for the conservation and management of all natural resources (Table 9.2). Although pollution occurs on a small scale, it nonetheless greatly impacts global climate change. Our study showed that, of the highly polluted mixed runoff that occurs on the island, 14.29% is from plastic packaging, 11.43% from vessel discharge, 5.71% from pet waste, all of which has a long-term impact on the ecosystem of St. Martin Island (Alam and Hassan, 1998; Kabir, 2006; IUCN, 2008).

9.4.3 Dumping Place of Wastes of St. Martin

The survey results (Table 9.3) also show that most of the respondents thought that fixed point dumping locations (25.71%) were the main source of pollutants in the study area. More tourism is good for business but bad for the environment. As their numbers increase, so does the amount of waste. St Martin's does not have a waste management mechanism in place to tackle this problem (Rabby, 2016). As a result, raw sewage and other waste are discharged in the open. This has a damaging impact on the island's ecology. The island's once picturesque sandy beaches are littered with debris. Most of the survey's respondents expressed the opinion that regular beach cleaning would help reduce trash (Roshid, 2015). Guidelines should be established to fix the number of tourists and address coral and shell collecting. The report noted that it was possible to earn a substantial amount of revenue through planned tourism without disturbing the ecological balance of the island.

From the above table we can see that wastes are the ultimate driver of one of the big pollution sources in any place all over the world. Particularly in St. Martin's Island it is of so much concern because of this place contain a huge amount of natural resources those are may can be on destruction (Islam, 2002; Sarwar, 2005). I have (6%) categorize wastes in some beside agricultural land as an organic fertilizer of core of chore of vegetables that are alarming to ecotourism and development. On the other hand (94%) wastes are not categorized that is so much concern be damaging to biodiversity and ecosystem health.

Implementation of administrative instructions is the ultimate author of the all kinds' good management (Smith et al., 2008). All governmental or nongovernmental organizations and individual research person those are find out the problems and recommendation to proper management that author is administration to implement. But unfortunately the implementation of governmental instructions most of the time slightly ignore that law to implement. Here is same job in St. Martin's Island. Here the implementation activities to environmental management are (23%) low degree while a high activity is (77%) that is very alarming to destroying ecosystem.

TABLE 9.3

Dumping Place of Wastes Identified by Stakeholders

Dumping Place	Respondents	Percentage (%)
Anywhere	9	25.71
Beside House	14	40
Fixed Point	9	25.71
Other	3	8.57
Total	35	100

TABLE 9.4

Responsibility for Environmental Mismanagement on
St. Martin's Island

Responsible for Mismanagement	Respondents	Percentage (%)
Administration	10	28.57
Hotel Owner	9	25.71
Local Businessman	9	25.71
Local Dweller	2	5.71
Tourist	2	5.71
Tourist Service Provider	3	8.57
Total	35	100

9.4.4 Environmental Mismanagement in St. Martin's Island

St. Martin's Island is the unique land in Bangladesh that provides lots of resources, pleasure, freshness and contribution to the economy of our country. But nowadays there are potential risks for environmental mismanagement (Islam et al., 2012b). Table 9.4 shows the rank of causes of environmental mismanagement. We see that highly responsible is (28.57%) local administration system and hotel owner and local businessman (25.71%) are the next responsible.

9.4.5 Major Drivers of Ecosystem and Water Quality

According to Hossain (2006), water samples from the inshore zone of St. Martin's Island were contaminated with faecal coliform bacteria up to 6 cfu/100 mL, and although this is below the intermediate risk level set by the World Health Organization, it is still an indicator of sewage contamination due to the presence of human and animal faeces (Rahman, 1999; Molony, 2006; Aziz et al., 2010; Thompson and Islam, 2010). Significantly the water also contained *Vibrio cholerae*, the cause of cholera, posing a health threat to local people and tourists (Hossain, 2006). About 116 ha of land on the island are cultivated, with homestead gardens occupying a further 7.4 ha, representing in total 37% of land use (Poush, 2006a). Farming mainly occurs in the north of the island (Uttar Para), with the main crops being chili and watermelon (Table 9.5). An indigenous small-bulb onion variety is also cultivated, and a small amount of maize is intercropped with chili.

A small amount of transplanted Aman rice is cultivated in the rainy season. Planted trees, particularly coconut, have replaced much of the original vegetation. Thus in 2006, 15,000 coconut palms were recorded on the island. Homestead coconut gardening is an important source of income (Alam and Hassan, 1998; Kabir, 2006; IUCN, 2008). Agriculture

TABLE 9.5

Major Drivers of Ecosystem and Water Quality

Driver	Pressure	State	Impact
Overconstruction	Soil, plants	Influx sediment, toxic chemical	Deforestation, erosion
Population growth	Agricultural land (high quantity of nutrients, pesticides, sediment)	6 cfu/100 mL (WHO) water contamination level	Damaging Zooxanthellae in coral, benthic, pelagic

is the cause of the ongoing destruction of habitats, especially the clearing of rocky land for cultivation and the filling in of lagoons. Additional problems are the cultivation of exotic and hybrid plants and the use of chemical pesticides and fertilizers.

9.4.5.1 Overbuilding

The construction of buildings and infrastructure causes direct physical damage to subtidal habitats and the adjacent coastline; it increases the influx of sediments from land-based and marine operations, introduces toxic chemicals, and has immediate physical impacts through extraction and trampling. Coastal current patterns have been altered as a result of jetty, breakwater and marina construction. Fragile shoreline habitat is also physically destroyed through recreational activities such as intertidal walking and boat anchoring.

9.4.5.2 Population Growth

Human settlement started on the island in the 1880s when several families migrated from what is now Myanmar to live on the island permanently. In the 1920s the hardwood trees of the island, reportedly mostly teak, were cut and sold to Myanmar (then Burma) (Rahman, 1999; Aziz et al., 2010). From the 1940s onwards it is reported that land was gradually converted to paddy cultivation, and from the 1960s onwards this involved converting the main lagoon to land for cultivation. In 1996 there was a population of around 3,700 people belonging to 535 families (MoEF, 2001b); in 2000 the population was 4,766 from 791 households (Islam, 2001), and in 2005 the population was 5,726 from 818 households (Poush, 2006a). See Figure 9.2.

FIGURE 9.2
Topography of St. Martin's Island, in Bangladesh.

Most of the inhabitants are ethnically Bengali and Muslim. This means that, by the late 2000s, the population density will likely be around 700 persons/km². By 2008 the island had the following significant and public buildings: 17 hotels, 12 restaurants, a government office, 2 mosques, 3 primary schools (including one that doubles as a cyclone shelter), one high school, a second cyclone shelter, a large new hospital, a lighthouse, a naval base and 2 village resource centres (Thompson and Islam, 2010).

9.4.5.3 Crowds of Tourists

The island has been a tourist destination for many years, but with recent developments in tourism infrastructure it has become one of Bangladesh's most popular tourist destinations. Tourism has increased steadily since it first began on the island. Official statistics on the number of tourists visiting the island are not available because there has been no systematic monitoring of visitor numbers. During a 45-day period in the period of December 1996–January 1997, between 150 and 200 visitors visited the island daily (Tomascik, 1997). According to the St. Martin's Island Project, the number of visitors for the whole tourist season for 2002–2003, 2003–2004 and 2005–2006 (2004–2005 figures are not available) was 62,520; 103,488; and 156,736 visitors, respectively. Tourism is concentrated in the winter, particularly December and January, when the island is most accessible, while the remainder of the year sees hardly any tourists (Thompson and Islam, 2010). A major problem resulting from tourism is uncontrolled and inadequate waste management. Untreated sewage is piped directly into the sea or stored in open ponds, adversely affecting marine and ground water quality (Thompson and Islam, 2010).

9.4.6 Resources That Promote Development in St. Martin's Island

Coral is one of the most important resources. An ecosystem-based management model could make a difference in its survival. Table 9.6 shows the strength of tourism on St. Martin's Island is growing robustly in Bangladesh (Table 9.6).

Tourism is the most important issue in relation to development not only on St. Martin's Island but in all of Bangladesh. People come to St. Martin's Island for many reasons, but mainly for its natural beauty and fresh air. This accounts for 43% of the tourism on the island, according to comments made by tourists themselves on surveys. Recreation and affordability are two more draws that bring people to the island. The barriers to tourism development on St. Martin's Island are shown in the Table 9.7.

It is true that tourism plays the most important role in the development of any country. In this island there is a huge amount of natural resources but at the same time there is some

TABLE 9.6

Strengths of Tourism Development

Attraction	Respondents	Percentage (%)
Affordable Service	2	5.71
Environmental Freshness	13	37.14
Natural Beauty	15	42.85
Other	2	5.71
Recreational Rides and Services	3	8.57
Total	35	100

TABLE 9.7

Barriers to Tourism Development

Barrier to Development	Respondents	Percentage (%)
Broker	3	8.57
High Expenses	10	28.57
Lack of Recreational Service	6	17.14
Other	1	2.85
Trash	12	34.28
Unfriendly Behaviour	3	8.57
Total	35	100

barrier to develop tourism. There are six types of barriers to developing tourism. Trash is the highest barrier to development followed by expensive stay.

9.5 Important Resources and Management Options in St. Martin's Islands

9.5.1 Land

All land on the island is privately owned, with the exception of 18.7 ha at Cheradia, which was recently purchased by the Ministry of Environment and Forest, and some small areas on which public buildings have been constructed. About 116 ha of land on the island are cultivated, with homestead gardens occupying a further 7.4 ha, representing in total 37% of land use (Rahman, 1999; Poush, 2006a; Aziz et al., 2010). Farming mainly occurs in the northern part of the island (Uttar Para), with the main crops being chili and watermelon. Shoreline erosion has been raised in a number of reports as an environmental problem. What seems not to be recognized is that coastal erosion is a natural cyclic process, part of the island's evolution (Chowdhury et al., 1997). Failure to recognize this important process as a natural phenomenon has resulted in a massive and ill-informed coastal works project that started in 1993. Huge quantities of intertidal boulders were removed from their original places and a rock wall was built along long stretches of the coastline (Alam and Hassan, 1998; Kabir, 2006; IUCN, 2008).

This has not only resulted in the destruction of important turtle nesting beaches but has accelerated erosion processes in other nearby stretches of coast, a phenomenon by now well recognized in other countries. Intertidal boulders have also been removed for use in road and house construction (Table 9.8).

The intertidal boulder reef is a natural barrier that protects the coastline from wave action and storm surges. Beach erosion is now evident in all stretches of coast where large numbers of boulders were removed. In addition, continual cutting of Pandanus beach vegetation for fuel wood by a number of poor families is a serious problem, resulting in soil and beach erosion (Thompson and Islam, 2010).

9.5.2 Water

The low salinity Bay of Bengal water into more saline Arabian Sea water and vice versa plays a crucial role in maintaining the freshwater–saltwater balance of the North Indian Ocean (Vinayachandran and Kurian, 2008). The massive inflow of freshwater and sediment from

TABLE 9.8

List of Major Resources in Saint Martin Islands in Bangladesh, Their Status, Current Scenarios and Possible Threats to These Resources at a Glance

Resources	Status	Current Scenarios and Possible Threats to the Resources of St. Martin Island
Land	8 km²	Erosion, pollution and degradation
Water	Marine (12 N. mile) Fresh water	Extraction (Fresh water) Contamination (Marine water)
Tourism	*St. Martin's Island* is very popular *tourist* spot	Over crowd, mismanagement and lacking of proper planning
Flora	Plants (144 species) marine algae (152)	Deforestation, Over construction and enhancing unplanned land use
Fauna	Crustacean (lobster, crabs, shrimp); Seaweeds; Molluscs	Harvest and export by local people
Coral	Coral; Huge amount of crude oil, plastic and other non-biodegradable waste are discharged in the water adjacent to the island	30,000 coral colonies are removed annually, for curio trade and human interferences.
Fish	Fish (225 species)	Increased number of fishing board
Birds	Birds (85 species)	Decreased number of birds due to poaching and human interferences
Others	Invertebrate (18)	lost and endangered

Source: Molony, L.-A. and National Project Professional Personnel. (2006). St. Martin's Island ECA Conservation Management Plan. Coastal and Wetland Biodiversity Management Project, Department of Environment, Dhaka, Bangladesh; Thompson, P.M. and Islam, M.A. (2010). Environmental Profile of St. Martin's Island Coastal and Wetlands Biodiversity Management Project: A Partnership between Department of Environment Ministry of Environment and Forest and UNDP, Bangladesh, and Bay of Bengal Large Marine Ecosystem project, 2015.

Note: Main resources those have existing on St. Martin's Island.

the Ganges and Brahmaputra Rivers, and locally from the Naf River, is also an important factor influencing the flora and fauna of the island. Thus, coral reef development is inhibited owing to low water salinity, high turbidity and the soft substrates present (Begum and Khondker, 2009).

Floods and heavy runoff during the rainy season introduce high quantities of sediments, nutrients and pesticides from poorly managed agricultural lands to inshore waters, and this has a negative impact on coastal ecosystems (Rahman, 1999; Aziz et al., 2010). Thus herbicides, even in low concentrations, interfere with the basic food chain by damaging corals and other primary producers, benthic or pelagic. Pesticides can selectively destroy zooplankton communities and larval stages of corals, while insecticides accumulate in animal tissues and interfere with physiological processes (Tomascik, 1997; Rahman, 1999; Aziz et al., 2010).

9.5.3 Flora

Since St. Martin's Island was originally a sedimentary continental island connected to the mainland of the Teknaf Peninsula as recently as 6,000–7,000 years ago, the flora of the island is similar to that of the mainland (Molony, 2006; Thompson and Islam, 2010). However, it has been significantly altered owing to human interventions since the island was first settled in the 1880s (Chowdhury et al., 2004). At that time the island most probably was covered with evergreen forest, reportedly with an abundance of teak trees (Tomascik,

1997; Islam et al., 2012a). Subsequently the loss of the original forest, continuing intensive agriculture and the recent increased number of tourists further changed the vegetation and landforms of the island, resulting in the loss of many of the flora and fauna species that once were abundant on the island. St. Martin's Island still has quite diverse vegetation because the remaining native species have been supplemented by a considerable number of cultivated and introduced species (Amin et al., 2014). Recent floral surveys recorded 260 plant species, including 150 herbs, 32 climbers, 25 shrubs and 53 trees, belonging to 58 families (Zaman, 2006). Aquatic vegetation has been less well studied, but recent surveys identified 151 species of benthic and drift algae, including a number of marine red algae (Aziz et al., 2008) and 18 species of bryophytes.

9.5.4 Fauna

9.5.4.1 Crustaceans (Lobster, Crabs, Shrimp)

Over 12 species of crab have been recorded from the island, including commercially important crab species such as the mangrove crab *Scylla olivacea*, which is widely distributed in the Bay of Bengal (Islam and Aziz, 1980; Roshid, 2015). Some of the other crab species recorded are the red egg crab *Atergatis integerrimus*, moon crab *Matuta lunaris*, flower moon crab *Matuta planipes*, crucifix crab *Charybdis feriatus*, flower crab *Portunus pelagicus*, three-spot swimming crab *Portunus sanguinolentus*, giant mud crab *Scylla serrata*, crenate swimming crab *Thalamita crenata*, soldier crab *Dotilla myctiroides*, horned ghost crab *Ocypode ceratophthalma* and horseshoe crab *Carcinoscorpius rotundicanda* (Islam, 2001; Ahmad et al., 2009). Crabs of the genus *Scylla* are strongly associated with mangrove areas throughout the Indian Ocean and form the basis of substantial fishery and aquaculture operations elsewhere, but not at St. Martin's Island (Thompson and Islam, 2010).

9.5.4.2 Seaweeds

A total of only nine species belonging to eight genera in four classes of the phylum Echinodermata have been identified to the species level from the island; these comprise four species of sea urchin, one species of sea star, three species of nudibranchs and one species of sea cucumber (Tomascik, 1997). There are also a number of species of brittle stars present, but they are cryptic and no collection or attempt to identify the species present has been made (Tomascik, 1997). The coral communities of the island are highly significant because there are only a few examples worldwide where coral communities dominate rock reefs as they do on St. Martin's (Chowdhury, 2006). The coral colonies are affected by many factors, both natural and anthropogenically influenced. St. Martin's Island can be classified as low in terms of low species dominance, meaning that no species dominates (Islam, 2001). So far, 66 coral species of 22 genera have been recorded. The genera *Porites*, *Favites*, *Goniopora*, *Cyphastrea* and *Goniastrea* are the most abundant (Alam and Hassan, 1998; Kabir, 2006; IUCN, 2008). In terms of coral coverage, *Porites* is by far the most important genus. In relative terms, almost all other coral genera, perhaps with the exception of *Acropora*, can be viewed as rare (Tomascik, 1997).

9.5.4.3 Molluscs

Marine molluscs are the most abundant large invertebrates found on the island; however, they are declining owing to unregulated harvesting. A total of 187 species of molluscs have been recorded from the island (MoEF, 2001b). Of these, 44 species are gastropods and the rest are bivalves. Numerically, the most abundant among the gastropod molluscs are Littorinidae (periwinkles), Neritidae (nerites), Trochidae (top shells), Cypraeidae (cowries), Muricidae (murex) and Conidae (cone shells). Tomascik (1997) reported the presence of some economically important gastropods, which at that time were abundant, e.g., *Conus striatus*, *Conust extile* and *Conus geographus*, and two economically important gastropods that are heavily depleted worldwide: *Trochus niloticus* and *Turbo marmoratus*.

9.5.4.4 Rocks

The entire terrestrial part of the island was once a rocky habitat, but this has gradually been altered through the removal of rocks and boulders for agriculture (Kaiser and Khadem, 2006). Now much of the land is cultivated and of very limited ecological and biodiversity interest. However, a small area of rocky land remains at Shil Bania, south of Dakhin Para Morong (lake) and west of the Coast Guard base (Amin et al., 2014). The majority of this area is covered by giant boulders similar to those of the intertidal zone, with some lowland pools (Alam and Hassan, 1998; Kabir, 2006; IUCN, 2008). This rocky land is the last remaining habitat for reptile species that are rare on the island, such as garden lizards of *Calotes* spp., two-banded monitor *Varanus salvator*, monocellate or Bengal cobra *Najakaouthia*, and birds of scrubby habitat and native herbs, shrubs and climbers. The rocky ground and shallow pools provide an excellent terrestrial microhabitat, especially during winter (Kaiser and Khadem, 2006). This 100 ha area is the last remaining rocky area on the island and has not yet been cleared, probably because the boulders are large and difficult to remove. However, local people are actively removing these rocks to improve the land for cultivation.

9.5.4.5 Coral

Commercial coral collection began in the 1960s and is now the professional activity of a few families. Of 332 family heads engaged in natural resource exploitation in 2000, almost 20% were coral collectors (Islam, 2001). The main threat to the future viability of coral communities comes from direct extraction of coral colonies. Until recently, *Acropora* was the main group exploited for the curio trade (Chowdhury et al., 2004). Most of the corals collected were sold in Cox's Bazar. The most recent data on the corals of St. Martin's Island are from a 1997 survey, which estimated that 30,000 coral colonies are removed annually, representing 24% of the existing population at that time. Coral removal has continued unabated since that time, so we can reasonably assume that the current status of coral at the site is very poor, and surveys of corals are an urgent priority (Thompson and Islam, 2010).

9.5.4.6 Fish

A total of 234 species of fish have been identified from the waters around the island, 89 of which are coral-associated species and only 16 of which are freshwater fish (Amin et al., 2014). Though coral reefs have not developed, the coral community supports fish fauna characteristic of coral reef environments (Ceballos-Lascurain, 1996). The most abundant coral- or reef-associated herbivores are the damselfish (Pomacentridae), parrotfish (Scaridae) and surgeonfish (Acanthuridae). Important coral- or reef-associated predators found here are Serranidae (groupers), Lutjanidae (snappers) and Lethrinidae (emperors) (Alam and Hassan, 1998; Kabir, 2006; IUCN, 2008). Five species of butterflyfish (Chaetodontidae) have been recorded from the island, as has one species of angelfish *Pomacanthus annularis*

(Pomocanthidae). Croakers (Sciaenidae) are also present. Other notable species that have been landed from deeper water by fishing boats operating from the island include the world's largest fish, the whale shark *Rhincodon typus*, a filter feeder on plankton that is considered to be globally vulnerable to extinction, and hammerhead shark *Sphyrna* sp., with its bizarre shaped head (Thompson and Islam, 2010).

9.5.4.7 Birds and Reptile

St. Martin's Island lies on the boundary or overlap zone of the East Asia–Australasian Flyway and the Central Asian Flyway and provides a stepping stone for a number of migratory wader or shorebird species (Molony, 2006; Thompson and Islam, 2010). A total of 85 species of birds have been confirmed from the island (35 resident species and 50 migratory species) (Islam, 2001). Mid-winter surveys conducted as part of the Asian Waterbird Census in 2008 and 2009 and other recent visits have recorded 43 water bird species; of these, the swift or greater crested tern *Sterna bergii* and lesser crested tern *Sterna bengalensis* are scarce in Bangladesh, and Pacific reef heron *Egretta sacra* is a rare vagrant, but the other species occur in other coastal areas of Bangladesh (Bird Life International, 2008). Species recorded include, for example, the ruddy shelduck *Tadorna ferruginea* (Islam, 2001).

9.6 Integration of Co-Management Strategy

Co-management or collaborative governance can be defined as 'a situation in which two or more social actors negotiate, define and guarantee amongst themselves a fair sharing of the management functions, entitlements and responsibilities for a given territory, area or set of natural resources' (Borrini-Feyerabend et al., 2000). This approach to managing natural resources and the conservation of biodiversity has become widespread internationally. In Bangladesh there is already considerable experience with community-based co-management in freshwater fisheries and wetlands, whereby local user communities have been strengthened and empowered by government with rights and responsibilities for managing specific areas within a framework of shared responsibilities and coordination with government (Thompson et al., 2003; Halder and Thompson, 2007).

9.6.1 Co-Management Framework

Coordination and cooperation among relevant authorities, organizations and agencies, along with other stakeholders from the private sector and local community, is even more important if there is to be biodiversity conservation and sustainable development on St. Martin's Island (Kaiser and Khadem, 2006; Chowdhury et al., 2011). Here the role and reach of government are limited; for example, land is a sensitive issue, and all land is considered to be private property, while government authority over the intertidal zone is unclear.

The key management issue is to achieve a balance between development, tourism in particular, and natural resource conservation, taking into account the environmental costs and benefits (Kaiser and Khadem, 2006; Rabby, 2016). A strong and effective mechanism and forum for cooperation among all stakeholders in development and conservation is needed on the island and at high levels.

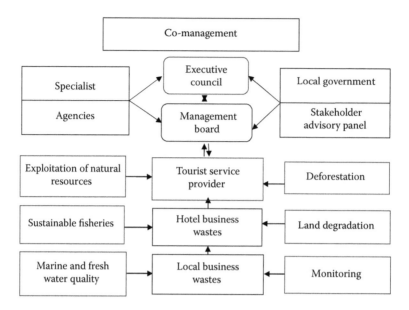

FIGURE 9.3
Co-management frameworks on St. Martin's Island.

This would most likely be in the form of a management committee, under the Environmental Clearance Certificate (ECA) framework/strategy being pursued through the Coastal and Wetland Biodiversity Management Project (CWBMP). Experience elsewhere in Bangladesh highlights the need for visible interventions and a meeting place/community centre for community participation in wetland management and conservation to be effective (Chowdhury et al., 2011). On St. Martin's Island an office building was constructed for the St. Martin's Biodiversity Conservation Project (SMBCP). The SMBCP now lies vacant, but it could easily be converted to serve multiple uses: as a collaborative management centre for regular community meetings and as an interpretive centre for visitors and as a facility for visiting researchers (Figure 9.3) (Islam et al., 2012b).

9.6.2 Zoning

The CWBMP has developed a proposal for zoning St. Martin's Island. The term *zoning* means dividing the island into logical units for management and conservation purposes, with the aim of defining and limiting uses and acceptable development in each zone (Amin et al., 2014). The purpose of developing a zoning system for conservation is to create a balance between biodiversity protection and economic development. The designation of a zoning system must reflect the natural and cultural values of the area as well as the current pattern of land use on the ground and the essential needs of local communities (Hebara, 2008).

A set of conservation zones has been developed by a Protected Areas Specialist through the CWBMP in direct consultation with local people and respecting both current land-use patterns and conservation needs (Chowdhury et al., 2011). The zoning plan was simplified to the maximum possible extent to make it pragmatic for implementation (Kaiser and Khadem, 2006). This zoning plan is in the process of being endorsed by government. It will then become the strategy of the Department of Environment in the management of the island. It is an important management tool developed by the competent authority to protect the core environmental values of the island in a pragmatic programme to encourage alternative

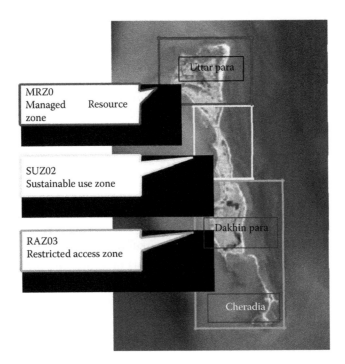

FIGURE 9.4
Zoning of St. Martin's Island, Bangladesh.

livelihoods for the island community (Hebara, 2008; Rabby, 2016). It is expected to provide a 'win-win' strategy between conservation and development because it divides the island into two zones: 50% will be dedicated to conservation and 50% will be designated for sustainable development. Some common conditions will be set covering the whole island (Figure 9.4).

An Environmental Clearance Certificate (ECC) based on an environmental impact assessment will be required for any development project on St. Martin's Island (Poush, 2006b). The approval of the ECC authority is mandatory for issuing an ECC for any project located within the island boundaries. Any activity that could result in the degradation of the natural conditions of any zone of the island is strictly forbidden (Siddiqui et al., 2008). Fishing activity should abide by environmental regulations; for instance, fishing with poisons or chemicals is strictly forbidden.

9.6.3 Managed Resource Zone

This zone covers the northern part of the island south to Golachipa and represents almost 25% of the total area of St. Martin's Island. It will function as a multiple-use zone where sustainable development is encouraged to ensure a sustainable flow of natural products and services for the local community without impinging upon the overall objective of the ECC. The key rules of this zone are as follows:

- It will be open for sustainable economic activities with particular emphasis on organic farming, traditional uses, ecotourism, cultural activities and art-and-crafts production.
- Agricultural production and planting of native species should have priority over exotic species (such as eucalyptus).

9.6.3.1 Sustainable-Use Zone

This zone covers the middle part of St. Martin's Island and represents 25% of the total area of the island. It starts from the southern border of the managed resource zone and continues southwards to end approximately 400 m north of the lagoon of Dakhin Para (MoEF, 2001a). This zone will form a buffer for the sensitive core zone in the southern part of the island and also will serve the community by providing sustainable flow of natural services and products through environmentally friendly projects such as organic farming, handcraft production and ecotourism. It also has a high carrying capacity? (Islam, 2001). The western beach will be protected as a nesting area for marine turtles. The key regulations for this zone are as follows:

- No more expansion of human settlements, development or infrastructure is allowed within this zone unless given special approval by the ECC authorities after due process through an ECC.
- Non-destructive marine sports, such as snorkeling, are allowed in this area.
- Only native plants may be planted here as part of a carefully planned ecosystem restoration programme.
- Alteration of natural features and landscape by the collection of natural components is strictly prohibited.
- Any activity that could change the natural processes of terrestrial or marine life is prohibited.
- Educational and public awareness activities are permitted provided they meet conditions in the management plan.
- Low-profile ecotourism is encouraged in this zone within the carrying capacity set by an integrated plan.

9.6.3.2 Restricted Access Zone

This zone covers the southern part of the island, known as Cheradia, and represents almost 50% of the total area of the island. This zone possesses the ecologically critical features for which the ECC was declared – coral–algal communities and coral-associated biodiversity – and thus deserve strict protection. Within this zone are also patches of mangrove and other natural vegetation including seaweeds (Steele, 1985; Barber and Chavez, 1986; Costanza et al., 1992; Powell et al., 1992. The sandy beaches are important nesting grounds for marine turtles, and the zone also contains spawning and nursery grounds for marine fishes and shrimps. A range of crabs, molluscs and echinoderms such as sea stars and sea anemones also occur here (Mahmud et al., 2013; Amin et al., 2014). There is little development in this area yet, but change threatens these fragile ecosystems. Hence it is expected that a so-called strict nature reserve or category I protected area will be established under the International Union for Conservation of Nature and Natural Resources (IUCN)'s classification, which is dedicated to enabling conservation and scientific research. The management strategy for this zone is to protect biodiversity and associated habitats, and the regulations for this zone are as follows:

- Human settlements and associated infrastructure and practices are not allowed.
- Alteration of natural features and landscape is strictly prohibited, including collection of natural components.

- Cutting of mangrove trees is strictly forbidden.
- Disposal of any pollutants, including solid wastes and oil, onto land, marine or estuarine water is strictly prohibited.
- Any activity that could change the natural process of terrestrial or marine life is prohibited.
- Hunting or disturbing wildlife in any way is strictly prohibited.
- A no-fishing zone extends for 500 m from the shore.

9.7 Recommendations

The following suggested guidelines would help to strengthen understanding and achieve better informed co-management between conservation authorities and the local community based on conservation, sustainable use and ecotourism:

- Regulate tourism businesses to bring greater benefits to local people – for example by requiring that the majority of employees be local people (islanders).
- Facilitate greater local employment in tourism by providing training to prospective local workers and entrepreneurs in both hospitality skills and sustainability/ biodiversity.
- Immediately minimize the negative impacts of tourism on local culture by providing visitors with appropriate literature, briefings, leading by example and corrective actions.
- Give opportunities to the local people to communicate with tourists, explaining their traditional perception of their environment.
- Carry out training programmes for tour operators who work with the local community in order to minimize culture shock and negative impacts.
- Encourage and facilitate financial support, especially microcredit schemes to community groups and individuals, and simplify forms and procedures for obtaining capital to start small appropriate tourism-related enterprises.
- Ensure that independent and financially sustainable community businesses are enabled and not penalized by government.
- Develop, with local co-managers, simple participatory monitoring of trends in natural resource use and visitation.
- Increase investments in capacity building and empowerment for local people and develop educational and environmental awareness campaigns and training programmes among the local community.
- Encourage research on understudied faunal groups and zones of the island.
- Create an enabling environment that gives priority to locals over development activities, provided these are consistent with sustaining biodiversity, for example encourage local employment through policies and incentives.

9.8 Conclusion

St. Martin's Island the only coral island in Bangladesh containing huge resources is now facing several potential risks. An ecosystem-based management model could make a difference in its survival. The island was once home to some unique vegetation and associated wildlife fauna on the southeastern coast of Bangladesh. However, the spectacular wildlife scenario has changed drastically over the last 15–20 years owing to habitat destruction in the name of tourism expansion. The apparently unproductive rocky area is highly degraded now. Local landowners sold almost all the land to outsiders, resulting in massive habitat degradation through, for example, the clearing of forests and displacement of natural rocks. The most damage has been caused by the anchoring of hundreds of boats that use very sharp metal anchors. As the anchors are pulled up, coral is broken into pieces. Fishermen also damage and disturb coral beds to save their nets, which get stuck during fishing. On the other hand, breaking and collecting coral for sale to tourists also endangers shallow water coral species. In addition, garbage strewn on shore or dumped into the sea by hotels and tourists is leading to pollution of the seabed. Many researchers have pointed out the detrimental effects of commercial tourism on St. Martin's and suggested that ecotourism be promoted in the area to preserve the coral reefs. It was further suggested that tourism be prohibited in certain areas of the island which are under severe threat from environmental degradation.

References

Ahmed, M. (1995). An overview on the coral reef ecosystem of Bangladesh. *Bangladesh Journal of Environmental Sciences*, 1, 67–73.

Ahmed, Z.U., Begum, Z.N.T., Hassan, M.A., Khondker, M., Kabir, S.M.H., Ahmad, M., Ahmed, A.T.A., Rahman, A.K.A., and Haque, E.U. (eds). (2009). *Encyclopedia of Flora and Fauna of Bangladesh*. Angiosperms; Dicotyledons. Asiatic Society of Bangladesh, Dhaka. Vol 6–12, 2007–2009.

Ahammed, S.S., Hossain, M.A., Abedin, M.Z., and Khaleque, M.A. (2016). A study of environmental impacts on the coral resources in the vicinity of the Saint Martin Island, Bangladesh. *International Journal of Scientific & Technology Research*, 5(1), 37–39.

Alam, M. (2003). Bangladesh country case study. Paper Presented at the National Adaptation Programme of Action Workshop, 9–11 September 2003, Bhutan.

Alam, M. and Hassan, M. (1998). The origin of beach rock of St. Martin's Island of Bay of Bengal, Bangladesh. *Oriental Geographer*, 42(2), 21–32.

Ali, S. (1975). Notes on a collection of shells from the St. Martin's Island. *Bangladesh Journal of Zoology*, 3, 153–154.

Ali, M.M. and Parvin, R. (2010). Strategic management of tourism sector in Bangladesh to raise gross domestic product: An analysis. *AIUB Business and Economic Paper Series*, July, American International University-Bangladesh, Dhaka, Bangladesh.

Amin, M., Roy, R. and Hasan, M. (2014). Modeling and optimization of decentralized microgrid system for St. Martin's Island in Bangladesh. *IJEIC*, 5(5), 1–12.

Aziz, A., Nurul Islam, A.K.M. and Jahan, A. (2002). Marine algae of St. Martin's Island, Bangladesh. IV. New records of red algae. *Bangladesh J. Bot.*, 31(2), 113–116.

Aziz, A., Nurul Islam, A.K.M. and Jahan, A. (2008). Marine algae of the St. Martin's Island, Bangladesh. VI. New records of species of the genus *Kallymenia* J. Ag. (Rhodophyta), *Bangladesh J. Bot.*, 37(2), 173–178.

Aziz, A., Islam, S. and Chowdhury, A.H. (2010). Marine algae of St. Martin's Island, Bangladesh. IX. New records of green algae (Chlorophyceae). *Bangladesh J. Plant Taxonomy*, 17(2), 193–198.

Aziz, A., Towhidy, S. and Alfasane, A. (2015). Sublittoral seaweed flora of the St. Martin's Island, Bangladesh. *Bangladesh J. Bot.*, 44(2), 223–236.

Banglapedia. (2008). Bay of Bengal. http://banglapedia.org/ht/B_0361.HTM (Downloaded on 27 January 2008).

Barber, R.T. and Chávez, F.P. (1986). Ocean Variability in Relation to Living Resources during the 1982/83 El Ni?o. *Nature*, 319, 279–285. http://dx.doi.org/10.1038/319279a0

Begum, M.A.H. and Khondker, M. (eds.) (2009). *Encyclopedia of Flora and Fauna of Bangladesh*. Vol. 27. Mammals. Asiatic Society of Bangladesh, Dhaka.

Biodiversity and Eco tourism project of St Martin's, Bangladesh. (2010). http://www.stmartinsbd.org

Bird Life International. (2008). BirdLife checklist. http://www.birdlife.org

Borrini-Feyerabend, G., Taghi Farvar, M., Nguinguiri, J.C. and Ndangang, V. (2000). *Co-management of Natural Resources: Organizing, Negotiating and Learning-by-doing*. GTZ and IUCN, Kasparek Verlag, Heidelberg, Germany.

Ceballos-Lascurain, H. (1996). *Tourism, Ecotourism and Protected Areas*. IUCN, Gland.

Chowdhury, A.H. (2006). Report on Marine Algae of St. Martin's Island. Conservation of Bio-Diversity, Marine Park Establishment and Ecotourism Development Project at St. Martin's Island, Department of Environment, Dhaka, Bangladesh.

Chowdhury, Md.S.M. (2006). Report on Invertebrates of St. Martin's Island. Conservation of Bio-Diversity, Marine Park Establishment and Ecotourism Development Project at St. Martin's Island, Department of Environment, Dhaka, Bangladesh.

Chowdhury, S.Q., Fazlul, A.T.M. and Hasan, H.K. (1997). Beachrock in St. Martin's Island, Bangladesh: Implications of sea level changes on beachrock cementation. *Mar. Geodesy.*, 20(1), 89–104. Sea Level Problems of Bangladesh.

Chowdhury S.Q., Hoque A.T.M., Hassan M.K. (1992). Coaatal Geomorphology of St. Martin's Islands, *Orient. Geogf.ap/i.*, 36(2), 30–43.

Chowdhury, M.I., Kamal, M., Alam, M.N., Aftabuddin, S. and Zafar, M. (2004). Environmental radio-activity of the St. Martin's Island of Bangladesh. *Radioprotection*, 39(1), 13–21.

Chowdhury, M.S.N., Hossain, M.S., Mitra, A. and Barua, P. (2011). Environmental functions of the Teknaf Peninsula mangroves of Bangladesh to communicate the values of goods and services. *Mesopot J. Mar. Sci.*, 26(1), 79–97.

Costanza, R., Norton, B. and Haskell, eds. (1992). Ecosystems Health: New Goals for Environmental Management. Inland Press, Washington, DC.

Fattah, Q.A. (1979). Protection of Marine Environment and Related Ecosystems of St. Martin's Island, *Proceedings of National Seminar on Protection of Marine Environment and Related Ecosystems*, Dhaka, 27–29 November, pp. 104–108.

Halder, S. and Thompson, P. (2007). Restoring wetlands through improved governance: community based co-management in Bangladesh. Technical Paper 1. Management of Aquatic Ecosystems through Community Husbandry, Winrock International, Dhaka, Bangladesh.

Hasan, M.M. (2009). Tourism and conservation of biodiversity: a case study of St. Martin's Island, Bangladesh. *Law, Social Justice and Global Dev. J.*, 1(13), 1–13.

Hebara, M.H. (2008). Proposed Zoning Plan for St. Martin Island ECA. Coastal and Wetland Biodiversity Management Project, Department of Environment, Bangladesh.

Hebara, S. (2009). Development of a Functional Zoning System for St. Martin's Island ECA. Coastal and Wetland Biodiversity Management Project, Department of Environment, Bangladesh

Hossain, M.M. (2006). Report on Aquatic Pollution of St. Martin's Island. Conservation of Bio-Diversity, Marine Park Establishment and Ecotourism Development Project at St. Martin's Island, Department of Environment, Dhaka, Bangladesh.

IPCC. (2004). http://www.ipcc.ch/

Islam, M.Z. (1999). Threats to sea turtle population in Bangladesh. *Technical Report*. MarineLife Alliance, 1998, 28.

Islam, M.Z. (2001). St. Martin Pilot Project, National Conservation Strategy (NCS) Implementation Project-1, Final Report, Ministry of Environment and Forest, Government of the Peoples Republic of Bangladesh, 2001, 119 pp.

Islam. (2001). Draft Final Report, St Martin Pilot Project, National Conservation Strategy Implementation Project-1, Ministry of Environment and Forests, Dhaka.

Islam, M.Z. (2002). Marine Turtle Nesting at St. Martin's Island, Bangladesh. *Mar. Turtle Newsl.*, 96, 19–21.

Islam, A.K.M.N. and Aziz, A. (1980). A marine angiosperm from St. Martin's Island, Bangladesh: Halodule uninervis (Forsskal) Ascherson. *Bangladesh J. Bot.*, 9(2), 177–178.

Islam, A.K.M.S., Rahman, M.M., Mondal, M.A.H. and Alam, F. (2012). Hybrid energy system for St. Martin Island, Bangladesh: An optimized model. *Procedia Eng.*, 49, 179–188.

Islam, M.N., Kitazawa, D., Kokuryo, N., Tabeta, S., Honma, T. and Komatsu, N. (2012a). Numerical modeling on transition of dominant algae in Lake Kitaura, Japan. *Ecol. Modell.*, 242, 146–163.

Islam, M.N., Kitazawa, D., Hamill, T. and Park, H.D. (2012b). Modeling mitigation strategies for toxic cyanobacteria blooms in shallow and eutrophic Lake Kasumigaura, Japan. *Mitigation Adaptation Strateg. Global Change*, 18(4), 449–470. Springer Publication.

IUCN. (2008). 2008 IUCN Red List of Threatened Species. www.iucnredlist.org (downloaded on 28 January 2009).

Kabir, S.M. (2006). Report on Soil Status and Rock and Sedimentations of St. Martin's Island. Conservation of Bio-Diversity, Marine Park Establishment and Ecotourism Development Project at St. Martin's Island, Department of Environment, Dhaka, Bangladesh.

Kaiser, M.S. and Khadem, S.K. (2006). Energy efficient system for St Martin's Island of Bangladesh. *Proc. J. Eng. Appl. Sci.*, 1, 93–97.

Khan, M.A.R. (1985). *St. Martins: a vanishing coral island of Bangladesh.* Tigerpaper (FAO/RAPA).

Khan, M.A.R. (1998). New Record of Six Marine Fishes From St. Martin's Coral Island, Bay of Bengal, in Bangladesh. *J. Bombay Nat. Hist. Soc.*, 95, 228–233.

Khan, M.H. (2008). The Wetlands: Our Ecological Heritage. Keynote paper presented at the Asian Wetland Convention of the Society of Wetland Scientists, Taipei, 23–26 October 2008

Mahmud, N. and Haider, S.M.B. (1992). A Preliminary Study of Corals of St. Martin's Island, Bangladesh. Institute of Marine Sciences, University of Chittagong, Bangladesh.

Mahmud, N., Hassan, A. and Rahman, M.S. (2013). Modelling and cost analysis of hybrid energy system for St. Martin Island using HOMER. *2013 International Conference on Informatics, Electronics and Vision (ICIEV)*, Dhaka, 2013, pp. 1–6.

MoEF. (2001a). Survey of Flora, National Conservation Strategy Implementation Project-1, Ministry of Environment and Forests, Dhaka.

MoEF. (2001b). Survey of Fauna, National Conservation Strategy Implementation Project-1, Ministry of Environment and Forests, Dhaka.

Molony, L.A. (2007). Conservation Management Plan of St. Martin's Island ECA. Coastal and Wetland Biodiversity Management Project, Department of Environment, Bangladesh.

Molony, L.-A. and National Project Professional Personnel. (2006). St Martin's Island ECA Conservation Management Plan. Coastal and Wetland Biodiversity Management Project, Department of Environment, Dhaka, Bangladesh.

National Geographic (nd). http://animals.nationalgeographic.com/animals.

Poush. (2006a). Land Use Survey Report. Coastal and Wetland Biodiversity Management Project, Department of Environment, Dhaka, Bangladesh.

Poush. (2006b). Reconnaissance Social Survey, Community Mobilisation for Biodiversity Conservation at Cox's Bazar. Coastal and Wetland Biodiversity Management Project, Department of Environment, Dhaka, Bangladesh.

Powell, R., Passaro, R.J. and Henderson, R.W. (1992). Noteworthy herpetological records from Saint [sic]Maarten, Netherlands Antilles. *Carib. J. Sci.* 28, 234–235.

Rabby, M.F. (2016). Study Report on the Environmental and Socio-economic Condition of Saint Martin's Island in Bangladesh, (a study tour report) to submitted at the Department of Geography and Environment, Jahangirnagar University, Savar, Dhaka, Bangladesh.

Rahman, S.M. (1999). Hydrology and Taxonomy of Some Seaweeds of the St. Martin's Island, Bangladesh, *Doctoral dissertation, M. Sc. Thesis.* Institute of Marine Science, University of Chittagong, Hathazari Upazila, Bangladesh.

Roshid, M. (2015). Modeling Ecosystems Based Management of the Saint Martin's Island in Bangladesh. A graduate research report has been to submitted at the Department of Geography and Environment, Jahangirnagar University, Savar, Dhaka, Bangladesh.

Sarwar, M.G.M. (2005). Impacts of sea level rise on the coastal zone of Bangladesh. *See* http://static.weadapt.org/placemarks/files/225/golam_sarwar.pdf.

Siddiqui, K.U., Islam, M.A., Kabir, S.M.H., Ahmad, M., Ahmed, A.T.A., Rahman, A.K.A., Haque, E.U. et al. (eds.). (2008). *Encyclopedia of Flora and Fauna of Bangladesh. Vol. 26 Birds.* Asiatic Society of Bangladesh, Dhaka.

Smith, B.D., Ahmed, B., Mowgli, R.M. and Strindberg, S. (2008). Species occurrence and distributional ecology of nearshore cetaceans in the Bay of Bengal, Bangladesh, with abundance estimates for Irrawaddy dolphins *Orcaella brevirostris* and finless porpoises *Neophocaenaphocaenoides. J. Cetacean Res. Manage.,* 10(1), 45–58.

Steele, J.E. (1985). Control of Metabolic Processes. In: G.A. Kerkut and L.I. Gilbert (eds.). *Comprehensive Insect Physiology, Biochemistry and Pharmacology.* Pergamon Press, Oxford. 99–145.

Thompson, P.M. and Islam, M.A. (2010). Environmental Profile of Environmental Profile of St. Martin's Island Coastal and Wetlands Biodiversity Management Project: A Partnership between Department of Environment Ministry of Environment and Forest and UNDP, Bangladesh.

Thompson, P.M., Sultana, P. and Islam, N. (2003). Lessons from community based management of floodplain fisheries in Bangladesh. *J. Environ. Manage.,* 69(3), 307–321.

Tomascik, T. (1997). Management Plan for Coral Resources of NarikelJinjira (St. Martin's Island). Final Report. National Conservation Strategy Implementation Project – 1, Dhaka, Bangladesh.

Vinayachandran, P.N. and Kurian, J. (2008). Modeling Indian Ocean circulation: Bay of Bengal fresh plume and Arabian Sea mini warm pool. *Proceedings of the 12th Asian Congress of Fluid Mechanics 18–21 August 2008,* Daejeon, Korea. http://www.afmc.org.cn/12thacfm/IL-1.pdf

Warrick, R.A., Bhuiya, A.H., Mitchell, W.M., Murty, T.S. and Rasheed, K.B.S. (1993). *Sea Level Changes in the Bay of Bengal.* Briefing Document No. 2. BUP, C.E.A.R.S. and UEA Norwich.

Zakaria, M.A. and Abdullah, R. (2002). Urban Drainage System, Malaysia. Dhaka: Don Publishers.

Zaman. (2006). Report on Floral diversity and angiospermic flora (coconut, pandanus and medicinal plants) plantation – in Saint Martin's Island. Conservation of Bio-Diversity, Marine Park Establishment and Ecotourism Development Project at St. Martin's Island, Department of Environment, Dhaka, Bangladesh.

10

Habitat Complexity of Tropical Coastal Ecosystems: An Ecosystem Management Perspective

Chandrashekher U. Rivonker, Vinay P. Padate, Mahabaleshwar R. Hegde and Dinesh T. Velip

CONTENTS

10.1 Introduction..264
    10.1.1 Ecosystem Concept..264
    10.1.2 Coastal Ecosystems..264
    10.1.3 Oceanography Off Central West Coast of India................................264
    10.1.4 Physiographic and Climatological Setting of Goa, India..................265
10.2 Coastal Ecosystems...266
    10.2.1 Coral Reefs..266
        10.2.1.1 Primary Productivity 267
        10.2.1.2 Calcification 267
        10.2.1.3 Nutrition..269
        10.2.1.4 Ecosystem Function...270
    10.2.2 Estuaries..270
        10.2.2.1 Types 271
        10.2.2.2 Salinity Adaptations (Osmosis)...271
        10.2.2.3 Biotic Structure..272
    10.2.3 Mangrove Wetlands 272
        10.2.3.1 Primary Production..273
        10.2.3.2 Heterotrophic Production..273
        10.2.3.3 Life Cycle...273
        10.2.3.4 Adaptations...274
    10.2.4 Sandy Shores 275
        10.2.4.1 Grain Size and Beach Profile 276
        10.2.4.2 Adaptations...276
        10.2.4.3 Biotic Communities 277
    10.2.5 Rock Patches 277
        10.2.5.1 Physical and Biological Factors 277
        10.2.5.2 Zonation Pattern..279
        10.2.5.3 Adaptation...279
        10.2.5.4 Species Diversity 279
10.3 Management of Coastal Resources 282
    10.3.1 Need for Management 282
        10.3.1.1 Natural Factors 282
        10.3.1.2 Anthropogenic Interference 282

10.3.2 Approach..282
    10.3.2.1 Public Awareness and Training.......................................283
    10.3.2.2 Legal Regulations and Implementation 283
Acknowledgements 284
References..284

10.1 Introduction

10.1.1 Ecosystem Concept

An ecosystem was originally defined by Tansley (1935) as 'a biotic community or assemblage and its associated physical environment in a specific place'. This concept finds relevance in budgetary approaches (Odum and Odum, 2000), studies of individual processes (Agren and Bosatta, 1996) and studies of the reciprocal interactions between disparate organisms and their effects in particular sites (Holling, 1995). It can be an analytic or a synthetic concept (Golley, 1993), and it can support an impressive variety of kinds of models (Ulanowicz, 1997). Ecosystem science, starting from the basic definition of Tansley (1935), has expanded to include many kinds of studies (Likens, 1992; Jones and Lawton, 1995; Pickett et al., 1997).

10.1.2 Coastal Ecosystems

Coastal ecosystems, although occupying a narrow fringe of the marine realm, account for the major share of productivity and biological diversity of the oceans. These ecosystems (particularly in the tropical Indo–Western Pacific regions) are highly productive owing to interplay between shallow bathymetry, habitat complexity, intrinsic hydrodynamics, meteorological conditions and terrestrial inputs, as compared to the almost barren (oligotrophic) open ocean environments. In addition, the high productivity supports large-scale fisheries and is a source of livelihood for millions.

10.1.3 Oceanography Off Central West Coast of India

Based on atmospheric forcing, the seasonality along the central west coast of India could be divided into four seasons: Southwest or summer monsoon (June–August), fall inter-monsoon (September–October), Northeast or winter monsoon (November–February) and spring inter-monsoon (March–May). This is also reflected in the spatial–temporal variations in productivity. The inter-monsoon periods are marked by higher sea surface temperature (SST) ($\sim$28°C), shallow mixed layer depths (MLD) (20–30 m) and strong stratification. Moreover, they are additionally marked by low primary production (14–21 m mol C m^{-2} day^{-1}), chlorophyll (ca. 45 m mol C m^{-2}) and undetectable levels of nutrients, especially nitrate in surface waters (Madhupratap et al., 2001). With the onset of the Southwest monsoon, southwesterly winds along the west coast cause movement of the surface waters away from the coast, thereby bringing about upwelling of colder, nutrient-rich and often oxygen-depleted waters from the subsurface by means of Ekman transport. This leads to blooms of mostly diatoms and dinoflagellates and increased productivity. The upwelling off Mangalore commences in June, propagates northwards up to 16°N and prevails until late August/early September, after which it subsides. Simultaneously, a low-level atmospheric jet known as the Findlater jet occurs along the northern part of this region and brings about open ocean upwelling, leading to shallower MLD and higher productivity and biomass (80 mmol C m^{-2} day^{-1}, 170 mmol C m^{-2}). Another unique feature, winter cooling, occurs north of 15°N, where cold, dry continental air blowing into the northern Arabian Sea causes cooling (SST ~24°C), densification of surface waters and sinking. This leads to deep MLDs (>100 m) despite weaker winds, and convective mixing injects nutrients into the surface layers (2–4 µM), generating higher production (40 mmol C m^{-2} day^{-1}) (Madhupratap et al., 2001).

10.1.4 Physiographic and Climatological Setting of Goa, India

Goa (Figure 10.1 shows a geographical map of Goa indicating various coastal ecosystems), with a coastline of about 105 km along NNW–SSE facing the Arabian Sea, supports diversified geological and ecological features and forms an integral part of the central west coast of India (Wagle, 1993). The seabed consists of silty clay up to 50 m and sandy silt from 50 to 100 m (Modassir and Sivadas, 2003), with an average slope of 1.50 m km^{-1} up to a depth of approximately 55 m, and the submarine contours are approximately parallel to the coastline (Veerayya, 1972). The bathymetry is intermittently interrupted by coral reefs (Rodrigues et al., 1998; Hegde and Rivonker, 2013) and submerged rocky patches, which extend from the cliffs and promontories along the adjacent rocky shores (Wagle, 1993). Coral reefs occur mostly in patchy forms around near-shore islands (Grande Island), in the vicinity of submerged rocks across and off estuarine mouths. The overlying waters perennially receive nutrient-rich freshwater influxes from the adjoining estuaries, with the Mandovi–Zuari estuarine complex (between 15°25′N and 15°31′N and between 73°45′E and 73°59′E) in particular being the most prominent, with a catchment area of 1700 km^2 (Qasim, 2003). The two major rivers, the Mandovi and Zuari, are connected to the Arabian Sea by Aguada and Mormugao Bays, respectively. Aguada Bay (4 km long) runs north–south (Shetye et al., 2007); Mormugao Bay (14 km long) runs in an east–west direction from the Western Ghats, and the rocky outcrops extending in a north–south line across the entrance of the bay separate it from the Arabian Sea (Rao and Rao, 1974). The other estuaries traversing the coastal region of Goa are Terekhol (26 km), Chapora (30 km), Sal (10 km), Talpona (9 km) and Galgibag (16 km) (Singh et al., 2004). Estuarine tides are of a semi-diurnal nature (Qasim and Sen Gupta, 1981) and carry seawater a considerable distance upstream.

This region experiences maximum precipitation during the Southwest monsoon, accompanied by stormy weather, while quieter conditions prevail during the rest of the year (Ansari et al., 1995). The intertidal estuarine marshy ecosystem is the transformation of the gentle slope of the near-shore banks of Mandovi and Zuari, which are filled with silt, clay and detritus transported by riverine influx from upper reaches, where mangrove vegetation occurs in high density. The marshy areas extend for a distance of 4 km and are inundated during high tide. The entire mudflats consist of loose muddy soil bordered by mangrove vegetation, making it highly productive for benthos, which support large numbers of economically important species (Ansari et al., 1995; Kulkarni et al., 2003). Moreover, rich mangrove vegetation exists along the banks of the Terekhol, Chapora and Sal estuaries (Singh et al., 2004).

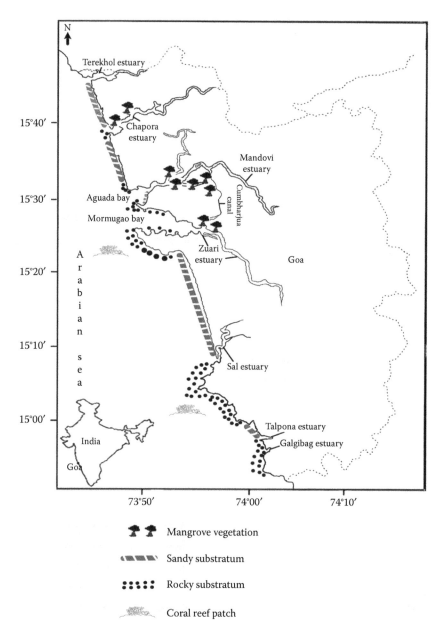

FIGURE 10.1
Map of Goa, central west coast of India, indicating various coastal ecosystems.

10.2 Coastal Ecosystems

10.2.1 Coral Reefs

Coral reefs are the most spectacular environments and are now being treated as endangered ecosystems. Biological diversity in these ecosystems is very high, with myriad varieties and an abundance of plant and animal life. They are ecologically distinct and are highly productive, with gross productivity of about 1800 g C m^{-2} year^{-1}. However, their surrounding waters are nutrient deficient, wherein the primary and secondary productivity is very small. Reef-building corals are anthozoan coelenterates of Class Scleractinia characterized by an ability to produce $CaCO_3$ as an external skeleton. They grow in a cumulative manner, giving rise to massive formations. Corals occur in subtropical and tropical environments in clear, transparent waters with temperatures above 18°C. The richest assemblages of coral reefs are known from Melanesian– South-east Asian areas representing about 700 species from 50 genera. On the other hand, the Caribbean Sea in the Atlantic supports 100 species from about 26 genera. Corals possess endosymbiotic zooxanthellae – photosynthetic organisms that are essential for the calcification process, as evident from the direct correlation between photosynthesis and calcification rate. The growth of corals is confined to a depth of 10 m, and as the depth increases, the growth declines, indicating a direct relationship with photosynthesis. These ecosystems also support a variety of reef fishes, mainly represented by herbivores, damselfish and sturgeon. Recent extensive biological sampling surveys (2005–2011) along Goa's coast have revealed 20 new records of rare reef species (Hegde et al., 2013), including *Caesio cuning* (Padate et al., 2010a) and *Temnopleurus decipiens* (Hegde and Rivonker, 2013), which are the first records of them outside their respective known geographical ranges (Figure 10.2 indicates the geographic distributional ranges of the two aforementioned species and their present occurrence along Goa's coast).

Moreover, the occurrence of reef fishes in bottom trawl hauls taken from the vicinity of submerged ships suggests that these structures also act as artificial reefs and enable the recruitment of reef fish larvae and subsequent inhabitation by adult fishes (Padate et al., 2010a).

10.2.1.1 Primary Productivity

Primary production in coral reef areas is made by all conceivable types of primary producer, which are highly diverse, ranging from zooxanthellae to seagrasses. The gross primary productivity in coral reefs ranges from 2 to 5000 g C m^{-2} year^{-1}. However, respiration by consumers is also of similar magnitude, indicating that most of what is being produced is consumed within the community. Therefore, zooxanthellae form one of the major primary producers of reef, fixing carbon at about 0.9 g C m^{-2} year^{-1}. The other components of primary producers in this ecosystem are mainly comprised of different types of benthic algae, coralline algae, seagrasses and filamentous algae attached to coral rubble.

10.2.1.2 Calcification

Individual coral polyps are measured in millimetres, yet coral reefs extend for hundreds of kilometres. These integrated structures are capable of withstanding cyclones and insidious effects of countless boring and grazing organisms. Reef-building corals require sunlight, warm water and zooxanthellae. However, the role of zooxanthellae has been the subject of great debate. Goreau and Goreau (1959) demonstrated very clearly using ^{45}Ca as tracer that zooxanthellae are essential in the calcification process. They demonstrated that growth rates of most corals (*Acropora prolifera*) in light were considerably higher than in the dark. They also observed that when the corals were held in the dark for several weeks, they extruded their zooxanthellae in the water, although polyps remained apparently healthy and active in food collection. Further, it was reported that the corals treated in this manner had a reduced rate of calcification even in light.

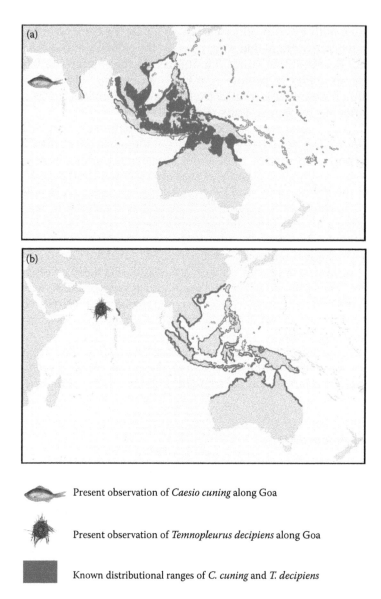

FIGURE 10.2
Range distribution maps of (a) *Caesio cuning* and (b) *Temnopleurus decipiens*, along with their respective photographs, indicating their occurrences outside these ranges.

The relationship of calcification to zooxanthellae was thought to be indirect since large apical polyps of some corals have very few zooxanthellae, yet calcification proceeds most rapidly at the tips of their branches. To gain a better understanding of these aspects, Pearse and Muscatine (1971) carried out ^{45}Ca experiments with staghorn coral (*Acropora cervicornis*) and hypothesized that zooxanthellae provided organic material for the construction of a skeleton matrix. Other studies demonstrating the role of zooxanthellae suggested that they provided glycolate, subsequently converted to glyoxylate, and then combined with urea to form allantoic acid. They proposed that allantoins served as the medium by which Ca and CO_2 were transported to sites of calcification.

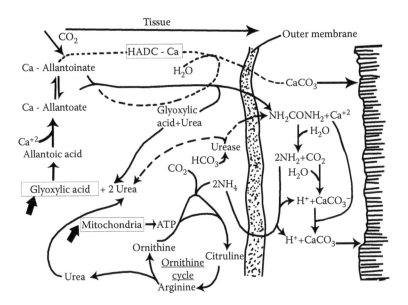

FIGURE 10.3
Diagrammatic representation of calcification mechanism in corals.

The mechanism of calcification (Figure 10.3 illustrates the calcification mechanism in corals) involves glyoxylic acid and mitochondria, both associated with zooxanthellae, as probable sites for the stimulation of the calcification process. Glyoxylic acid combines with urea to form allantoic acid, which under the influence of calcium from seawater forms calcium allantoate. This, in combination with CO_2, forms calcium salts of hydroxyl-acetylene diureide carboxylic acid, which serves as a medium for the deposition of $CaCO_3$ on the skeleton. On the other hand, the nitrogenous (excretory) products released by corals under the influence of CO_2 and urease enter the ornithine cycle, initially forming citruline, then arginine and ornithine. The propagation of the ornithine cycle is mainly mediated through energy supply as ATP by mitochondria. This is then converted to urea, which then combines with glyoxylic acid.

10.2.1.3 Nutrition

Corals meet their nutritional requirements from several sources, namely through extracellular products of zooxanthellae; by capturing particles in the water, especially zooplankton, bacteria, phytoplankton and particulate matter scavenged from the substrate and possibly dissolved organic matter. The mechanism of capture among these species includes raptorial use of tentacles bearing nematocysts, the use of mucus as a trap, ciliary currents to carry the trapped particles to their mouth and the extrusion of mesenterial filaments through the mouth. However, coral reef ecosystems are highly transparent clear waters that support zooplankton and other particles. To assess the role of zooplankton in coral nutrition, Johannes and Tepley (1974), using time-lapse photography, concluded that no more than 10% of a coral reef's energy requirements are derived from this source, suggesting that zooxanthellae are the primary source of carbon for corals. On the other hand, zooplankton communities in coral reefs are mainly represented by a damsel population that hides within the reef during the day and emerges at night. Hence, random collections of zooplankton from such ecosystems may greatly underestimate their true abundance.

From the foregoing discussion, it appears that zooplankton may be a major source of diet of highly voracious carnivorous corals. However, it is very unlikely that they provide more than a fraction of the metabolic needs of the whole community.

10.2.1.4 Ecosystem Function

Coral reefs primarily act as sinks for CO_2, which is accumulated as $CaCO_3$ that forms colossal substrates over millions of years and serve as a substrate for the settlement of myriad varieties of marine flora and fauna. Moreover, these structures act as a buffer for other coastal ecosystems and entire shorelines as they drastically reduce (by up to 95%) the insidious effects of waves and surges. Corals are filter feeders that enhance the water quality of near-shore waters by consuming large quantities of suspended particulate matter. Coral-associated symbiotic bacteria convert molecular nitrogen in water to nitrogenous products that are taken up by reef-associated plants, which are grazed upon by herbivores, thereby allowing N_2 to enter the food web (Figure 10.4 illustrates key components of a coral reef ecosystem). Corals themselves serve as food for invertebrates and fishes, whose calcareous faecal pellets either settle at the bottom or are carried off by waves and currents to adjacent ecosystems, where they contribute to the beach sands. Coral reefs are havens for commercially important fish and invertebrates and thus support large-scale artisanal as well as mechanized fisheries. Moreover, corals are exploited for limestone, jewellery and medicinal purposes. Coral reefs are tourist attractions and thus make it possible to generate valuable foreign exchange.

10.2.2 Estuaries

Pritchard (1967) defined an estuary as 'a semi-enclosed coastal body of water which has a free connection with the open sea and within which seawater is measurably diluted with fresh water derived from land drainage'. Estuaries are one of the most complex and dynamic ecosystems, with high variability leading to stressful environments. These environments act as buffer zones between freshwater and seawater, subjected to the dilution of seawater depending upon the density gradient and basin morphology. Estuaries form an excellent area for nutrient traps and are highly productive. These are enriched by sewage effluents and agricultural runoff.

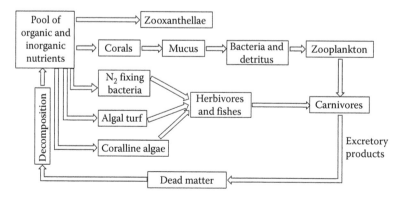

FIGURE 10.4
Schematic diagram of various abiotic and biotic constituents of a coral reef ecosystem.

Biotic communities in estuarine ecosystems are mainly euryhaline and are referred to as transitional/opportunistic species. The number of species living in estuarine environments/ecosystems is significantly less than in marine or freshwater habitats. Further, most truly estuarine organisms are derived from marine regimes owing to fluctuating salinity, which requires a certain kind of physiological specialization to survive; however, most estuaries do not have a sufficiently long geological history to permit the development of completely estuarine fauna.

10.2.2.1 Types

10.2.2.1.1 Mixed Estuaries

In mixed estuaries, the tidal flow is moderate (partially mixed) to strong (well mixed), and salinity generally decreases away from the mouth. Based on tidal amplitude, Goan estuaries exhibit partially mixed behaviour during the non-monsoonal period (Unnikrishnan and Manoj, 2007).

10.2.2.1.2 Stratified Estuaries

Stratified estuaries are formed when riverine flow is stronger than the incoming tide. Goan estuaries exhibit this behaviour at the mouth during the Southwest monsoon season, when freshwater input greatly exceeds tidal flow, resulting in the formation of sand banks at estuarine mouths (Qasim, 2003).

10.2.2.2 Salinity Adaptations (Osmosis)

Estuarine fauna exhibit two different approaches, namely osmoconformation (isoosmotic body fluids) and osmoregulation (use of specialized organs), to overcoming salinity fluctuations in their ambient environment. In lower animal phyla such as Cnidaria, Ctenophora and Echinodermata, the internal osmolarity of body fluids is almost equal to the surrounding seawater. In contrast, other invertebrates, such as annelids, crustaceans and molluscs, possess specialized organs that regulate the internal salt concentration. In shelled molluscs, the mantle cavity is hermetized (waterproofed), while at the same time changes occur in protein and RNA synthesis, and osmotic and volume regulation is carried out by intracellular amino acids and inorganic ions. Moreover, exposure to extreme salinity triggers the closure of shells and burrowing in the substratum (Berger and Kharazova, 1997). In crustaceans, osmoregulation is triggered by neuro-endocrinal secretions that control filtration by specialized organs such as metanephridia. Moreover, their embryos develop in pouches and cyst envelopes, and eggs are exposed to the external environment to become acclimatized to ambient salinity (Charmantier and Charmantier-Daures, 2001). Several sharks employ both osmoconformation and ion regulation through their rectal gland. Marine teleostean fishes possess hypotonic internal fluids; hence ion concentration is regulated through the removal of salt ions by mitochondria-rich chloride cells located in gills. Their kidneys produce concentrated urine, and the gastrointestinal tract compensates for salt loss through continuous uptake by drinking seawater and feeding. Estuarine gobies and mudskippers possess well-vascularized skin that undertakes active ion transport (Marshall and Grosell, 2006). Moreover, they are highly mobile and undertake three types of migration to overcome salinity fluctuations in estuaries, namely offshore release of eggs, larval migrations to inshore nursery areas and regular non-reproductive migrations (shelter and foraging) between freshwater and seawater (Pittman and McAlpine, 2001).

10.2.2.3 Biotic Structure

Most of the estuarine fauna is mainly derived from the seawater regime and categorized as euryhaline, i.e., they tolerate salinity down to 5 parts per thousand (ppt). Truly estuarine species are found in a salinity range of 15–18 ppt and are not found either in marine or freshwater. Estuarine fauna are dominated by polychaetes, oysters, clams, crabs and shrimps. Goan estuaries are known to harbour diverse assemblages of benthic, epibenthic and pelagic fauna (Parulekar et al., 1980; Ansari et al., 1995), including several rare species (Padate et al., 2010b; Hegde et al., 2013). A recent study by Padate (2010) reported 134 taxa of epibenthic and pelagic fauna, including 5 elasmobranchs, 100 teleosts, 2 gastropods, 24 crustaceans, 2 sea snakes and 1 cnidarian from the mouth regions of the Mandovi and Zuari estuaries. Among these, some of the estuarine genera may be limited to the seaward side not by physiological tolerances but by biological interactions such as competition and predation. Therefore, the species composition of any given estuary may not be easily defined. Much of these species have their origin in freshwater and cannot tolerate salinity above 5 ppt, thereby limiting themselves to the upper stretch of the estuary. These include the freshwater puffer *Tetraodon fluviatilis* (Padate et al., 2013a) and the shrimps of the palaemonid genus *Macrobrachium* (Padate, 2010). Transitional opportunistic species include migratory species those crossed over in an estuary either for breeding or nursery purposes (e.g., elasmobranchs, eels, ariid catfishes, snappers, sciaenids) (Padate, 2010). This also includes a few organisms that spend part of their life in the estuary, particularly crabs of the genus *Scylla* (Padate et al., 2013b).

Much of the estuarine productivity in recent times has been strongly influenced by eutrophication, mainly caused by increased use of artificial nitrogen fertilizers and anthropogenic activities associated with shoreline development. Goan estuaries are prone to ore spillage originating from iron ore mining in the hinterland and barge traffic along estuarine channels. These activities generate excessive amounts of human waste, leading to water-quality deterioration, whereas the activities pertaining to shoreline development, particularly removing natural vegetation, lead to the replacement of natural vegetation by weedy algae.

10.2.3 Mangrove Wetlands

Mangroves are morphologically and physiologically diversified, highly evolved plant communities that cover 60%–75% of Earth's coastlines. The unique features of a mangrove community are shallow root systems, thick leaves and aerial biomass. Overall, 53 true species of mangroves have been reported worldwide, and the criteria adopted to distinguish these communities are complete fidelity to a mangrove environment, possession of morphological specializations such as aerial roots and viviparity, ability to establish in a wide range of substrates and synchronization with the local hydrological regime. These communities are distributed in tropical and subtropical ecosystems where the water temperature exceeds 24°C. The best luxuriant growth of these species occurs in the Asian region, particularly along the Indo-Malaysian range. Besides these, Sunderbans in India and part of Bangladesh also form ideal sites for the establishment of these species.

In India, mangroves occupy an area of about 6740 km² representing about 7% of global mangrove cover. The east coast of India, including Sunderbans, Bhitarkanika wildlife sanctuary, Andaman and Nicobar Islands and the estuarine belts of Mahanadi, Godavari, Krishna and Kaveri, account for about 80% of India's total mangrove cover. In contrast, the west coast of India from Kachchh to Kerala contributes only 20%. The limited distribution of mangroves along the west coast is mainly due to the peculiar coastal structure and the nature of estuaries formed by non-perennial rivers (except Narmada and Tapti). The afore-mentioned conditions do not support the establishment of these communities.

Mangrove vegetation occurs from the highest level of spring tide to mean tide level. These are protected areas or sheltered shores formed by reduced wave action. Additionally, mangroves occur in fully saline conditions as well as banks of estuaries. Crabs, molluscs and other invertebrates are permanent inhabitants, whereas migratory shrimp and fishes move in and out with the tides. The upper canopy of mangroves supports a rich diversity of insects, insectivorous fauna and piscivorous birds.

10.2.3.1 Primary Production

High incident solar radiation and the ability of mangroves to take up freshwater from sea-water are the major factors that enable primary production in mangrove areas. The role of phytoplankton biomass in such aquatic ecosystems is believed to be of lesser significance. In mangrove ecosystems, the gross primary productivity is approximately $8 \text{ g C m}^{-2} \text{ day}^{-1}$. However, total respiration is almost of a similar level. It has been estimated that the export of particulate matter from such ecosystems is around $1.1 \text{ g C m}^{-2} \text{ day}^{-1}$. Miller (1972) pre-pared a detailed model of leaf production in terms of solar radiation, associated tem-perature, transpiration, respiration and gross and net photosynthesis in illuminated and shaded zones and found that leaf water stress induced stomatal closure at the top of the canopy on a bright clear sunny day such that maximum production occurred in the mid-dle of the canopy, and they further emphasized that mangroves are of great importance compared to phytoplankton biomass in some nutrient-deficient tropical waters.

10.2.3.2 Heterotrophic Production

In a mangrove ecosystem, about 5% of the mangrove leaf is directly consumed by poten-tial grazers, and the remaining 95% enters the aquatic environment as debris (Figure 10.5 depicts pathways of energy flow through the mangrove trophic web). An estimation of the biochemical composition of a leaf on a tree suggests that it contains 6.1% protein on a dry weight basis, whereas a fallen leaf has 3.1% protein on a dry weight basis. This leaf that has fallen in seawater, over a period of time, is subject to decomposition, and the detrital particles get enriched with organic matter, and after a period of about 12 months, the pro-tein content increases to about 22%. This forms a nutritious food source for the majority of inhabitants mediated through bacterial protoplasm (Heald, 1969). Further, to explain the importance of detritus in fish and invertebrate diets, Heald analyzed the stomach contents of 10,000 individuals and defined a detritus consumer as a species whose digestive tract contains on average 20% of vascular plant detritus by volume on an annual basis. Using these criteria, he reported that about 33% of the species represented mainly by fishes, poly-chaete worms, crabs, chironomid worms and crustaceans such as cumaceans, copepods and mysids were detritus consumers.

10.2.3.3 Life Cycle

Mangroves exhibit a unique reproductive strategy wherein seeds germinate on trees and drop into the water. Seedlings are dispersed by water, and the embryo continuously devel-ops, grows with the parent tree and may even fall after 3–4 years. The floats on the seed act as a buoy, and therefore the seed floats upright when water levels are high due to tidal inundation. As the tide recedes, the upright seed, owing to the decrease in water level,

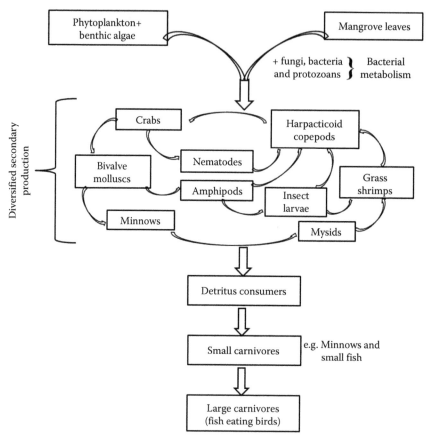

FIGURE 10.5
Schematic diagram of heterotrophic production in a mangrove ecosystem.

touches the bottom and, over a period of time, develops roots and leaves and continues to grow (Figure 10.6 illustrates different stages in the life cycle of a mangrove species).

10.2.3.4 Adaptations

10.2.3.4.1 Salinity

Mangroves are facultative halophytes, which means that salt is not essential for their growth. However, growth in seawater is advantageous owing to a lack of competition. It is pertinent that only a limited number of vascular plants have invested evolutionary energy to adapt to intertidal areas. Although an adaptation, it is noteworthy that these species of vascular plants might be expending some energy to overcome stress in these areas such that they derive an advantage of 'no competition' in such environments. These species exhibit various approaches to and methods of withstanding these environmental anomalies. Mangrove genera such as *Rhizophora* and *Bruguiera* prevent much of the salt from entering body tissue through filtration at the root level. About 90% of the salt is excluded through roots, as evidenced by the high concentration (up to 97%) of salts near the roots. In *Sonneratia*, the plants prevent water loss through the closure of stomatal openings. This is noticed in areas with reduced availability of freshwater coupled with a changing portion

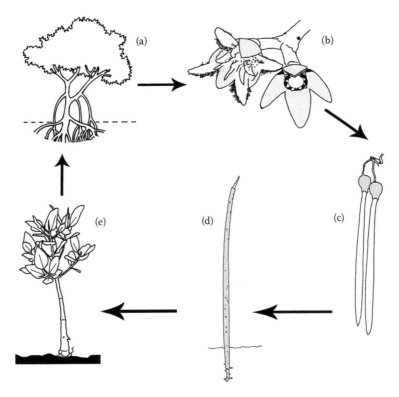

FIGURE 10.6
Life cycle of mangrove species *Rhizophora mucronata*: (a) tree; (b) flowers; (c) fruits and seedlings; (d) germination; (e) young plant.

of the leaf with respect to solar radiation. In *Acanthus*, the plants concentrate salt in the old leaf or bark by forming crystals on the leaf, which are carried and then dropped.

10.2.3.4.2 Water-logged Conditions

In the normal tidal rhythm, the partial pressure of oxygen (O_2) falls when air roots are submerged in the water. The lenticels are hydrophobic under submergence and are effectively closed, preventing entry of air and water. On the other hand, respiration takes up oxygen from air spaces and releases carbon dioxide (CO_2). CO_2 thus released is readily soluble in water and is not effective at replacing the volume of O_2 removed; in this way, the gas pressure within the roots is reduced. This was confirmed by direct measurement of the gas composition in submerged *Avicennia* spp. Consequently, once the roots are covered by the tide, the O_2 level therein falls and the CO_2 level does not increase to compensate for the pressure fall because it is readily soluble in water. For this reason, a negative pressure develops in the air spaces during submergence. Therefore, when the tide recedes and the lenticels are reopened, there is a relatively rapid inhalation of air as gaseous pressure is equalized through lenticels.

10.2.4 Sandy Shores

Sandy shores, intertidal sand flats and protected sand flats are common along the world's coastlines and are better known to human population as sites for recreation. Exposed sandy beaches are devoid of macroscopic life because the environmental conditions are adverse and do not support such species. On the other hand, protected sand flats are densely populated with large numbers of macro-organisms.

10.2.4.1 Grain Size and Beach Profile

Grain size on a sandy beach is affected by wave action and varies with the beach and the season. In tropical environments, the profile of a beach depends on the season. During light wave action particles are fine, whereas heavy wave action leads to coarser particles or gravel beds. The importance of particle size to organism abundance and distribution depends on water retention capacity and suitability for burrowing. The finer particle size through capillary action retains water in the interstitial spaces when the tide recedes, whereas coarser particles allow the water to drain quickly. Protected sand flats are seasonally variable and consist of fine grain sand. In such ecosystems, wind is a function of the size of the water body, and therefore the effect of wind and wave action is less important.

Sand and gravel particles are small and unstable. As a wave strikes the shore, it picks up certain particles, keeps them in suspension and redeposits them elsewhere. Hence, the particles are constantly removed and redeposited, causing wave-induced substrate movement. During high wave action, coarser particles dominate because most of the finer particles are kept in suspension and carried away, allowing coarser particles to settle. Most beaches at the high-water mark support coarser particles, whereas the low-water mark is dominated by finer particles. This suggests that the wave energy on a beach is closer to the high-water mark where waves break and is farther from the low-water mark where the increasing depth reduces the impact of wave action. Hence the substrate itself is in the motion regulated by wave action, so any change in the wave intensity would change not only the grain size but also the profile of the beach.

10.2.4.2 Adaptations

Biological communities occurring in such unstable environments are bestowed with a variety of adaptations categorized under two major types:

1. Burrowing deep in the substrate such that the depth of the sediment is not affected by passing waves (e.g., clam, *Tiavela sturdoran*). Deep-burrowing organisms possess heavy shells that help them to retain in sediments and possess long siphon tubes that enable feeding.
2. Ability to burrow very quickly such that a passing wave does not remove the animal from the substrate (e.g., annelid worms, sand crabs). Among these organisms, the limbs are highly modified to dig wet sand quickly. They burrow quickly before wave motion carries them offshore.

Other adaptations among sandy shore organisms are smooth shells (clams) that reduce resistance to burrowing in sand. A few gastropod shells (e.g., *Trochus* spp.) possess special ridges that make it possible to grip in the sediments, whereas in some onshore echinoderms the spines are reduced. A special adaptation noticed in the sand dollar (*Dendraster excentricus*) is an iron compound that accumulates in a special area of the digestive tract referred to as a 'weight belt' that enables the organism to stay down during wave action.

10.2.4.3 Biotic Communities

Much of the biota present along sandy shores is mainly composed of algae (e.g., *Ulva, Enteromorpha, Sargassum*), which are seasonally abundant and form clumps. Benthic diatoms of different forms are found attached to sand particles, whereas in protected sand flats diverse microflora, benthic diatoms, dinoflagellates and blue green algae form brownish or greyish film on sediments.

Sessile organisms like mussels, barnacles or oysters do not inhabit exposed sandy shores because there is no firm substrate for attachment. Moreover, they do not have access to food sources, as evidenced by the fact that there is little or no primary productivity generated in this ecosystem. Hence, most of the animal communities in this zone depend for food on phytoplankton brought by seawater, organic debris brought by waves and the consumption of other beach animals.

However, intertidal sand flats support a variety of polychaete worms, nematodes, bivalves, molluscs and crustaceans. Primary productivity is confined to microfilms and epiphytes. However, 90%–95% of epiphytes add to the detritus. The distribution of carnivores is relatively smaller and is mainly dominated by suspension or detritus feeders (e.g., crabs, bivalves). On the sandy shores of Goa harbour only a few brachyura exist such as *Ocypoda* and *Dotilla* and Anomura such as *Emerita*.

10.2.5 Rock Patches

The rocky intertidal zone, with its hard substratum, is densely populated by microbes and shows a great diversity of plants and animals, in contrast to sandy and muddy shores, which appear to be barren.

10.2.5.1 Physical and Biological Factors

Zonation along the rocky shore is mainly determined by physical and biological factors.

1. *Physical factors*:
    a. *Tidal exposure/amplitude*: This is exclusively determined by the tide, reflecting the tolerance of organisms to increasing exposure, the resultant desiccation and temperature extremes. However, the disadvantage is that the rise and fall of a tide follows a smooth pattern with no sudden breaks, and therefore it is the tide level that determines the extent of occurrence of an organism near the high-water mark. The critical tide level is defined as the maximum change in exposure time with very short vertical movement. However, this hypothesis does not apply at low tide levels, owing mainly to the diverse topography and variations in exposure time. This suggests that the upper limit for an organism to occur in a particular location is set by physical factors such as temperature and desiccation. The temperature is found to act as a synergist along with desiccation, causing mortality of the biological community. The role of solar radiation, although not very clearly understood, suggests that ultraviolet radiation has deleterious effects on living cells. Light, on the other hand, also regulates the distribution of intertidal algae because the absorption spectra of light vary with depth. However, under natural conditions, a mixture of intertidal algae occur, suggesting that their distribution is mainly regulated by the interaction of other factors and by the physiology of algae.

2. *Biological factors*: The role of biological factors in regulating the biological communities along rocky shores tends to be more complex, and those factors are closely related to each other.

   a. *Competition*: Intertidal rocky shores constitute only 0.003% of the world's total marine ecosystems, suggesting that one of the limited resources in this ecosystem is a restricted area, and therefore, there is increased competition for space, causing densely populated habitats. Along intertidal rocky shores, primarily based on the time of the spawning season, the recruitment of a particular species is likely to occur. Among these, the small barnacle *Chthalamus stellatus* is known to be recruited first and to occupy the highest zone, followed by *Balanus balanoides*, which occupies the mid-intertidal zone. The reason for the disappearance of *C. stellatus* from mid-littoral region is mainly due to competition from *B. balanoides*, which either overgrows or uplifts or crushes the young *Chthalamus* species. In the higher zone, however, *B. balanoides* cannot evict *Chthalamus* outright because it cannot tolerate the high temperatures in the upper zone – a case of partial function of biological competition. A complex case of competition is seen among mussels and several species of barnacles – a dominant competitor for space. Given enough time and freedom from potential predators, these species overgrow and outcompete all other organisms and take over the complete substrate throughout the intertidal zone. However, this is a slow process. The availability of empty space facilitates rapid colonization by barnacles, which persist until mussels enter. Once the mussels enter, they outcompete and destroy barnacles by settling on top of them because they are competitively superior. As long as they remain in the intertidal zone, they control the space. However, because these species are filter feeders/suspension feeders, they survive and grow well in subtidal environments. But these species do not occur in subtidal ecosystems, and this fact is explained by another biological factor.

   b. *Predation*: The dominant species in the intertidal zone are mussels followed by barnacles. Barnacles, despite being competitively inferior to mussels, occur mainly because of predatory sea stars, *Pisaster ochraceus*, which prefer preying upon mussels, thereby preventing them from completely outgrowing barnacles and preventing mussels from occupying the entire space. Simultaneously, barnacles occur as individuals or in clumps in the intertidal zone owing to predation by predatory gastropods (*Nucella* spp.), which regulate the population in the narrow bend of the upper subtidal zone. However, they do not proceed towards higher water mark because their movement is restricted by excessive desiccation. Regulation of *Nucella* (and mussels) is also carried out by *P. ochraceus*, which acts as the top predator, controlling and regulating the structure of the entire community; hence, they are often referred to as the keystone species. Along the upper intertidal zone, no predation is seen mainly because both *P. ochraceus* and *Nucella* feed only during high tide and need long periods of submergence to attack their prey. Moreover, they cannot withstand the high temperatures and desiccation problems in the upper intertidal zone. The main theme of intertidal ecology is that wherever predation is absent, competition is greater.

   c. *Grazing*: The animal communities comprised of gastropods, molluscs, crustaceans, sea urchins and fishes exclusively graze upon intertidal algae and create open space and altered physical conditions those affect algal zonation, species diversity, patchiness and succession through recolonization.

d. *Larval settlement*: The recruitment of invertebrate communities in these ecosystems is a function of algal film and favourable environmental conditions, and therefore the zonation pattern is structured by larval choice. Larval recruitment and settlement in such environments vary with space and time. Under favourable conditions, heavy recruitment leads to increased competition, predation and probably changes in adult population and community structure. Therefore, the same area may harbour communities that not only differ in composition over time but also change in relative numbers of individuals. Hence, these factors play an important role in determining community structure in such ecosystems.

10.2.5.2 Zonation Pattern

Zonation along rocky shores is mainly characterized by colour, morphology or a combination of both. The vertical extent of zonation is mainly regulated by the slope of the shore, tidal range and exposure to wave action. However, zonation reported from different places mainly depends upon region and local topography.

10.2.5.3 Adaptation

Algal populations subject to increased grazing intensity have developed a few defence mechanisms. A few algae have developed an ability to deposit $CaCO_3$ in their tissues (*Halimeda* spp.), whereas others develop woody tissue (*Egregia* spp.) upon maturity that reduces palatability. Further, a few species have also developed an ability to defend themselves by chemical means, accumulating toxic compounds. *Desmarestia* species from Pacific coast accumulates sulfuric acid that is sufficient to erode the $CaCO_3$ teeth of a potential grazer, *Strongylocentrotus franciscanus*. A few other algal species have developed an ability to produce alkaloids, phenolic compounds and halogenated metabolites. A few species have also developed the ability to overcome grazing pressure through natural history. A few species (*Microcystis* spp.) are known to have a low crust with a slow growth rate and high grazing resistance, whereas a few species have an upright frondose form with a high growth rate and low grazing resistance (seagrasses).

Rocky-shore organisms are intermittently exposed and therefore have developed strategies to avoid desiccation. This includes shell closure by shelled molluscs and barnacles during low tide, burrowing by crabs and other invertebrates in the soft substratum and hiding beneath rocks, between crevices or within rock cavities. These strategies also protect these organisms from predation and secondarily from wave action that causes them physical damage by dislodging them from the substratum and even harming essential vital organs.

10.2.5.4 Species Diversity

Rocky shores are among the most complex ecosystems and have diverse floral and faunal assemblages. Thin microalgal films support diverse microbial flora, which serve as food for a wide array of larval invertebrates (Roper et al., 1984). Primary consumers include filter feeding sedentary invertebrates (barnacles) and grazers (sea urchins). Predators such as brachyuran crabs (families Eriphiidae, Pilumnidae, Xanthidae, Grapsidae and Portunidae) and birds constitute the higher tier of the rocky-shore food web, whereas scavengers such as holothurians enable recycling of decomposed matter into the food web (see Table 10.1).

TABLE 10.1

New Records of Coastal Macrofaunal Species along with Their Habitats from Goa, West Coast of India

Sr. No.	Species	Habitat
1	*Charybdis (Charybdis) goaensis* (Padate et al., 2010)[a,e]	Sandy
2	*Callionymus sublaevis* (McCulloch, 1926)[b,f]	Rocky/coral reef
3	*Thysanophrys armata* (Fowler, 1938)[b,g]	Sandy/silt
4	*Hydatina velum* (Gmelin, 1791)[b,f]	Sandy/rocky
5	*Raphidopus indicus* (Henderson, 1893)[b,g]	Soft muddy/sandy
6	*Scylla olivacea* (Herbst, 1796)[g]	Mangrove/muddy
7	*Charybdis (Charybdis) variegata* (Fabricius, 1798)[b,g]	Sandy
8	*Hexapus estuarinus* (Sankarankutty, 1975)[b,h]	Sandy bottom
9	*Caesio cuning* (Bloch, 1791)[c,g]	Coral reef
10	*Stomopneustes variolaris* (Lamarck, 1816)[b,g]	Rocky
11	*Haustellum (Vokesimurex) malabaricus* (Smith, 1894)[b,h]	Sandy bottom
12	*Morula anaxeres* (Kiener, 1835)[d,h]	Sandy/rocky
13	*Trigonostoma scalarifomis* (Lamarck, 1822)[b,h]	Sandy bottom
14	*Cistopus indicus* (Orbigny, 1840)[b,h]	Muddy bottom
15	*Parapenaeopsis maxillipedo* (Alcock, 1905)[b,h]	Muddy/mangrove
16	*Macrobrachium equidens* (Dana, 1852)[b,h]	Sandy/estuary
17	*Thalassina anomala* (Herbst, 1804)[b,h]	Muddy/mangrove
18	*Diogenes miles* (Fabricius, 1787)[b,h]	Muddy bottom
19	*Clibanarius infraspinatus* (Hilgendorf, 1869)[b,h]	Sandy/soft silt
20	*Diogenes alias* (McLaughlin and Holthuis, 2001)[b,h]	Sandy/muddy/coral reef
21	*Albunea symmysta* (Linnaeus, 1758)[b,h]	Sandy bottom
22	*Harpiosquilla raphidea* (Fabricius, 1798)[b,h]	Sandy/soft clay
23	*Philyra globus* (Fabricius, 1775)[b,h]	Sand/silt
24	*Schizophrys aspera* (H. Milne Edwards, 1834)[b,h]	Rocky
25	*Himantura walga* (Müller and Henle, 1841)[b,h]	Sandy bottom
26	*Himantura gerrardi* (Gmelin, 1789)[b,h]	Sandy/rocky/coral reef
27	*Himantura marginata* (Blyth, 1860)[b,h]	Sandy reef
28	*Neotrygon kuhlii* (Müller and Henle, 1841)[b,h]	Sandy/rocky/coral
29	*Aetobatus flagellum* (Bloch and Schneider, 1801)[b,h]	Sandy/rocky
30	*Rhinobatos obtusus* (Müller and Henle, 1841)[b,h]	Sandy/muddy
31	*Ilisha sirishai* (Rao, 1975)[b,h]	Pelagic/euryhaline
32	*Thryssa setirostris* (Broussonet, 1782)[b,h]	Sandy/rocky/seagrass
33	*Thryssa mystax* (Bloch and Schneider, 1801)[b,h]	Sandy/rocky/coral
34	*Hyporhampus limbatus* (Valenciennes, 1847)[b,h]	Muddy/mangrove/estuary
35	*Hippocampus kuda* (Bleeker, 1852)[b,h]	Mangrove/rocky/estuary
36	*Apogon fasciatus* (White, 1870)[b,h]	Sandy/muddy bottom
37	*Archamia bleekeri* (Günther, 1859)[b,h]	Muddy/clay
38	*Scomberoides commersonnianus* (Lacepede, 1801)[b,h]	Pelagic coral
39	*Trachinotus mookalee* (Cuvier, 1832)[b,h]	Coral reef
40	*Heniochus acuminatus* (Linnaeus, 1758)[b,h]	Rocky/coral
41	*Drepane longimana* (Linnaeus, 1758)[b,h]	Sandy/rocky/coral
42	*Platax teira* (Forsskål, 1775)[b,h]	Coral reef
43	*Gerres erythrourus* (Bloch, 1791)[b,h]	Sandy
44	*Gerres longirostris* (Lacepede, 1801)[b,h]	Estuary/mangrove
45	*Gazza minuta* (Bloch, 1795)[b,h]	Sandy/silt
46	*Leiognathus brevirostris* (Valenciennes, 1835)[b,h]	Sandy/rocky

(Continued)

TABLE 10.1 (*Continued*)

New Records of Coastal Macrofaunal Species along with their Habitats from Goa, West Coast of India

Sr. No.	Species	Habitat
47	*Monodactylus argenteus* (Linnaeus, 1758)[b,h]	Estuary/mangrove
48	*Upeneus tragula* (Richardson, 1846)[b,h]	Coral reef
49	*Nemipterus bipunctatus* (Valenciennes, 1830)[b,h]	Rocky/coral reef
50	*Parascolopsis townsendi* (Boulenger, 1901)[b,h]	Sandy/soft bottom
51	*Pempheris molucca* (Cuvier, 1829)[b,h]	Coral reef/rocky
52	*Dendrophysa russelii* (Cuvier, 1829)[b,h]	Rocky
53	*Johnius amblycephalus* (Bleeker, 1855)[b,h]	Muddy/soft bottom
54	*Johnius carutta* (Bloch, 1793)[b,h]	Muddy/estuary
55	*Johnius coitor* (Hamilton, 1822)[b,h]	Muddy/estuary
56	*Epinephelus coioides* (Hamilton, 1822)[b,h]	Coral/sandy/mangrove
57	*Epinephelus erythrurus* (Valenciennes, 1828)[b,h]	Rocky/coral reef
58	*Sparidentex hasta* (Valenciennes, 1830)[b,h]	Rocky/coral reef
59	*Pomadasys furcatus* (Bloch and Schneider, 1801)[b,h]	Soft bottom/coral reef
60	*Plectorhinchus gibbosus* (Lacepede, 1802)[b,h]	Rocky/coral reef
61	*Plectorhinchus schotaf* (Forsskal, 1775)[b,h]	Rocky/coral reef
62	*Yongeichthys criniger* (Valenciennes, 1837)[b,h]	Muddy/coral reef
63	*Parachaeturichthys polynema* (Bleeker, 1853)[b,h]	Muddy/coral reef
64	*Oxyurichthys paulae* (Pezold, 1998)[b,h]	Muddy/coral reef
65	*Callionymus japonicus* (Houttuyn, 1782)[b,h]	Sandy/coral reef
66	*Callionymus sagitta* (Pallas, 1770)[b,h]	Muddy/mangrove/estuary
67	*Eurycephalus carbunculus* (Valenciennes, 1833)[b,h]	Muddy bottom
68	*Cynoglossus dispar* (Day, 1877)[b,h]	Muddy bottom
69	*Synaptura albomaculata* (Kaup, 1858)[b,h]	Muddy bottom
70	*Brachirus orientalis* (Bloch and Schneider, 1801)[b,h]	Coral reef/sandy bottom
71	*Acreichthys hajam* (Bleeker, 1851)[b,h]	Coral reef
72	*Odonus niger* (Rüppell, 1836)[b,h]	Coral reef
73	*Diodon hystrix* (Linnaeus, 1758)[b,h]	Coral reef
74	*Lactoria cornuta* (Linnaeus, 1758)[b,h]	Rocky/coral/seagrass
75	*Triacanthus nieuhofii* (Bleeker, 1852)[b,h]	Sandy bottom
76	*Takifugu oblongus* (Bloch, 1786)[b,h]	Estuary/coral reef
77	*Arothron immaculatus* (Bloch and Schneider, 1801)[b,h]	Seagrass
78	*Tetraodon fluviatilis* fluviatilis (Hamilton, 1822)[b,h]	Estuary/muddy bottom
79	*Arius subrostratus* (Valenciennes, 1840)[b,h]	Muddy bottom
80	*Nemapteryx caelata* (Valenciennes, 1840)[b,h]	Muddy bottom
81	*Netuma bilineata* (Valenciennes, 1840)[b,h]	Muddy bottom
82	*Muraenesox bagio* (Hamilton, 1822)[b,h]	Estuary/mangrove
83	*Gymnothorax pseudothyrsoideus* (Bleeker, 1853)[b,h]	Coral reef/muddy bottom
84	*Trachinocephalus myops* (Forster, 1801)[b,h]	Sandy bottom/coral reef

Source: Hegde, M. R. et al. *Indian Journal of Geo-marine Sciences*, 42, 900–901, 2013. With permission.

[a] Padate et al., 2010a.
[b] Present study.
[c] Padate et al., 2010b.
[d] Kumbhar and Rivonker, 2012.
[e] New to science.
[f] New to Indian waters.
[g] New to west coast of India.
[h] New to Goa coast.

10.3 Management of Coastal Resources

10.3.1 Need for Management

Goa, with a coastline of 105 km, is marked by varied habitats including reefs (Rodrigues et al., 1998), mangroves, mudflats, estuaries (Shetye et al., 2007) and sandy and rocky shores that support rich, diversified demersal assemblages (Rivonker et al., 2008). The importance of the Goa region with respect to demersal fishery potential has been emphasized by Rao and Dorairaj (1968).

10.3.1.1 Natural Factors

During the Southwest monsoon, the upwelled subsurface off the west coast of India brings hypoxic and even anoxic water over the shelf. But this water is generally prevented from surfacing owing to the presence of a thin (<10 m), warm, fresher layer that forms as a result of intense rainfall in the coastal zone. Off Goa, near-bottom oxygen concentrations reach suboxic levels in August, and complete denitrification is followed by the sulphate reduction in September. The quality and quantity of primary production are affected when suboxic waters ascend to the euphotic zone (Naqvi et al., 2009). The emigration of demersal fish from the shallow suboxic zone and frequent episodes of fish mortality, presumably caused by the surfacing of O_2-depleted water, affect fishery resources (Naqvi et al., 2009).

Moreover, enhanced productivity during the Southwest monsoon season often triggers blooms of phytoplankton, including several harmful algal species. The proliferation, development and subsequent senescence of these blooms result in water-quality deterioration with hazardous implications for the coastal and estuarine biota. Moreover, violent storms during this season also threaten some of the fragile coastal ecosystems.

10.3.1.2 Anthropogenic Interference

However, these coastal ecosystems are under constantly increasing threats from various anthropogenic inputs. The near-shore fishing grounds are subjected to intensive exploitation by mechanized and traditional fishing throughout the year, except a brief period during the Southwest monsoon. Discharge of untreated domestic sewage, synthetic fertilizers and industrial effluents, as well as incidental spillage of mineral ore through barge accidents into the bay-estuarine waters, has led to water-quality deterioration, resulting in an increased frequency of harmful algal blooms. Accidental oil spills and sinking of cargo vessels also cause widespread pollution of coastal habitats and interfere with ecosystem functioning. Recent development activities such as the expansion of existing port and jetty facilities, as well as rampant sand mining along estuarine channels, directly threaten the structural integrity of the estuarine habitats.

10.3.2 Approach

Modifications of the natural ecosystems, particularly those of anthropogenic origin, are deleterious for the health and functioning of these fragile ecosystems. This situation requires a holistic approach to acquiring deeper insight into the role of ecological processes that govern demersal species populations and, ultimately, regulate species distributions. This necessitates the creation of baseline data to understand the composition and nature of biological assemblages (animals, plants and microbes) within an ecosystem. Such an enterprise would involve long-term and intensive field surveys incorporating the collection of biological (primary production, species richness and abundance) and physico-chemical parameters (temperature, salinity, pH, dissolved oxygen, nutrient concentration). Analyses of these data and their comparison with existing published literature or global (and regional) databases would make it possible to make inferences about the variations in geochemical and biological processes regulating and, therefore, influencing the health of the concerned ecosystems. There is a need to study coastal processes that overcome the deterioration effect of changing land-use patterns and monsoonal sequences, thereby diluting their deleterious impacts on the diverse faunal assemblages. However, these constitute only the preliminary stages of a holistic strategy in the management of coastal resources and serve as guidelines to spread awareness among both the general public and government agencies to sensitize people to certain complex environmental issues as well impart specialized training in framing and implementing environmentally friendly regulations.

10.3.2.1 Public Awareness and Training

General awareness among the general public, particularly coastal human communities, regarding the complexity of coastal ecosystems and the resultant ecosystem function is vital. This will not only sensitize people to the fragility of the impacted coastal environments but also educate them about the benefits of environmentally friendly development for their means of earning a livelihood.

Awareness programmes conducted by local government departments and non-governmental organizations (NGOs) are the most common and effective means of imparting information (particularly NGOs) to the public from far-flung rural areas. These include thematic training workshops, seminars, group discussions and street exhibitions, as well as through print and television media. Among these, print and television media may be used to educate the general public about the perils of environmental deterioration on their livelihoods. Training workshops, seminars and group discussions take things a step further towards capacity-building measures aimed at the direct involvement of the general public in the conservation and sustainable management of environmental resources.

10.3.2.2 Legal Regulations and Implementation

Numerous legal regulations such as the Wildlife (Protection) Act of 1972, the Forest (Conservation) Act of 1980, the Environmental (Protection) Act of 1986 and the Biological Diversity Act of 2002 have been promulgated to forward the cause of conservation of endangered natural habitats. Furthermore, the Ministry of Environment and Forests, Government of India, issued special Coastal Regulation Zone notifications in 1991 to impose restrictions on changes in land-use patterns, including urbanization, industrialization and mining in ecologically sensitive coastal regions, as well as prohibit any activities endangering coastal ecosystems. To protect larger areas of natural marine and coastal habitats inclusive of human communities, the government has declared three ecologically sensitive areas – the Gulf of Mannar, Sundarbans and the Nicobar Islands – to be Marine Biosphere Reserves that fall under Category V protected areas of the International Union for Conservation of Nature (IUCN). In addition, there are four marine national parks (two on Andaman and Nicobar Islands and one each in Gujarat and West Bengal) that fall under Category II of the IUCN.

Implementation of the aforementioned legislation is achieved through various central and state government agencies such as the Ministry of Environment and Forests, the respective state forest departments and biodiversity boards that strive not only to conserve coastal living resources but also protect indigenous knowledge of these resources. The state biodiversity boards operate at regional and local levels for the regulation and conservation of local resources with the aim of attaining sustainable utilization. The state biodiversity boards are mainly governed by national guidelines imposed from time to time.

Acknowledgements

The authors are grateful to the Ballast Water Management Programme, India, carried out by the National Institute of Oceanography (CSIR), India, for the Directorate General of Shipping, Ministry of Shipping, Government of India. MRH and DTV acknowledge the University Grants Commission and Council of Scientific and Industrial Research (CSIR) respectively for funding their research. Thanks are also due to the registrar, Goa University, for providing the necessary facilities.

References

Agren, G. I. and Bosatta, E. 1996. *Theoretical Ecosystem Ecology: Understanding Element Cycles.* New York: Cambridge University Press.

Ansari, Z. A., Chatterji, A., Ingole, B. S., Sreepada, R. A., Rivonkar, C. U. and Parulekar, A. H. 1995. Community structure and seasonal variation of an inshore demersal fish community at Goa, west coast of India. *Estuarine, Coastal and Shelf Science* 41:593–610.

Berger, V. J. and Kharazova, A. D. 1997. Mechanisms of salinity adaptations in marine molluscs. *Hydrobiologia* 355:115–126.

Charmantier, G. and Charmantier-Daures, M. 2001. Ontogeny of osmoregulation in crustaceans: The embryonic phase. *American Zoologist* 45:1078–1089.

Golley, F. B. 1993. *A History of the Ecosystem Concept in Ecology: More Than the Sum of the Parts.* New Haven/London: Yale University Press.

Goreau, T. F. and Goreau, N. I. 1959. The physiology of skeleton formation in corals. II. Calcium deposition by hermatypic corals under various conditions in the reef. *The Biological Bulletin* 117:239–250.

Heald, E. J. 1969. The production of organic detritus in a South Florida estuary. *PhD Diss.*, University of Miami, Florida.

Hegde, M. R. and Rivonker, C. U. 2013. A new record of *Temnopleurus decipiens* (de Meijere, 1904) (Echinoidea, Temnopleuroida, Temnopleuridae) from Indian waters. *Zoosystema* 35:97–111.

Hegde, M. R., Padate, V. P., Velip, D. T., and Rivonker, C. U. 2013. An updated inventory of new records of coastal macrofauna along Goa, west coast of India. *Indian Journal of Geo-marine Sciences* 42:898–902.

Holling, C. S. 1995. What barriers? What bridges? In: *Barriers and Bridges to the Renewal of Ecosystems and Institutions*, ed. L. H. Gunderson, C. S. Holling, and S. S. Light, 3–34. New York: Columbia University Press.

Johannes, R. E. and Tepley, L. 1974. Examination of feeding of the reef coral *Porites lobata* in situ using time lapse photography. *Proceedings of the Second International Coral Reef Symposium* 2:127–131.

Jones, C. G. and Lawton, J. H. 1995. *Linking Species and Ecosystems.* New York: Chapman and Hall.

Kulkarni, S. S., Rivonker, C. U., and Sangodkar, U. M. X. 2003. Role of meio-benthic assemblages in detritus based food chain from estuarine environment of Goa. *Indian Journal of Fisheries* 50:465–471.

Kumbhar, J. V. and Rivonker, C. U. 2012. A new record of *Morula anaxares* with a description of the radula of three other species from Goa, central West coast of India (Gastropoda: Muricidae). *Turkish Journal of Fisheries and Aquatic Sciences*, 12:189–197. doi: 10.4194/1303-2712-v12_1_22

Likens, G. E. 1992. *Excellence in Ecology, 3: The Ecosystem Approach: Its Use and Abuse.* Oldendorf/Luhe (Germany): Ecology Institute

Madhupratap, M., Nair, K. N. V., Gopalakrishnan, T. C., Haridas, P., Nair, K. K. C., Venugopal, C., and Gauns, M. 2001. Arabian Sea oceanography and fisheries of the west coast of India. *Current Science* 81:355–361.

Marshall, W. S. and Grosell, M. 2006. Ion-transport, osmoregulation, and acid-base balance. In: *The Physiology of Fishes Third Edition*, ed. D. H. Evans and Claiborne, J. B., 179–196. Boca Raton: CRC Press.

Miller, P. C. 1972. Bioclimate, leaf production and primary production in red mangrove canopies in South Florida. *Ecology* 53:22–45.

Modassir, Y. and Sivadas, S. 2003. Conservation and sustainable management of molluscan resources along Goa coast, Central West coast of India. In: *Recent Advances in Environmental Science*, ed. K. G. Hiremath, 399–421. New Delhi: Discovery Publishing House.

Naqvi, S. W. A., Naik, H., Jayakumar, A., Pratihary, A. K., Narvenkar, G., Kurian, S., Agnihotri, R., Shailaja, M. S., and Narvekar, P. V. 2009. Seasonal anoxia over the western Indian continental shelf. In: *Indian Ocean Biogeochemical Processes and Ecological Variability*, ed. J. D. Wigget, R. R. Hood, S. W. A. Naqvi, K. H. Bink, and S. L. Smith, 333–345. Washington, D. C. American Geophysical Union.

Odum, H. T. and Odum, E. C. 2000. *Modeling for all Scales: An Introduction to System Simulation.* San Diego, California: Academic Press.

Padate, V. P. 2010. Biodiversity of demersal fish along the estuarine–shelf regions of Goa. *Ph.D. Thesis*, Goa University.

Padate, V. P., Rivonker, C. U., and Anil, A. C. 2010a. A note on the occurrence of reef inhabiting, red-bellied yellow tail fusilier, *Caesio cuning* from outside known geographical array. *Marine Biodiversity Records* 3:1–6.

Padate, V. P., Rivonker, C. U., Anil, A. C., Sawant, S. S., and Venkat, K. 2010b. A new species of portunid crab of the genus *Charybdis* (De Haan, 1833) (Crustacea: Decapoda: Brachyura) from Goa, India. *Marine Biology Research* 6:579–590.

Padate, V. P., Rivonker, C. U., Anil, A. C., Sawant, S. S., and Venkat, K. 2013a. First record of the freshwater puffer *Tetraodon fluviatilis fluviatilis* from the coastal waters of Goa, west coast of India. *Indian Journal of Geo-marine Sciences* 42:466–469.

Padate, V. P., Rivonker, C. U., and Anil, A. C. 2013b. A new record of *Scylla olivacea* (Decapoda, Brachyura, Portunidae) from Goa, central west coast of India – A comparative diagnosis. *Indian Journal of Geo-marine Sciences* 42:82–89.

Parulekar, A. H., Dhargalkar, V. K., and Singbal, S. Y. S. 1980. Benthic studies in Goa estuaries. Part 3. Annual cycle of macrofaunal distribution, production and trophic relations. *Indian Journal of Marine Sciences* 9:189–200.

Pearse, V. B. and Muscatine, L. 1971. Role of symbiotic algae (zooxanthellae) in coral calcification. *The Biological Bulletin* 141:350–363.

Pickett, S. T. A., Burch, W. R. Jr., Foresman, T. W., Grove, J. M., and Rowntree, R. 1997. A conceptual framework for the study of human ecosystems in urban areas. *Urban Ecosystems* 1:185–199.

Pittman, S. J. and McAlpine, C. A. 2001. Movements of marine fish and decapod crustaceans: Process, Theory and Application. *Advances in Marine Biology* 44:205–294.

Pritchard, D. W. 1967. What is an estuary: physical viewpoint. In: *Estuaries*, ed. G. F. Lauf, 3–5. Washington DC: American Association for the Advancement of Science Publication.

Qasim, S. Z. 2003. *Indian Estuaries*. Mumbai: Allied Publication Pvt. Ltd.

Qasim, S. Z. and Sen Gupta, R. 1981. Environmental characteristics of the Mandovi – Zuari estuarine system in Goa. *Estuarine, Coastal and Shelf Science* 13:557–578.

Rao, K. V. and Dorairaj, K. 1968. Exploratory trawling off Goa by the Government of India fisheries vessels. *Indian Journal of Fisheries* 15:1–13.

Rao, D. G. and Rao, T. C. S. 1974. Textural characteristics of sediments of Mormugao Bay, Central West coast of India. *Indian Journal of Marine Sciences* 3:99–104.

Rivonker, C. U., Padate, V. P., Sawant, S. S., Venkat, K., and Anil, A. C. 2008. Spatio-temporal variations in demersal fish community along the fishing grounds of Goa, West coast of India, *Proceedings of International Conference on Bio-fouling and Ballast Water Management*, 5–7 February, 2008 held at National Institute of Oceanography, Dona Paula, Goa India.

Rodrigues, C. L., Caeiro, S., and Raikar, S. V. 1998. Hermatypic corals of the Goa coast, west coast of India. *Indian Journal of Marine Sciences* 27:480–481.

Roper C. F. E., Sweeney M. J., and Nauen C. E. 1984. FAO Species Catalogue. Vol. 3. Cephalopods of the world. An annotated and illustrated catalogue of species of interest to fisheries. *FAO Fisheries Synopsis No.* 125(3):277p. Rome: FAO.

Shetye, S. R., Shankar, D., Neetu, S., Suprit, K., Michael, G. S., and Chandramohan, P. 2007. The environment that conditions the Mandovi and Zuari estuaries. In: *The Mandovi and Zuari estuaries*, ed. S. R. Shetye, M. Dileep Kuma, and D. Shanka, 3–28. Bangalore: Lotus Printers.

Singh, I. J., Singh, S. K., Kushwaha, S. P. S., Ashutosh, S., and Singh, R. K. 2004. Assessment and Monitoring of estuarine mangrove forests of Goa using satellite remote sensing. *Photonirvachak – Journal of the Indian Society of Remote Sensing* 32:167–174.

Tansley, A. G. 1935. The use and abuse of vegetational concepts and terms. *Ecology* 16:284–307.

Ulanowicz, R. E. 1997. *Ecology, the Ascendant Perspective*. New York: Columbia University Press.

Unnikrishnan, A. S. and Manoj, N. T. 2007. Numerical models. In: *The Mandovi and Zuari estuaries*, ed. S. R. Shetye, M. Dileep Kumar, and D. Shankar, 39–48. Bangalore: Lotus Printers.

Veerayya, M. 1972. Textural characteristics of Calangute beach sediments, Goa Coast. *Indian Journal of Marine Sciences* 1:28–44.

Wagle, B. G. 1993. Geomorphology of Goa and Goa coast – A review. *Giornale di Geologia serie 3* 55:19–24.

11

Biomonitoring Ecosystem Health: Current State of Malaysian Coastal Waters

W. O. Wan Maznah, Khairun Yahya, Anita Talib,
M. S. M. Faradina Merican and S. Shuhaida

CONTENTS

11.1 Introduction to Coastal Ecosystem Health and Dynamics 287
11.2 Environmental Issues of Malaysian Coastal Waters...289
11.3 Biomonitoring to Assess Coastal Environmental Status in Malaysia.......................290
    11.3.1 Basic Concepts in the Use of Indicator Species 290
    11.3.2 Eutrophication...291
    11.3.3 Heavy Metal 292
11.4 Ecological Modelling as a Tool to Characterize Coastal Ecosystem Health 294
    11.4.1 Introduction of Modelling Technology in Biomonitoring...............................294
    11.4.2 Modelling of Macrobenthos Communities along Penang Coastal
             Waters: A Case Study 294
             11.4.2.1 Macrobenthic Communities on Rocky Habitats 294
             11.4.2.2 Macrobenthic Communities on Soft-Bottom Habitats 296
11.5 Conclusion 299
Acknowledgement..300
References...300

11.1 Introduction to Coastal Ecosystem Health and Dynamics

Coastal zone is defined as the geomorphologic area in which the interaction between the marine and land parts occurs in the form of complex ecological and resource systems made up of biotic and abiotic components coexisting and interacting with human communities and relevant socio-economic activities (UNEP, 2008). The coastal areas of a tropical marine environment, which encompass ecosystems from upland, freshwater portions of river basins through an interface estuary, mangrove and coral reefs until they reach the open ocean (Stiling, 1996), are known to be the most productive zones of marine ecosystems (Amirrudin et al., 2004).

The various conditions on the coastline, such as tidal exposure, coastal current, temporal and spatial changes, monsoon, freshwater and nutrient input, represent their unique characteristics. This near-shore area is also a highly natural productive area that provides habitats for a diversity of flora and fauna. Thus, these habitats play an important role in the global economy and biodiversity (Smith and Hollibaugh, 1993). The coastal zone is the border between land and water; therefore, the physical and biological processes of the coastal zones are influenced by the land. The particular combinations of various environmental factors may contribute to the area-specific characteristic of an ecosystem and ultimately determine the specific ecological attributes of the system.

In Malaysia, the term 'coastal zone' has yet to be conclusively defined in a legal context, but an attempt was made by the Economic Planning Unit in 1992 (Kamaruddin, 1998) to define a coastal zone as a strip of land extending 5 km inland from the coastline/shoreline or, if the coastal zone is lined with mangrove/nipah swamps, extending 5 km from the inner boundary of the swamp; in the case of peat swamps, the coastal zone starts from the shoreline to the inner boundary of the swamp, which could be more than 5 km wide. The seaward limit is 200 nm (nautical miles) from the coastline covering the Exclusive Economic Zone (EEC). This definition makes the coastal zone in Malaysia an area 5 km in width in the coastal areas of all states except for Federal Territory, which has no coastline. According to Kamaruddin (1998), the term coastal zone would include all islands within Malaysia's EEC. Malaysia has a long coastline that varies from scenic bays flanked by rocky headlands to shallow mudflats lined with mangrove forests.

Major ecosystems in tropical coastal areas are constituted by mangroves, estuaries, coral reefs, seagrass beds and beaches. Mangroves are tide-influenced wetlands made up of mangrove forests, estuaries and associated habitats along the coastlines (Ong, 1995; Kathiresan and Bingham, 2001; Bagarinao and Primavera, 2005). It serves as a productive ecosystem and supports a high abundance and a diverse variety of marine organisms. Estuaries are partially enclosed water bodies where freshwater from a river merges with the ocean (Telesh and Khlebovich, 2010). Mangroves and estuaries provide a feeding, breeding and nursery ground for juvenile fish and crustaceans (Beck et al., 2001; Kostecki et al., 2010; Chew and Chong, 2011). In addition, mangroves also act as buffer regions against tidal erosion to protect ecosystems (Bagarinao and Primavera, 2005).

Coral reefs and seagrass beds are ecosystems that support marine biodiversity and sustain human activities (Chua et al., 2000). They contribute to fisheries, marine tourism, shoreline protection, nursery and feeding grounds (Arshad et al., 2011). Coral reef and seagrass bed ecosystems serve as an important habitat for sea turtles (Jackson, 1997) and the highly endangered dugong (Fortes, 1990; Mansor et al., 2005). In Malaysia, in the areas of the Strait of Malacca, the South China Sea, fishing grounds, islands, estuaries, coral reefs, mangrove swamps and lagoons, activities such as fishing, wildlife conservation, aquaculture, trade and commerce flourish (Kamaruddin, 1998). These areas are also marked as the most bioproductive areas and provide gateways for world commercial and inland trade, port facilities, industry, oil wells and power plants (Sharifah Mastura, 1992).

The Strait of Malacca is one of the world's busiest straits; it connects the Indian Ocean to the South China Sea and is sandwiched in by the west coast of Peninsular Malaysia and the east coast of the island of Sumatra in Indonesia. The strait is surrounded by heavily urbanized areas with extensive industrial and agricultural development and receives significant freshwater run-off from watersheds (Chua et al., 2000). This makes the strait vulnerable to pollution because it subsequently receives discharge from land-based and sea-based activities (Hii et al., 2006). Such threats by human activities to coastal ecosystems are observable and there is evidence showing that point source (sewage discharge, manufacturing, agro-based industries and animal farms) and non-point source (agricultural activities and surface runoffs) pollution are inevitable and might lead to environmental degradation (Hii et al., 2006).

11.2 Environmental Issues of Malaysian Coastal Waters

Economic development and population growth have degraded the pristine coastal zones around the globe. Modification of biological, chemical and physical conditions within coastal zones as a result of pollution and rapid exploitation of natural resources has increased global concern about the fate of these ecosystems.

In Malaysia, growing human population and coastal development have resulted in widespread land-based and sea-based sources of pollution (Abdullah et al., 1999). Shipping activities and discharge, land reclamation and coastal power plants, domestic and industry discharges, aquaculture and fisheries and tourism have brought with them environmental pollution to coastal regions (Chua et al., 2000; Omar, 2003). Sharifah Mastura (1992) briefly described the issues related to coastal management in Malaysia: (1) indiscriminate cutting of coastal mangrove forests for aquaculture, agriculture and tourism development projects disrupts the ecological functioning of coastal forests; diminished mangrove forests also cause coastal flooding and coastal erosion; (2) fisheries are an important sector in the exploitation of coastal resources, and overfishing, especially along the west coast of Peninsular Malaysia, has reduced the volume of landing, increased trash fish components and reduced species diversity; (3) the stability of the coast is also being impacted by mass tourism development and its associated facilities, which are not adequately planned; (4) pollution of water courses such as in streams, channels and public drains has caused organic and inorganic pollution to enter rivers and coastal waters. River mouths in Malaysia also face siltation problems, which reduces the water depth for fishing access, hindering the development of the fishing industry; (5) sea-level rise and its associated problems, which include coastal flooding and erosion, are an increasing threat. Coastal erosion is one of the main problems in Malaysia that threaten the livelihood and property of coastal communities. Erosion is caused by natural events, such as major storms during high tide, and human activity. Human activities that contribute to the root cause of coastal erosion include the construction of ports/harbours, marinas and bridges, for example. Sand mining activities that remove materials from the coastal system lead to human-induced erosion, which interrupts and reduces sediment transport in the coastal system (Sharifah Mastura, 1992).

Coastal areas are subject to various environmental conditions; every factor plays an important role in the ecosystem. Hence, an environmental change such as a variation in seawater quality may affect a natural ecosystem. Marine organisms in the ecosystem are sensitive to environmental changes (Suthers et al., 2009). Therefore, information about water-quality changes in coastal areas plays a crucial role in detecting natural environmental changes in coastal waters. In addition, it helps to distinguish changes caused by natural processes from those resulting from human activities. Such information could also serve as background data for future monitoring and providing an overview of the existing trends of environmental quality.

Coastal areas are under increasing pressure from human activities and anthropogenic pollution, resulting in reduced water quality (Ghazali, 2006). Such pollution has had a severe impact on the local ecology and economy. According to Yusoff et al. (2000) and Omar (2003), river water quality deteriorated as a result of agricultural and industrial waste discharge into river systems, and this has had negative effects on populations within the river system in the downstream area. One of the prevailing effects of anthropogenic coastal degradation is the blooming of jellyfish, which causes problems in the tourism industry and poses threats to the operation of coastal power plants (Masilamoni et al., 2000; Purcell et al., 2007). Excessive nutrient input caused by human activity is the most likely contributor to jellyfish blooms. Land-based human activities might be the culprit in the presence of other nutrients in coastal waters, which leads to jellyfish blooms. The frequency of jellyfish blooms in waters of the Strait of Malacca could act as an indicator of marine environmental degradation stemming from human activities.

Hence, water quality has become an important parameter in determining pollution and environmental conditions (Maiti, 2004; Suthers et al., 2009). Therefore, proper environmental management is needed to monitor and manage coastal zones. Laws and regulations should be revised and be more strictly enforced in order to reduce the human impact on coastal areas.

11.3 Biomonitoring to Assess Coastal Environmental Status in Malaysia

11.3.1 Basic Concepts in the Use of Indicator Species

Human activities have caused an unprecedented rise in pollutants and contaminants in the coastal regions of the world, with increased levels of environmental degradation with negative effects on the ecology and socio-economic features of coastal ecosystems. The science of assessing the state of coastal environments has progressed from measuring physical parameters and chemical assays to including biological responses from organisms, referred to as bioindicators (Holt and Miller, 2010). In the context of coastal systems, qualitative and quantitative approaches are used in biomonitoring including bioassessments, bioaccumulations, toxicity and behavioural bioassays and community health (Roux et al., 1993).

Bioindicators provide evidence of contaminant exposure generally assessed by their presence or absence, frequency of occurrence and physiological or behavioural variation (Rainbow, 1995; Villares et al., 2002). The method of using bioindicators includes the use and understanding of biological processes, species or communities to assess the quality of the environment and how it changes over time in response to natural and anthropogenic disturbance (Holt and Miller, 2010). These bioindicators reflect the bioavailability of contaminants and disturbances in different coastal regions and periods of time (Conti and Cecchetti, 2001). Since the deployment of bioindicators as effective and rapid tools for environmental monitoring in the 1960s (Conti and Cecchetti, 2001), several criteria have been identified as ensuring suitable biota will be selected for biomonitoring (Conti and Cecchetti, 2003).

Suitable criteria of useful bioindicators should be comprehensible without requiring expert knowledge so that environmental conditions at present and in the future can be understood by the general public and implemented by decision makers (Hakanson and Blenckner, 2008). These criteria involve organisms that (Conti and Cecchetti, 2001; 2003; Hakanson and Blenckner, 2008)

- Are sensitive to contaminants, however to the extent organisms could accumulate these contaminants without causing death to the target organism;
- Are preferably sedentary, or less mobile; this is attributed to their specific distribution, which could be reflective of the contaminants or disturbance at a particular location;

- Contain the same contaminant level in tissues, which correlates with their surrounding environment;
- Can be identified and sampled easily;
- Are widely abundant and distributed across multiple spatial and temporal scales;
- If necessary, could survive laboratory conditions as well as be usable in laboratory bioassays of contaminant absorption.

Considering the multidimensional nature of coastal ecosystems and different responses of each system to various levels of contaminants, no single species can adequately indicate every type of disturbance or stress in all environments (Holt and Miller, 2010). Therefore, it is also important to predict the responses of multi-taxa assemblages to disturbances, because changes in assemblages are detected more readily than changes in the abundance of individual indicator species (Terlizzi et al., 2005).

The use of bioindicators is somewhat limited by the environment in which they live, so not all species can serve as successful bioindicators. Benthic organisms are among the most dominant and widely abundant species in coastal ecosystems (Carmichael et al., 2004; Cardoso et al., 2007) and are often classified as good indicators of the status of an environment. Macrofaunal assemblages are sensitive to changes in the environment due to their taxonomic diversity, wide range of physiological tolerance to stress and the variability of life history strategies that allow benthic macrofauna to respond to a wide range of environmental changes and different types of disturbances (Austen and Widdicombe, 2006; Patricio et al., 2012). These benthic fauna inhabit ecosystems at scales that reflect localized conditions (Chapman, 1998). The usefulness of some species as indicators include the advantage that they provide an early warning of degradation in an ecosystem, indicate only pollutants that are bioavailable and have the most important effect on specific groups of animals, and have cumulative biological responses that can be observed after the event that caused these responses are over (Linton and Warner, 2003). In Malaysia, the use of bioindicators has been incorporated into a number of detailed studies pertaining to coastal ecosystem health. The following subsection therefore focuses on two sources of disturbances, eutrophication and heavy metal pollution, that could potentially become major threats to Malaysian coastal ecosystems and the response from potential bioindicators inhabiting the environment.

11.3.2 Eutrophication

Eutrophication is categorized as a change occurring in a water body as a result of increased inputs of nutrients (Nixon, 1995). The eutrophication process is a major problem caused by increased inputs of inorganic nitrogen and phosphorus into coastal ecosystems (Gray, 1992; Nixon, 1995; Cloern, 2001). The magnitude of change in a community can indicate the severity of the impact of increased nutrients (Gray et al., 2002), while a return to predisturbance conditions could indicate recovery (Kingston, 2002; Kraufvelin et al., 2006). Native species will be replaced by opportunistic species that are more adaptable to conditions of high nutrient content (Duarte, 1995; Cummins et al., 2004). Therefore, species that can withstand highly enriched conditions may increase in abundance when eutrophication occurs, while the opposite may occur with species that benefit from decreased nutrient concentrations (Cardoso et al., 2004, 2010). For example, the major presence and increased abundance of polychaetes were found in eutrophied seagrass beds (Cardell et al., 1999; Cai et al. 2001; Mendez, 2002; Cardoso et al., 2004), while, in contrast, eutrophication negatively influenced the growth rate and biomass of gastropods (Cardoso et al., 2005). Changes in feeding groups have also been observed in response to eutrophication, from populations dominated by mostly suspension feeders to surface and subsurface deposit feeders (Pearson and Rosenberg, 1978; Sarda et al., 1996).

In Malaysia, Chung (2012) observed the occurrence of dense jellyfish populations throughout the year correlated with high nutrient levels in the coastal waters of Penang. Nutrient inputs from the nearby river are suspected to contribute to excessive nutrient levels in the coastal area (Nurul Ruhayu, 2010; Nurul Ruhayu and Khairun, 2013). A high number of jellyfish in Penang coastal waters might imply that jellyfish are sensitive to nutrient fluctuations, and slight changes might be sufficient to cause a notable impact on jellyfish populations (Purcell et al., 2007). Meanwhile, another study by Elham (2013) on jellyfish *Chrysaora* spp. detected that the spawning and peak reproductive season for these species was in July. Therefore, the frequency of jellyfish blooms in waters of the Strait of Malacca could act as an indicator of marine environment degradation as a result of human activities.

Aside from invertebrate population increase, algal blooms are also indicative of high nutrient inputs; in Sabah, for example, algal blooms have been linked to organic matter and inorganic matter effluents from aquaculture farms (Anton et al., 2000). A similar pattern in the abundance of dinoflagellates was also identified in Malacca, where a high abundance of dinoflagellates was found in coastal waters in eutrophied areas exposed to sewage outfalls, with 10 out of 35 species potentially becoming harmful algal blooms (HABs) (Normawaty et al., 2007). Recorded cases of these 10 species in the production of HABs are scattered from Australia throughout Asia, while in Malaysia *Noctiluca scintillans* have caused red tides in Penang and the Straits of Johor in Peninsular Malaysia (see review in Table 2 from Normawaty et al., 2007).

The algae are versatile as they are also perfect indicators of short-term impact thanks to their rapid reproduction rates and very short life cycles. Sampling techniques are simple and economical and create minimal impact on resident biota (Rott, 1991). Standard methods exist for the assessment of functional and non-taxonomic structural characteristics of the community, enhancing their versatility for biomonitoring (Rott, 1991). There are a total of 377 species of marine algae recorded in Malaysia (Phang et al., 2005). High diversity and wide distribution of the group are important characteristics that serve as reliable biological monitors of the Malaysian coastal areas. Algae communities inhabiting these water bodies are able to provide evidence of the environmental history of the water by displaying different sensitivities at the interspecific level and recovery rates to substances in the water or by accumulation of substances in their cells (Lowe and Pan, 1996). These algae are suitable indicators reflecting changes that occur over time in the environment that they inhabit.

11.3.3 Heavy Metal

Over the years, the rise of the manufacturing sector in Malaysia has contributed significantly to heavy metal contamination in the coastal ecosystem (Ong and Kamaruzzaman, 2009). Industries including metal finishing processes (Rahman and Surif, 1993), electronics and semiconductors (Hamid and Sidhu, 1993) are major contributors of waste containing high concentrations of metals, including Cd, Cu, Zn, Ni, Fe, Al, Zn and Mn in Malaysian coastal waters. Other less substantial contributors include pig farming (Ismail and Rosniza, 1997), mining (Yusof et al., 2001) and shipping activities

(Abdullah et al., 1999). These anthropogenic influences tend to elevate the levels of contaminants present in the water to potentially toxic concentrations that could be detrimental to human health (Agusa et al., 2005; Hajeb et al., 2009a,b). The ability of heavy metals to penetrate the food web via direct consumption of water or organism, consequently affecting humans through the food web, spurred interest in the issue among local scientists. Studies on bioindicators to monitor the impact of heavy metal contaminants in Malaysian coastal areas are widespread. However, the majority of these studies are conducted on the west coast of Peninsular Malaysia, where most manufacturers are located.

Nearly 50% of bioindicator studies deal with bivalves (Yap, 1999; Yap et al., 2002a,b, 2003, 2007; Yusof et al., 2004; Ismailm, 2006; Edward et al., 2009; Yunus et al., 2010; Timur, 2014). The green-lipped mussel, *Perna viridis*, is by far the most commonly used in Malaysia. *P. viridis* has been aquacultured since the 1970s (Sivalingam, 1977) and was used as a bioindicator for heavy metals in the coastal waters of Penang in the 1980s (Sivalingam and Bhaskaran, 1980; Sivalingam et al., 1982; Sivalingam, 1985). Its wide geographical distribution, sedentary lifestyle, low species variability and high accessibility with the capacity to accumulate pollutants in its soft tissue favour its selection as a reliable bioindicator of heavy metal (Yap et al., 2004). In their laboratory ecotoxicology studies, Yap et al. (2003) found that heavy metal levels in the soft tissue of their specimens increased 10–30 times following exposure to the substance. This indicates the specimen's capacity to accumulate pollutants. Information on the background levels of heavy metals present in the soft tissue of *P. viridis* are available in the literature, which provides good baseline information for future research (Ismail et al., 2000; Yap et al., 2003). Yap et al. (2003) also found traces of Cd and Pb in the shell material they examined, which correlates with the presence of these metals in the environment. This indicates the possibility of utilizing the shell as a reliable heavy metal indicator. Apart from bivalves, the response of other macrofauna, including arthropods (Ismail et al., 1991; Hajeb et al., 2009a,b; Zulkifli et al., 2012), polychaetes (Gholizadeh et al., 2012) and fish (Kamaruzzaman et al., 2010; Bashir et al. 2013), to heavy metal pollution has also been studied. However, these only constitute nearly 40% of the bioindicator studies conducted in Malaysia. The least explored groups are the macroflora comprised of seaweeds and other algae. Less than 5% of these have been studied.

Although a number of extensive studies have been conducted on the coastal ecosystem of Malaysia, the response of organisms to heavy-metal pollution is still not widely understood. For instance, Hashmi et al. (2002) concluded that the use of tiger prawn, *Penaeus monodon*, as bioindicators of heavy-metal concentrations in an aquaculture pond did not yield accurate results because high concentrations of heavy metals were not detected readily in muscle tissue, which could be attributed to the considerable interspecific variability in the rate of uptake of individual metals. In this context, if the selection of bioindicators were made using only single species, the actual condition regarding heavy-metal contamination might not be ascertained, which could lead to harmful effects on human health (Jarup, 2003). To determine the amount of heavy metal contaminants in the environment, Cuong et al. (2005) collected mangrove macrofauna comprising two taxonomic groups, molluscs and crabs in mangroves, located in the Strait of Johor. The authors found heavy metals in the environment which were identified in tissues of both molluscs and crabs; however, the highest concentrations were found in gastropods, *Thais gradata* and *Telescopium telescopium*, highlighting the importance of these species as indicator species among the other fauna inhabiting mangroves and potentially other coastal ecosystems such as seagrasses, mudflats and coral reefs.

11.4 Ecological Modelling as a Tool to Characterize Coastal Ecosystem Health

11.4.1 Introduction of Modelling Technology in Biomonitoring

Machine learning is concerned with the design and development of algorithms and techniques that allow computers to 'learn'. The major focus of machine learning research is to extract information from data automatically using computational and statistical methods. The most common task in modelling through machine learning techniques is to find relations of a set of input parameters to one or more output parameters. This kind of analysis is called supervised because the goal of the analysis is given by the user. In case no explicit goal parameter is available, the analysis is called unsupervised. These are data-driven modelling techniques, where the algorithms will try to find patterns in data sets.

We shall consider the use of unsupervised modelling techniques to illustrate the machine learning techniques such as neural networks respectively to model the complex non-linear relationships inherent in environmental data sets using two case studies of macrobenthos on rocky shores and soft-bottom habitats. A non-supervised artificial neural network (ANN) will be used in ecological modelling where a self-organizing map (SOM) is applied to order the data by similarity and to cluster the same input variables into groups of similar input (Gevrey et al., 2006). A neural network was used to classify the species into their respective zonations and subsequently help to determine which species belong to which intertidal or other coastal zonations. A non-supervised neural network application developed by Vesanto (1999) distinguishes between similar and dissimilar features of normalized input data, which can be mapped as clustered inputs, where the features are expressed by Euclidian distances, which are calculated between the inputs and can be visualized as a unified distance matrix (U-matrix) and as a partitioned map (K-means).

Regional or national studies of coastal water ecosystems provide large volumes of site-specific data, which may carry some valuable information that can be used to derive certain spatial patterns of biological communities on different scales (i.e., from a local to a regional area). This has been a key issue in defining community patterns on a large scale for sustainable ecosystem management. There have been many attempts to classify coastal waters based on the distribution patterns of aquatic organisms and pollution (Bakalem et al., 2009; Borja et al., 2004).

11.4.2 Modelling of Macrobenthos Communities along Penang Coastal Waters: A Case Study

11.4.2.1 Macrobenthic Communities on Rocky Habitats

Benthic communities are used extensively as indicators of environmental quality status because many studies have demonstrated that benthic organisms respond predictably to many kinds of natural stress and the possible effects of anthropogenic stress.

Barnacles are sessile crustaceans that cause biofouling (Southward, 1987), unwanted and undesirable attachment and growth of organisms that readily occur on clean, submerged, unprotected artificial surfaces. Biofouling begins with the adhesion of microfoulers, such as bacteria, diatoms and microalgae, to a surface, which causes a biofilm to form, followed by the settlement of soft and hard macrofoulers composed of barnacles, mussels and tubeworms (Southward, 1987). Their distribution is affected by a few environmental factors, which include biotic interactions like predation, seasonal changes in environmental conditions, i.e., air temperature, and spatially variable characteristics, i.e., rock topography (Blythe, 2008). In this case study, barnacles are monitored during the period 2012–2013 at various sampling stations and the data are modelled using SOM to study their distribution around various coastal areas of Penang in connection with anthropogenic effects.

Figure 11.1 shows the classification of intertidal zonation as mapped on a U-matrix using SOM from MATLAB 6.5 software. The colouring in the U-matrix map indicates the relative distances between neighbouring data of the input data space, which ranged from light to dark colours. The light colours represent neighbouring data with the smallest distances belonging to a cluster, while the darker colours visualize the greatest distance between neighbouring data and signify borders between clusters. The K-means algorithm partitions the input data space into three clusters based on the U-matrix.

Data collected from Penang National Park by Tay (2008) and Omar et al. (2012) showed the dominance of *Chthamalus* spp. and *Balanus* spp. According to Stephenson and Stephenson (1972), the distribution pattern of intertidal barnacles is vertically zoned. Their vertical zonation is influenced by different factors. The results of the SOM analysis (Figure 11.1) showed that *B. amphitrite* was the most dominant species in the lower zone compared with *Chthamalus* spp. According to Newman and Stanley (1981), competitive exclusion is the principal factor leading to the decline of chthamaloids below the upper intertidal zone owing to competition with balanoids. Balanoids are competitively superior to chthamaloids in competition for space, and this restricted the distribution of chthamaloids from accommodating or reaching lower intertidal zone (Stanley and Newman, 1980).

Our results, however, revealed a higher abundance of *Balanus* spp. on the eastern side of Penang. This could be partly due to the rapid development in Penang that degraded its natural environment, especially in the eastern part of the island. *Chthamalus* is observed

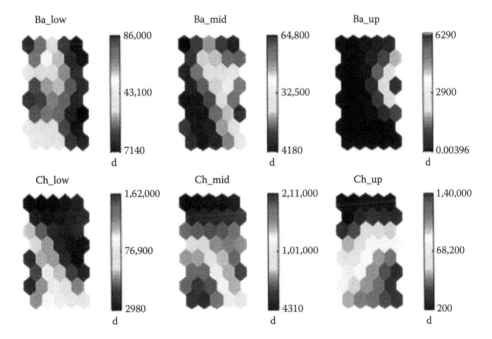

FIGURE 11.1
Distribution of intertidal zonations (lower, middle and upper zones) of *Balanus* and *Chthamalus* across the eastern, northern and western coastal regions of Penang Island based on U-Matrix and K-means clustering using SOM analysis.

from this study to be more abundant in the less developed areas of Penang, i.e., on the northern and western areas of Penang. The possible sources of pollution in Penang coastal regions are from the direct discharge of sewage into the sea (Sivalingam, 1984) and pollutants from river outflow. Apart from sewage, coastal reclamation and aquaculture activities in the state have also contributed to the destruction of natural coastal ecosystems. This modelling case study has highlighted the possibility that human activities may also have an impact on the distribution and abundance of barnacle communities.

11.4.2.2 Macrobenthic Communities on Soft-Bottom Habitats

At present, studies on the ecological distribution of macrobenthos inhabiting soft-bottom habitats are scarce. In this case study, we monitored the composition of the shallow soft-bottom communities along the Penang National Park coastal waters (from June 2010 to April 2011 at 24 sampling stations). Molluscs were the most abundant family, comprising more than 60% of all individuals. The macrobenthic communities were assigned to ecological groups based on the AZTI Marine Biotic Index (AMBI) and Shannon's Diversity Index (Table 11.1).

TABLE 11.1

Most Abundant Macrobenthic Organisms at Each Station and Their Assignment to Ecological Groups Based on AMBI and Shannon's Diversity Index.

Site	Station	Ecological Group (AMBI)	Biotic Index	Biotic Coefficient	Shannon's Diversity (H')	Ecological Status (H')	Dominating Ecological Group	Benthic Community Health
T.B.	S1	1.23	2	0.107	1.93	Poor	III	Unbalanced
	S2	0.45	1	0.076	1.86	Poor		Impoverished
	S3	0.07	0	0.06	2	Moderate		Normal
	S4	0.39	1	0.33	1.84	Poor	I	Impoverished
	S5	0.1	0	0.096	2.48	Moderate		Normal
	S6	0.26	1	0.27	1.99	Poor		Impoverished
T.A.	S1	0.31	1	0.05	2.06	Moderate		Impoverished
	S2	0.24	1	0.155	2.13	Moderate		Impoverished
	S3	0.17	0	0.109	2.27	Moderate	I	Normal
	S4	0.64	1	0.5	2.02	Moderate		Impoverished
	S5	0.15	0	0.13	1.93	Poor		Normal
	S6	0.41	1	0.203	2.4	Moderate		Impoverished
T.K.	S1	0.14	0	0.33	3.08	Good		Normal
	S2	0.12	0	0.26	2.66	Moderate		Normal
	S3	0.25	1	0.21	2.64	Moderate	I	Impoverished
	S4	0.07	0	0.059	2.25	Moderate		Normal
	S5	0.13	0	0.12	1.81	Poor		Normal
	S6	0.24	1	0.2	2.01	Moderate		Impoverished
P.A.	S1	0.65	1	0.29	2.74	Moderate		Impoverished
	S2	0.6	1	0.47	2.43	Moderate		Impoverished
	S3	0.53	1	0.39	2.59	Moderate		Impoverished
	S4	0.38	1	0.026	1.97	Poor	I	Impoverished
	S5	0.18	0	0.2	2.14	Moderate		Normal
	S6	0.2	0	0.17	2.22	Moderate		Normal

Note: T.B. = Teluk Bahang; T.A.= Teluk Aling; T.K. = Teluk Ketampi; P.A. = Pantai Acheh.

The outcome of this grouping is the dendrogram of the cluster analysis with the Ward's linkage method, as shown in Figure 11.2.

The non-supervised ANN algorithm separated the ecological quality status of the coastal waters of Penang National Park into clusters visualized by the U-matrix and K-means and macrobenthic abundances, as shown in Figure 11.3. SOM units were classified into two main clusters based on the dendrogram of the cluster analysis using the Ward's linkage method (Figure 11.2) and, as shown in Figure 11.3, the clusters were further divided into three subclusters at different levels of the Euclidean distance. The clusters accordingly reflected ecological status differences among the study locations. It recognized the top cluster of the U-matrix as being related to unpolluted status and the bottom-left cluster related to slightly polluted status. The plane quality was reasonable, with a quantization error of 0.106 and a topographic error of 0.003.

The SOM accordingly classified the sample sites into three groups based on ecological conditions. The classification showed that spatial variation was the main factor in characterizing macrobenthic communities collected on north-western Penang Island. The sampling points nearest to the coastal area were grouped in cluster 2, showing high abundance (Figure 11.3). Samples from the disturbed areas due to rehabilitation and constructions were grouped together in cluster 3, showing extremely low abundance. This is commonly observed in physically disturbed areas with poor colonization by aquatic organisms. Cluster 1 is composed of other sampling stations in the coastal waters of Penang National Park. Samples from deeper habitats were separated from samples from shallow habitats.

Clear spatial changes in the macrobenthic assemblage composition related to the hydrodynamics and morphodynamics were found along the studied depth gradient (Dolbeth et al., 2007). Macrobenthic communities differ between shallow depths (stations near the coastal), which are subjected to hydrodynamics acting on the seabed, and deeper areas, which experience no significant wave energy effect at the seabed (till station 1200 m

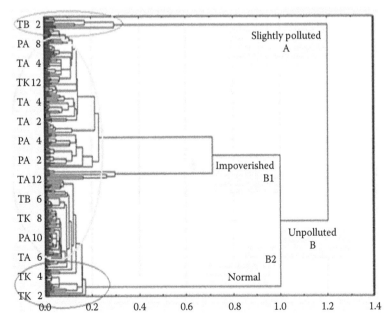

FIGURE 11.2
Ordination and clustering of ecological quality status in Penang National Park as defined in Table 11.1.

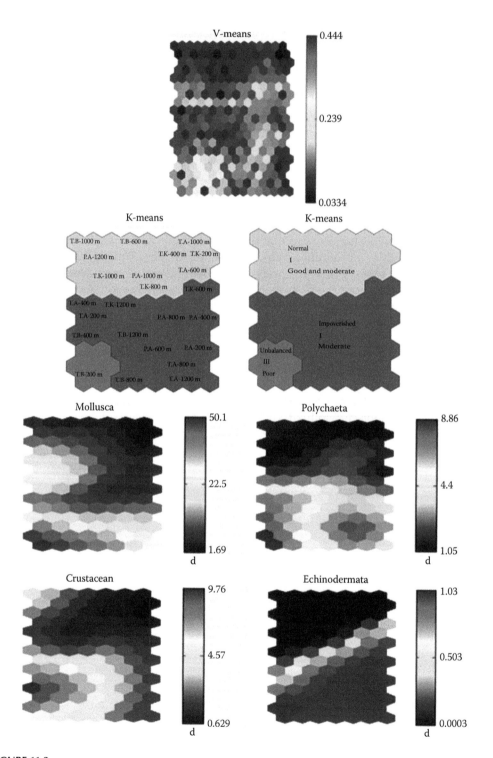

FIGURE 11.3
Ordination and clustering of Mollusca, Polychaeta, Crustacean and Echinodermata patterns according to different sampling stations in Penang National Park by means of non-supervised ANN.

seaward) (Dolbeth et al., 2007). The dominant feeding guilds of the macrobenthic assemblage at the shallowest stations (stations at 200 to 400 m) were mainly predator and carnivore. These feeding guilds were made up of several families, mainly polychaetes, molluscs and amphipods of crustaceans. Most of these polychaetes were predators and were distributed in coarser sand at some of the sampling sites, which was also observed by Gaston (1987). Their presence, coexisting with other functional feeding guilds, is assumed to play an important and structuring role in the community (Wieking and Kroncke, 2005; Levinton and Kelaher, 2004). Apart from polychaetes, molluscs were also dominant in the benthic communities over the coastal water of Penang National Park of Malaysia. These macrobenthic communities, particularly the dominant polychaetes and molluscs in the shallow parts of the coastal waters of north western Penang Island, could be affected by the sewage discharged by the establishments and habitations along the coast. Solid and liquid waste (including pig wastes) from local communities was being illegally discharged from the rivers to the sea.

These case studies show how ecological modelling techniques can be applied as tools to understand our natural coastal ecosystems and processes and to characterize our coastal ecosystem health. Statistical and neural network models were used to classify soft-bottom habitat macrobenthos based on their spatial distribution and neural network model of SOM applied to study the spatial distribution of rocky-shore barnacle species in their respective zonations across Penang. Our results showed the possible effects of pollution on the ecology of these macrobenthic communities and our study may serve to integrate the recent pollution history of different kinds of pollutants in the sediment. The potential of macrobenthic communities from both soft-bottom and rocky shores as good indicators of the health of our ecological or environmental quality status are highlighted through both modelling case studies.

Modelling is an important tool in ecological studies because it can improve our understanding of how species are distributed in relation to land use and how species could serve as indicators of the overall coastal ecosystem health.

11.5 Conclusion

Coastal areas in Malaysia are under increasing pressure from human activities and anthropogenic pollution. Since no one group of organisms is always best suited for detecting and assessing the environmental disturbance associated with human activities, indicators derived from several groups of organisms should be included in water-quality-monitoring programmes to provide a good signal of ecosystem change. The use of various taxonomic groups, however, should not be narrowed to a specific range of organisms one should consider when selecting bioindicators of an environmental condition or anthropogenic disturbance; rather, it shows the potential of considering different groups of taxa in coastal ecosystems as effective ecological indicators. This means more emphasis should be placed on understanding not only the composition of species but also the responses from these species in the face of environmental change brought about by anthropogenic disturbance. In our search for indicators of coastal ecosystem health, biomonitoring and data-based modelling that is knowledge-driven is needed for the transition of knowledge from ecological research to ecosystem management.

Acknowledgement

We wish to thank the publisher for granting us the opportunity to contribute to the knowledge base on problems and issues associated with the coastlines of Malaysia and the biomonitoring that can be done to achieve sustainability and prevent disastrous consequences for precious natural resources. We would also like to express our thanks to the Dean of School of Biological Sciences and the Director of the Centre for Marine and Coastal Studies at Universiti Sains Malaysia for their continued support in fostering the intellectual atmosphere needed to produce scientific literature for the benefit of academia and the general public. We would also like to give special recognition to university faculty, staff and students for their assistance in this research and publication. Last but not least, we would like to acknowledge various funding bodies, including USM Grants 1001/PBIOLOGI/815052 and 203/PPANTAI/6711215, for the data that made this book chapter possible.

References

Abdullah, A. R., Tahir, N. M., Loong, T. S., Hoque, T. M., Sulaiman, A. H. 1999. The GEF/UNDP/IMO Malacca straits demonstration project: sources of pollution. *Marine Pollution Bulletin* 39: 229–233.

Agusa, T., Kunito, T., Yasunaga, G., Iwata, H., Subramanian, A., Ismail, A., Tanabe, S. 2005. Concentrations of trace elements in marine fish and its risk assessment in Malaysia. *Marine Pollution Bulletin* 51: 896–911.

Amirrudin, A., Yusri, Y., Siti Azizah, M., Ahyaudin, A. 2004. The fish fauna of Pantai Acheh Forest Reserve in Penang, Malaysia. Species composition, local distribution and inter-site relationship. *Journal of Bioscience* Universiti Sains Malaysia Press, 15(1): 49–61.

Anton, A., Alexander, J., Ogata, T., Fukuyo, Y. 2000. The relationship between PSP toxin produced by Pyrodinium bahamensevar compressum in shell fish and the plankton in Sabah, Malaysia. *Proceeding of the 9th International Conference on Toxic Phytoplankton*, Tasmania.

Arshad, A., Ara, R., Amin, S., Effendi, M., Zaidi, C. C., Mazlan, A. G. 2011. Influence of environmental parameters on shrimp post-larvae in the Sungai Pulai seagrass beds of Johor Strait, Peninsular Malaysia. *Scientific Research and Essays* 6: 5501–5506.

Austen, M. C., Widdicombe, S. 2006. Comparison of the response of meio- and macrobenthos to disturbance and organic enrichment. *Journal of Experimental Marine Biology and Ecology* 330: 96–104.

Bagarinao, T. U., Primavera, J. H. 2005. Code of practice for sustainable use of mangrove ecosystems for aquaculture in Southeast Asia. SEAFDEC Aquaculture Department, Tigbauan, Iloilo, Philippines.

Bakalem, A., Ruellet, T., Dauvin J. C. 2009. Benthic indices and ecological quality of shallow Algeria fine sand community. *Ecological Indicators* 9: 395–408.

Bashir, F. H., Othman, M. S., Mazlan, A. G., Rahim, S. M., Simon, K. D. 2013. Heavy metal concentrations in fishes from the coastal waters of Kapar and Mersing, Malaysia. *Turkish Journal of Fisheries and Aquatic Sciences* 13: 375–382.

Beck, M. W., Heck Jr, K. L., Able, K. W., Childers, D. L., Eggleston, D. B., Gillanders, B. M., Halpern, B., Hays, C. G., Hoshino, K., Minello, T. J. 2001. The identification, conservation, and management of estuarine and marine nurseries for fish and invertebrates. *BioScience* 51: 633–641.

Blythe, J. N. 2008. Recruitment of the intertidal barnacle Semibalanus balanoides; metamorphosis and survival from daily to seasonal timescales. *Doctoral Dissertation*. Degree of Doctor of Philosophy in Biological Oceanography.

Borja, A., Franco, J., Muxika, I. 2004. The biotic indices and the water framework directive: the required consensus in the new benthic monitoring tools (correspondence). *Marine Pollution Bulletin* 48: 405–408.

Cai, L. Z., Lin, J. D., Li, H. M. 2001. Macroinfauna communities in an organic-rich mudflat at Shenzhen and Hong Kong, China. *Bulletin of Marine Science* 69: 1129–1138.

Cardell, M. J., Sarda, R., Romero, J. 1999. Spatial changes in sublittoral soft-bottom polychaete assemblages due to river inputs and sewage discharges. *Acta Oecologica International Journal of Ecology* 20: 343–351.

Cardoso, P. G., Bankovic, M., Raffaelli, D., Pardal, M. A. 2007. Polychaete assemblages as indicators of habitat recovery in a temperate estuary under eutrophication. *Estuaries Coastal and Shelf Science* 71: 301–308.

Cardoso, P. G., Brandao, A., Pardal, M. A., Raffaelli, D., Marques, J. C. 2005. Resilience of *Hydrobiaulvae* populations to anthropogenic and natural disturbances. *Marine Ecology Progress Series* 289: 191–199.

Cardoso, P. G., Leston, S., Grilo, T. F., Bordalo, M. D., Crespo, D., Raffaelli, D., Pardal, M. A. 2010. Implications of nutrient decline in the seagrass ecosystem success. *Marine Pollution Bulletin* 60: 601–608.

Cardoso, P. G., Pardal, M. A., Lillebo, A. I., Ferreira, S. M., Raffaelli, D., Marques, J. C. 2004. Dynamic changes in seagrass assemblages under eutrophication and implications for recovery. *Journal of Experimental Marine Biology and Ecology* 302: 233–248.

Carmichael, R. H., Shriver, A. C., Valiela, I. 2004. Changes in shell and soft tissue growth, tissue composition, and survival of quahogs, *Mercenaria mercenaria*, and softshell clams, *Mya arenaria*, in response to eutrophic-driven changes in food supply and habitat. *Journal Experimental Marine Biology and Ecology* 313: 75–104.

Chapman, M. G. 1998. Relationships between spatial patterns of benthic assemblages in a mangrove forest using different levels of taxonomic resolution. *Marine Ecology Progress Series* 162: 71–78.

Chew, L. L., Chong, V. 2011. Copepod community structure and abundance in a tropical mangrove estuary, with comparisons to coastal waters. *Hydrobiologia* 666: 127–143.

Chua, T. E., Gorre, I. R. L., Adrian Ross, S., Bernad, S. R., Gervacio, B., Corazon Ebarvia, M. 2000. The Malacca Straits. *Marine Pollution Bulletin* 41: 160–178.

Chung, C. C. 2012. Species composition, abundance and distribution of jellyfish in the coastal waters of Penang National Park and Manjung. *MSc thesis*. Universiti Sains Malaysia, Penang, Malaysia.

Cloern, J. E. 2001. Our evolving conceptual model of the coastal eutrophication problem. *Marine Ecology Progress Series* 210: 223–253.

Conti, M. E., Cecchetti, G. 2001. Biological monitoring: lichens as bioindicators of air pollution assessment - a review. *Environmental Pollution* 114: 471–492.

Conti, M. E., Cecchetti, G. 2003. A biomonitoringstudy: trace metals in algae and molluscs from Tyrrhenian coastal areas. *Environmental Research* 93: 99–112.

Cummins, S. P., Roberts, D. E., Zimmerman, K. D. 2004. Effects of the green macroalga *Enteromorpha intestinalis* on macrobenthic and seagrass assemblages in a shallow coastal estuary. *Marine Ecology Progress Series* 266: 77–87.

Cuong, D. T., Bayen, S., Wurl, O., Subramaniam, K., Wong, K. K. S., Sivasothi, N., Obbard, J. P. 2005. Heavy metal contamination in mangrove habitats of Singapore. *Marine Pollution Bulletin* 50: 1713–1744.

Dolbeth, M., Ferreira, O., Teixeira, H., Marques, J. C., Dias, J. A., Pardal, M. A. 2007. Beach morphodynamics impact on a macrobenthic community along a subtidal depth gradient. *Marine Ecology Progress Series* 352: 113–124.

Duarte, C. M. 1995. Submerged Aquatic Vegetation in Relation to Different Nutrient Regimes. *Ophelia* 41: 87–112.

Edward, F. B., Yap, C. K., Ismail, A., Tan, S. G. 2009. Interspecific variation of heavy metal concentrations in the different parts of tropical intertidal bivalves. *Water, Air, and Soil Pollution* 196(1–4): 297–309.

Elham, M. 2013. Molecular taxonomy, population genetics and reproductive studies of the jellyfish (Scyphozoa) in selected areas of Malaysian waters. *PhD thesis*. Universiti Sains Malaysia, Penang, Malaysia.

Fortes, M. D. 1990. *Seagrasses: a resource unknown in the ASEAN region*, International Specialized Book Services. ICLARM, Manila, Philippines, p 46.

Gaston, G. R. 1987. Benthic Polychaeta of the middle Atlantic bight: feeding and distribution. *Marine Ecology Progress Series* 36: 251–262.

Gevrey, M., Worner, S., Kasabov, N., Pitt, J., Giraudel, J. L. 2006. Estimating risk of events using SOM models: a case study on invasive species establishment. *Ecological Modelling* 197: 361–372.

Ghazali, N. H. M. 2006. Coastal erosion and reclamation in Malaysia. *Aquatic Ecosystem Health and Management* 9: 237–247.

Ghoziladeh, M., Yahya, K., Talib, A., Ahmad, O. 2012. Distribution of macrobenthic polychaete families in relation to environmental parameters in North West Penang, Malaysia. *World Academy of Science, Engineering and Technology* 72, 673–678.

Gray, J. S. 1992. Eutrophication in the sea. In: Colombo, G., Ferrarri, I., Ceccherelli, V. U., Rossi, R. (Eds.). *Marine eutrophication and population dynamics*, pp. 3–16. Olsen and Olsen, Fredensborg, Denmark.

Gray, J. S., Wu, R. S. S., Or, Y. Y. 2002. Effects of hypoxia and organic enrichment on the coastal marine environment. *Marine Ecology Progress Series* 238: 249–279.

Hajeb, P., Christianus, A., Ismail, A., Zadeh, S. S, Saad, C. R. 2009a. Heavy metal concentration in horseshoe crab (*Carcinoscorpius rotundicauda* and *Tachypleusgigas*) eggs from Malaysian coastline. In: *Biology and conservation of horseshoe crabs*, pp. 455–463. Springer, Boston, Massachusetts, US.

Hajeb, P., Jinap, S., Ismail, A., Fatimah, A. B, Jamilah, J., Rahim, M. A. 2009b. Assessment of mercury level in commonly consumed marine fishes in Malaysia. *Food Control* 20(1): 79–84. doi: 10.1016/j.foodcont.2008.02.01

Hakanson, L., Blenckner, T. 2008. A review on operational bioindicators for sustainable coastal management—Criteria, motives and relationships. *Ocean and Coastal Management*. 51(1): 43–72.

Hamid, M. Y. A., Sidhu, H. S. 1993. Metal finishing wastewater: characteristics and minimization. In: Yeoh, B. G., Chee, K. S., Phang, S. M., Isa, Z., Idris, A., Mohamed, M. (Eds.). *Waste management in Malaysia: current status and prospects for bioremediation*, pp. 41–49. Ministry of Science, Technology and the Environment, Kuala Lumpur, Malaysia.

Hashmi, M. I., Mustafa, S., Tariq, S. A. 2002. Heavy metal concentrations in water and tiger prawn (*Penaeusmonodon*) from grow-out farms in Sabah, North Borneo. *Food Chemistry* 79: 151–156.

Hii, Y. S., Law, A. T., Shazili, N. A. M., Abdul Rashid, M. K., Lokman, H. M., Yusoff, F. M., Ibrahim, H. M. 2006. The straits of Malacca: hydrological parameters, biochemical oxygen demand and total suspended solid. *Journal of Sustainability Science and Management* 1(1): 1–14.

Holt, E. A., Miller, S. W. 2010. Bioindicators: using organisms to measure environmental impacts. *Nature Education Knowledge* 3(10): 8. https://www.nature.com/scitable/knowledge/library/bioindicators-using-organisms-to-measure-environmental-impacts-16821310

Ismail, A. 2006. The use of intertidal molluscs in the monitoring of heavy metals and organotin compounds in the West coast of Peninsular Malaysia. *Coastal Marine Science* 30(1): 401–406.

Ismail, A., Badri, M. A., Ramlan, M. N. 1991. Heavy metal contamination in fiddler crabs (*Ucaannulipes*) and hermit crabs (*Clibanarius* sp.) in a coastal area of northern Peninsular Malaysia. *Environmental Technology* 12(10): 923–926.

Ismail, A., Rosniza, R. 1997. Trace metals in sediments and molluscs from an estuary receiving pig farms effluent. *Environmental Technology* 18: 509–515.

Ismail, A., Yap, C. K., Zakaria, M. P., Tanabe, S., Takada, H., Ismail, A. R. 2000. Green-lipped Mussel Pernaviridis (L.) as a biomonitoring agent for heavy metals in the west coast of Peninsular Malaysia. In: Shariff, M., Yusoff, F. M., Gopinath, N., Ibrahim, H. M., Mustapha, A. N. (Eds.). *Towards sustainable management of the Straits of Malacca, technical and financial options*, pp. 553–559. Malacca Straits Research and Development Centre (MASDEC), University Putra, Malaysia.

Jackson, J. B. C. 1997. Reefs since Columbus. *Coral Reefs* 16: 23–32.

Jarup, L. 2003. Hazards of heavy metal contamination. *British Medicine Bulletin* 68: 167–182.

Kamaruddin, H. 1998. Coastal zone management in Malaysia – pollution control. *Malaysian Journal of Law and Society* 2: 46–55.

Kamaruzzaman, B. Y., Ong, M. C., Rina, S. Z., Joseph, B. 2010. Levels of some heavy metals in fishes from Pahang River Estuary, Malaysia. *Journal of Biological Sciences* 10(2): 157–161.

Kathiresan, K., Bingham, B. L. 2001. Biology of mangroves and mangrove ecosystems. *Advances in Marine Biology* 40: 81–251.

Kingston, P. F. 2002. Long-term environmental impact of oil spills. *Spill Science & Technology Bulletin* 7:53–61.

Kostecki, C., Le Loc'h, F., Roussel, J. M., Desroy, N., Huteau, D., Riera, P., Le Bris, H., Le Pape, O. 2010. Dynamics of an estuarine nursery ground: the spatio-temporal relationship between the river flow and the food web of the juvenile common sole (*Soleasolea*, L.) as revealed by stable isotopes analysis. *Journal of Sea Research* 64: 54–60.

Kraufvelin, P., Moy, F. E., Christie, H., Bokn, T. L. 2006. Nutrient addition to experimental rocky shore communities revisited: delayed responses, rapid recovery. *Ecosystems* 9: 1076–1093.

Levinton, J., Kelaher, B. 2004. Opposing organizing forces of deposit-feeding marine communities. *Journal of Experimental Marine Biology and Ecology* 300: 65–82.

Linton, D. M., Warner, G. F. 2003. Biological indicators in the Caribbean coastal zone and their role in integrated coastal management. *Ocean Coast Management* 46: 261–276.

Lowe, R. L., Pan, Y. 1996. *Benthic algal communities as biological monitors. Algal ecology: freshwater benthic Ecosystems*, pp. 705–739. Elsevier, San Diego, California.

Maiti, S. K. 2004. *Handbook of methods in environmental studies: water and wastewater analysis*, pp. 307. ABD Publishers, Jaipur. ISBN 81-8577-34-0.

Mansor, M., Ali, A., Baharuddin, M. H. M., Sah, S. A. M., Othman, A. F. 2005. A Preliminary Survey: Distribution and Habitat Suitability of Dugong (Dugong Dugon) in Peninsular Malaysia, 1–13.

Masilamoni, J. G., Jesudoss, K. S., Nandakumar, K., Satpathy, K. K., Nair, K. V. K., Azariah, J. 2000. Jellyfish ingress: a threat to the smooth operation of coastal power plants. *Current Science* 79(5): 567–569.

Mendez, N. 2002. Annelid assemblages in soft bottoms subjected to human impact in the Urias estuary (Sinaloa, Mexico). *Oceanologica Acta* 25: 139–147.

Newman, W. A., Stanley, S. M. 1981. Competition wins out overall: reply to Paine. *Paleobiology* 7(4): 561–569.

Nixon, S. W. 1995. Coastal marine eutrophication - a definition, social causes, and future concerns. *Ophelia* 41: 199–219.

Normawaty, M. N., Anton, A, Nakisah, M. A. 2007. Biodiversity of dinoflagellates in the coastal waters off Malacca, Peninsular Malaysia. Borneo Science. *The Journal of Science and Technology* 21(1): 12–18.

Nurul Ruhayu, M. R. 2010. The Impact of Anthropogenic Activities in Pinang River, Balik Pulau, Penang. *M. Sc. Thesis*, Universiti Sains Malaysia, Penang, Malaysia.

Nurul Ruhayu, M. R., Khairun, Y. 2013. Trends of Sediment Loading in Catchment Areas of Pinang River in Malaysia. APCBEES Procedia, Elsevier, Vol. 5, 128–133.

Omar., A., Tay, P. F., Khairun, Y. 2012. Distribution of intertidal organisms in the shore of Teluk Aling, Penang, Malaysia. *Publications of the Seto Marine Biological Laboratory* 41: 51–61.

Omar, I. 2003. *National Report of Malaysia on the formulation of a transboundary diagnostic analysis and preliminary framework of a Strategic Action Programme for the Bay of Bengal*. BOBLME Report. pp. 108.

Ong, J. E. 1995. The ecology of mangrove conservation and management. *Hydrobiologia* 295: 343–351.

Ong, M. C., Kamaruzzaman, B. Y. 2009. An assessment of metals (Pb and Cu) contamination in bottom sediment from South China Sea Coastal Waters, Malaysia. *American Journal of Applied Sciences* 6(7): 1418–1423.

Patricio, J., Adao, H., Neto, J. M. et al. 2012. Do nematode and macrofauna assemblages provide similar ecological assessment information? *Ecology Indicators* 14: 124–137.

Pearson, T. H., Rosenberg, R. 1978. Macrobenthic succession in relation to organic enrichment and pollution of the marine environment. *Oceanography and Marine Biology Annual Review* 16: 229–311.

Phang, S. M., Wong, C. L., Lim, P. E., Yeong, H. Y., Chan, C. X. 2005. Seaweed diversity of the Langkawi islands with emphasis on the northeastern region. *Malaysian Journal of Science* 24: 77–94.

Purcell, J. E., Uye, S., Lo, W. 2007. Anthropogenic causes of jellyfish blooms and their direct consequences for humans: a review. *Marine Ecology Progress Series* 350: 153–174.

Rahman, R. A., Surif, S. 1993. Metal finishing wastewater: characteristics and minimization. In: Yeoh, B. G., Chee, K. S., Phang, S. M., Isa, Z., Idris, A., Mohamed, M. (Eds.). *Waste management in Malaysia: current status and prospects for bioremediation*, pp. 3–7. Ministry of Science, Technology and the Environment, Malaysia.

Rainbow, P. S. 1995. Biomonitoring of heavy metal availability in the marine environment. *Marine Pollution Bulletin* 31: 183–192.

Rott, E. 1991. Methodological aspects and perspectives in the use of periphyton for monitoring and protecting rivers. In: Whitton, B. A., Rott, E., Friedrich, G., (Eds.) *Use of algae for monitoring rivers*, pp. 9–16. Institutfür Botanik, Universitat Innsbruck, Innsbruck, Austria.

Roux, D. J., van Vliet, H. R., van Veelen, M. 1993. Towards integrated water quality monitoring: assessment of ecosystem health. *Water SA* 19(4): 275–280.

Sarda, R., Valiela, I., Foreman, K. 1996. Decadal shifts in a salt marsh macroinfaunal community in response to sustained long-term experimental nutrient enrichment. *Journal of Experimental Marine Biology and Ecology* 205: 63–81.

Sharifah Mastura, S. A. 1992. The coastal zone in Malaysia. Processes, issues and management plan. Background paper, Malaysian National Conservation Strategy. Economic Planning Unit, Kuala Lumpur.

Sivalingam, P. M. 1977. Aquaculture of the green mussel, *Mytilus viridis* Linnaues, in Malaysia. *Aquaculture* 11: 297–312.

Sivalingam, P. M. 1984. Ocean disposal and land reclamation problems of Penang, Malaysia. *Conservation Recycling* 7(24); 85–98.

Sivalingam, P. M. 1985. An overview of the Mussel Watch monitoring programme in conjunction with the Westpac programme. *Journal of Asian Environment* 6: 39–50.

Sivalingam, P. M., Allapitchay, I., Kojima, H., Yoshida, T. 1982. Mussel watch of PCBs and persistent pesticides residues in *Pernaviridis* Linnaeus from Malaysian and Singapore waters. *Applied Geography* 2: 231–237.

Sivalingam, P. M., Bhaskaran, B. 1980. Experimental insight of trace metal environmental pollution problems 1in mussel farming. *Aquaculture* 20: 291–303.

Smith, S., Hollibaugh, J. 1993. Coastal metabolism and the oceanic organic carbon balance. *Reviews of Geophysics* 31(1): 75–89.

Southward, A. J. 1987. *Pollution and fouling Crustacean issues 5: Barnacles biology*, pp. 405–431. A. A. Balkema, Rotterdam, Netherlands.

Stanley, S. M., Newman, W. A. 1980. Competitive exclusion in evolutionary time: the case of acorn barnacles. *Paleobiology* 6: 173–183.

Stephenson, T., Stephenson, A. 1972. *Life between tidemarks on rocky shores*. WH Freeman, San Francisco.

Stiling, P. D. 1996. *Ecology: Theories and applications*, pp. 539. 2nd Edition. Prentice Hall, Upper Saddle River, NJ.

Suthers, I., Bowling, L., Kobayashi, T., Rissik, D. 2009. Sampling methods for plankton. In: Suthers, I. M., Rissik, D. (Eds.). *Plankton: a guide to their ecology and monitoring for water quality*, pp. 73–114. Criro Publishing, Collingwood, Victoria, Australia.

Tay, P. F. 2008. The distribution of intertidal organisms around the shore of Teluk Aling, Pulau Pinang. Bachelor Science in Biology, Universiti Sains Malaysia, Pulau Pinang.

Telesh, I. V., Khlebovich, V. V. 2010. Principal processes within the estuarine salinity gradient: a review. *Marine Pollution Bulletin* 61: 149–155.

Terlizzi, A., Scuderi, D., Fraschetti, S., Anderson, M. J. 2005. Quantifying effects of pollution on biodiversity: a case study of highly diverse mollus can assemblages in the Mediterranean. *Marine Biology* 148: 293–305.

Timur, P. 2014. Trace metals in Thais clavigera along coastal waters of the east coast of peninsular Malaysia. *Sains Malaysiana* 43(4): 529–534.

UNEP, 2008. *Protocol on Integrated Coastal Zone Management in the Mediterranean.* UNEP 2008 Annual Report. pp. 106.

Vesanto, J. 1999. SOM-based data visualization methods. *Intelligent Data Analysis* 3: 111–126.

Villares, R., Puente, X., Carballeira, A. 2002. Seasonal variation and background levels of heavy metals in two green seaweeds. *Environmental Pollution* 119: 79–90.

Wieking, G., Kroncke, I. 2005. Is benthic trophic structure affected by food quality? The Dogger Bank example. *Marine Biology* 146: 387–400.

Yap, C. K. 1999. Accumulation and distribution of heavy metals in green-lipped mussel Pernaviridis (Linnaeus) from the west coast of Peninsular Malaysia (Doctoral dissertation, Universiti Pertanian Malaysia).

Yap, C. K., Edward, F. B., Tan, S. G. 2007. Determination of heavy metal distributions in the green-lipped mussel *Pernaviridis* as bioindicators of heavy metal contamination in the Johore straits and Senggarang, Peninsular Malaysia. *Trends in Applied Sciences Research* 2(4): 284–294.

Yap, C. K., Ismail, A., Tan, S. G. 2003. Can the byssus of green-lipped mussel *Pernaviridis* (Linnaeus) from the west coast of Peninsular Malaysia be a biomonitoring organ for Cd, Pb and Zn? Field and laboratory studies. *Environment International* 29(4): 521–528.

Yap, C. K., Ismail, A., Tan, S. G. 2004. Biomonitoring of heavy metals in the west coastal of Peninsular Malaysia using the green-lipped mussel *Pernaviridis*, Present status and what next? *Pertanika Journal of Tropical Agriculture Science* 27(2): 151–161.

Yap, C. K., Ismail, A., Tan, S. G., Omar, H. 2002a. Correlations between speciation of Cd, Cu, Pb and Zn in sediment and their concentrations in total soft tissue of green-lipped mussel *Pernaviridis* from the west coast of Peninsular Malaysia. *Environment International* 28(1): 117–126.

Yap, C. K, Ismail, A., Tan, S. G., Omar, H. 2002b. Occurrence of shell deformities in green-lipped mussel *Pernaviridis* (Linnaeus) collected from Malaysian coastal waters. *Bulletin of Environmental Contamination and Toxicology* 69(6): 0877–0884.

Yunus, K., Suhaimi, M., Zahir, M. et al. 2010. Determination of some heavy metal concentrations in razor clam (Solenbrevis) from Tanjung Lumpur coastal waters, Pahang, Malaysia. *Pakistan Journal of Biological Sciences* 13(24): 1208–1213.

Yusof, A. M., Mahat, M. N., Omar, N., Wood, A. K. H. 2001. Water quality studies in an aquatic environment of disused tin-mining pools and in drinking water. *Ecological Engineering* 16: 405–414.

Yusof, A. M., Yanta, N. F., Wood, A. K. 2004. The use of bivalves as bio-indicators in the assessment of marine pollution along a coastal area. *Journal of Radio Analytical and Nuclear Chemistry* 259(1): 119–127.

Yusoff, M. K., Sulaiman, W. N. A., Yaziz, M. I. 2000. Impact of piggery waste on river water quality in the vicinity of pig farm in Malaysia. In: Shariff, M., Yusoff, F. M., Gopinath, N., Ibrahim, H. M., Nik Mustapha, R. A. (Eds.). *Towards sustainable management of the straits of Malacca*, pp. 571–578. Malacca Straits Research and Development Centre (MASDEC), Universiti Putra Malaysia, Serdang, Malaysia.

Zulkifli, S. Z, Mohamat-Yusuff, F., Ismail, A. 2012. Bioaccumulation of selected heavy metals in soldier crabs, *Dotilla myctiroides* (Decapoda: Ocypodidae) from Bagan Lalang, Selangor, Malaysia. *Acta Biologica Malaysiana* 1(3): 94–100.

12

Conservation and Management of Saltwater Crocodile (Crocodylus porosus) in Bhitarkanika Wildlife Sanctuary, Odisha, India

Lakshman Nayak, Satyabrata Das Sharma and Mitali Priyadarsini Pati

CONTENTS

12.1 Introduction 308
    12.1.1 Distribution of Mangroves 308
12.2 Distribution of Saltwater Crocodile 311
    12.2.1 Worldwide Distribution of Saltwater Crocodiles... 311
    12.2.2 Distribution of Saltwater Crocodiles in India 311
    12.2.3 Distribution of Saltwater Crocodiles in Odisha (Bhitarkanika) 311
12.3 Biology of Saltwater Crocodile 312
    12.3.1 Habitat 312
    12.3.2 Morphological Characteristics 313
    12.3.3 Appearance... 313
    12.3.4 Movement 314
    12.3.5 Food and Feeding Habits... 314
    12.3.6 Communication and Behaviour 314
    12.3.7 Reproduction and Parental Care 315
        12.3.7.1 Mating... 315
12.4 Why Should We Conserve the Saltwater Crocodile?.. 315
    12.4.1 Status 315
    12.4.2 Reasons for Conservation.. 315
    12.4.3 Reasons to Conserve Crocodiles 316
12.5 Conservation and Management of Saltwater Crocodile in Bhitarkanika
    Conservation Area... 316
12.6 Present Status of Saltwater Crocodile in Bhitarkanika 317
    12.6.1 Saltwater Crocodile Project at Dangmal-Bhitarkanika National Park 317
    12.6.2 Attacks on Humans by Wild Crocodiles... 318
    12.6.3 Attack on Domestic Livestock by Wild Crocodiles 318
12.7 Conclusion 319
References... 320

12.1 Introduction

Mangroves are salt-tolerant, complex and dynamic ecosystems that occur in tropical and subtropical intertidal regions. 'Mangrove' has been variously defined in the literature. Several researchers have opined that plants growing in between the highest and the lowest tidal limits may be considered mangroves (Aubreville, 1964; Blasco et al., 1975; Blasco, 1977; Clough, 1982; Grzimek et al., 1976; MacNae, 1968; Naskar and Guha Bakshi, 1987; Tomlinson, 1986). Mangroves derive their physical, chemical and biological characteristics from the sea as well as from inflowing freshwater from upland forests. Once occupying 75% of tropical coastland inlets, today mangroves are restricted to just a few pockets. Among the two mangrove distribution zones recognized, the eastern zone (East African coast, Pakistan, India, Myanmar, Malaysia, Thailand, the Philippines, southern Japan, Australia, New Zealand, south eastern archipelago) harbours a greater percentage of genera and species than the western zone (Atlantic coasts of Africa and Americas and the Galapagos Islands).

Around 70% of mangroves occur along the east coast of India and they include (1) Ganga Delta (Sunderbans), (2) Mahanadi Delta (Bhitarkanika), (3) Krishna Delta, (4) Godavari Delta and (5) Cauvery Delta (Badola and Hussain, 2003). The Bhitarkanika Mangrove Ecosystem, the second largest mangrove forest in mainland India, harbours the highest diversity of Indian mangrove flora, the largest rookery of olive ridley sea turtles in the world, the last of the three remaining populations of saltwater crocodiles in India, the largest known population of king cobras and water monitor lizards, one of the largest heronries along India's east coast and one of the highest concentrations of migratory water fowl (Badola and Hussain, 2003). The largest diversity of estuarine and mangrove obligate fishery resources and prawns is found in the Bhitarkanika Conservation Area (BCA). The mangrove and the associated forests provide the subsistence requirements of timber, fuel wood, tannin, honey, roof thatch and fodder for local communities (Chadah and Kar, 1999). Thus, BCA supports a wider range of flora and fauna, plays a greater ecological and economic role and is therefore of greater conservation value than other similar areas.

12.1.1 Distribution of Mangroves

Mangroves are distributed in the tropical and subtropical zones between the Tropic of Cancer and Tropic of Capricorn, with the Malaysian and Indonesian region supporting the greatest diversity of mangrove communities.

Chapman (1976) has reported 90 species of mangroves from around the world. Saenger et al. (1983) recorded 83 species, UNDP/UNESCO (1986) reported 65 species, and Tomlinson (1986) mentioned only 48 mangrove species, of which 40 were considered true mangroves found in the Old World Tropics (Indo-West Pacific Region) and 8 true mangroves from the New World Tropics. According to the statistics of the forest department, the mangrove forest coverage area during 1995 was 4533 km² and during 1999 it was about 4871 km² of the country's territory. It increased significantly, by 615 km², from 1991 to 1999. It showed an increase of 338 km² in the period 1995–1999. The total area according to the Fishery Survey of India (FSI) report in 1995 was 4533 km².

Mangroves are typically tropical coastal vegetation found in the intertidal regions of river deltas and backwater areas. Extensive mangrove forests occur in major river deltas such as the mouths of the Budhabalanga, Subarnarekha, Brahmani, Baitarani

(Bhitarkanika), Mahanadi and Devi River on the Odisha coast extending over 480 km along the Bay of Bengal. Adequate protection was given to its rich floral and faunal diversity in 1975 when it was declared a wildlife sanctuary. Out of the total sanctuary area of 672 km², the core area, comprised of 145 km², was declared a national park in 1998 (Thatoi, 2004).

Mangrove forests serve as ecotones between land and sea, and elements from both are stratified horizontally and vertically, between the forest canopy and subsurface soil. In India, mangroves occur in two groups, those of the west coast and those of the east coast. India has a coastline that is approximately 8129 km long and that can be divided into east and west coasts and island chains. The east coast covers the maritime states like Tamil Nadu, Andhra Pradesh, Odisha, West Bengal and Andaman-Nicobar Islands. The west coast extends from Kerala, Karnataka, Goa, Maharashtra and Gujarat and also includes the coral atolls of Lakshadweep Islands. The total mangrove area along the Indian coast is estimated to be approximately 700,000 ha. The mangroves along the east coast of India are larger (80%) than those of the west coast (20%) because the terrain – plains – of the east coast has a gradual slope compared with the steep gradient along the west coast.

The state of Odisha is located on the east coast of the Indian peninsula, which is quite rich in natural resources. It is home to several biodiversity hot spot areas of the Indian subcontinent and has the richest, most biodiverse regions in South-East Asia. It has seven major river deltas of various sizes and shapes formed by the rivers Subarnarekha, Budhabalanga, Baitarani, Brahmani, Mahanadi, Rushikulya and Bahuda. The natural boundaries of the Bhitarkanika Sanctuary are defined by Dhamara River to the north, Maipura River to the south, Brahmani River to the west and the Bay of Bengal in the east. Bhitarkanika Mangroves are a location of rich, abundant green energetic ecosystems lying in the estuarine region of Brahmani–Baitarani in the north-eastern corner of the Kendrapara district of Odisha. The Bhitarkanika Mangroves forest is situated at 20°-30′ N to 20°-48′ N and 86°-45′ to 87°-03′E. It is the second largest mangrove of India and has an ecological importance because it is internationally famous for its estuarine crocodile conservation. It covers mangrove forests, rivers, creeks, estuaries, backwaters, accreted land and mudflats. Mangroves flourish in the deltaic region formed by rich alluvial deposits of the Brahmani, Baitarani, Maipura and Dhamra Rivers (Gopi and Pandav, 2007). The climate of the area is characterized by three seasons: summer (March–June), monsoon (July–October) and winter (November–February). The annual temperature varies from 15 to 30°C, the average rainfall is 1670 mm, and relative humidity is 75% to 80%. The deltaic slopes of Bhitarkanika Wildlife Sanctuary (BKWS) are extremely low lying and subject to regular tidal inundations. The average elevation above mean tide level is between 1.5 and 2 m and extends up to 3.4 m (Dani and Kar, 1999). The mangrove soils are fine-grained silt or clay formed by the sedimentation of the Mahanadi and Brahmani Rivers (Chadha and Kar, 1999).

A small area of the Bhitarkanika mangrove forest was declared a national park in 1988. The park encompasses an area of 672 km² of the Bhitarkanika Mangroves. This mangrove swamp lies in the river delta of the Brahmani, Baitarani, and Dhamra Rivers. The national park is surrounded by the BKWS and was designated a Ramsar site in 2002 (second in the state). Wild animals can be seen, such as the leopard cat, fishing cat, jungle cat, hyena, wild boar, spotted deer, porcupine, dolphin, saltwater crocodile including the partially white (sankhua) crocodile, python, king cobra, water monitor lizard, terrapin, marine sea turtle, kingfisher, woodpecker, hornbill, bar-headed geese, brahminy duck, pintail white-bellied sea eagle, tern, sea gull, waders and a large variety of resident and migratory birds. In addition, the Indian government established 18 biosphere reserves in India (categories roughly corresponding to International Union for Conservation of Nature (IUCN) Category

V Protected Areas), 166 national parks and 515 animal sanctuaries (IUCN Category IV Protected Area).

Within the Odisha state of eastern India, saltwater crocodiles persist only within Bhitarkanika National Park and adjacent waterways within the Kendrapara district. Human–crocodile conflict has become an issue in recent years, with dozens of human fatalities reported within the past decade. Saltwater crocodiles persist in small numbers only within the Sundarbans portion of West Bengal. The saltwater or estuarine crocodile (*C. porosus*) is the largest of all living reptiles. It is found in suitable habitat throughout South-East Asia and northern Australasia. The park is home to the endangered saltwater crocodile (*C. porosus*), white crocodile, Indian python, king cobra, black ibis, darters and many other species of flora and fauna. Bhitarkanika National Park was created in September 1998 from the core area of the BKWS, which was created in 1975. Areas of 145 km² have been designated as composing Bhitarkanika National Park through Notification 19686/F & E dated 16 September 1998 from the Forests and Environment Department, Government of Odisha. The sanctuary is the second largest mangrove ecosystem in India.

In Odisha, the estuarine or saltwater crocodile (*Crocodylus porosus*) or Baula kumbhira, Kuji Khumbhiora (in the Oriya language) is restricted to the mangrove swamps of the Brahmani – Baitarani Delta of the north-eastern portion of the state, which falls within Bhitarkanika National Park. Around the mid-1970s, the population of these saltwater crocodiles had fallen to a critical level, leaving only a small viable population in the main Bhitarkanika River and a few adjoining creeks. The decline of the population was mostly due to overexploitation, poaching and indiscriminate hunting. To save these greatly endangered species from extinction, a conservation programme was launched by the state government through the Forest Department.

The saltwater crocodile has a vast geographical range that extends from Cochin on the west coast of India to the Sunderbans in West Bengal and to the Andaman Islands. Single individuals can be found at quite large distances from their usual range because they can travel long distances (over 1000 km) by sea. The species is seriously endangered as a result of hunting and largely from loss of habitat, particularly breeding sites. Surveys in the Andamans have brought to light the precarious position of the animal on the Andaman and Nicobar Islands. It is now extinct in Kerala and Tamil Nadu. A sanctuary for the species was established on Bhitarkanika Island and adjacent areas in the Brahmani Baitarni River estuary in Odisha.

Currently, *C. porosus* is not listed as threatened under the Environmental Protection and Biodiversity Conservation Act 1999, but the future of the species is uncertain. The primary source of the decline of this species is habitat loss and continuous hunting by humans, who highly value its leather products, with skins fetching prices in the thousands of dollars. The largest population of *C. porosus* occurs in the Northern Territory of Australia (100,000 to 150,000). Other populations of *C. porosus* around the world do not have such high numbers, reaching only into the hundreds or a few in areas like India, Sri Lanka and Thailand. Populations in some areas have declined drastically over the years owing to habitat loss from the development of coastal areas. *Crocodylus porosus* is the largest and one of the most powerful and intelligent reptile species living today. Crocodiles, as a whole, have existed on Earth for millions of years, demonstrating that their unique physiology and hunting abilities have made them very successful at surviving changing environmental conditions over time. Their importance to biodiversity lies in the key role that they play as predators at the top of the food chain, and sometimes they pose a risk to human lives. Owing to increasing habitat destruction, the coastal areas where *C. porosus* is normally found are disappearing, resulting in the dispersion of the species in search of new homes.

12.2 Distribution of Saltwater Crocodile

12.2.1 Worldwide Distribution of Saltwater Crocodiles

Saltwater crocodile numbers have been severely depleted throughout the world in the vast majority of their range, with sightings in areas such as Thailand, Cambodia, Laos and Vietnam becoming extremely rare, and the species may in fact even be extinct in one or more of these countries. It is presumed that the least crocodilian to become globally extinct due to their wide distribution and almost precolonial population sizes in Northern Australia and New Guinea. In India this crocodile is extremely rare in most areas but very common in the north-eastern part of the country (mainly Odisha and the Sunderbans). The population is sporadic in Indonesia and Malaysia, with some areas harbouring large populations (Borneo for example) and others with very small, so-called at-risk populations (the Philippines). The saltwater crocodile is also present in very limited portions of the South Pacific, with an average population in the Solomon Islands, a very small and soon-to-be-extinct population in Vanuatu (where the population officially only stands at three) and a decent but at-risk population in Palau (possibly rebounding).

12.2.2 Distribution of Saltwater Crocodiles in India

In India, the estuarine crocodile is restricted in its distribution to the tidal estuaries, marine swamps, coastal brackish water lakes and lower reaches of the larger rivers. The saltwater crocodile has a vast geographical range that extends from Cochin on the west coast of India to the Sunderbans in West Bengal and to the Andaman Islands.

It is known by various names in different parts of the country:

English: Saltwater crocodile, estuarine crocodile

Oriya: Baula kumbhira, Kuji Khumbhiora

Hindi, Gujarat, Marathi: Mugger

Bengali: Kuhmir

Kannada: Mossalay

Tamil: Muthalai

Telugu: Moseli

Malayalam: Muthala, Cheengkani

12.2.3 Distribution of Saltwater Crocodiles in Odisha (Bhitarkanika)

In Odisha, the estuarine or saltwater crocodile (*C. porosus*) is restricted to the mangrove swamps of the Brahmani–Baitarani Delta of the north-eastern portion of the state, which falls within the Bhitarkanika National Park. Around the mid-1970s, the population of these saltwater crocodiles had dropped to a critical level, leaving only a small viable population in the main Bhitarkanika River and a few adjoining creeks. The decline of the population was mostly due to overexploitation, poaching and indiscriminate hunting. To save these greatly endangered species from extinction, a conservation programme was launched by the state government through the Forest Department. This project was started with active assistance from the Food and Agriculture Organisation of the United Nations Development Programme. The major achievement of this project in its first phase has been to rear and rehabilitate the saltwater crocodile. Apart from rearing and releasing 1717 crocodiles in nature, 26 captive reared crocodiles have been supplied to other state projects.

12.3 Biology of Saltwater Crocodile

Systematic Position

Kingdom: Animalia

Phylum: Chordata

Class: Sauropsida

Order: Crocodilia

Family: Crocodylidae

Subfamily: Crocodylinae

Genus: *Crocodylus*

Species: *porosus*

Common Oriya Name: Baula Kumbhira (Figure 12.1)

12.3.1 Habitat

Saltwater crocodiles generally spend the tropical wet season in freshwater swamps and rivers, moving downstream to estuaries in the dry season, and sometimes travelling far out to sea. Crocodiles compete angrily with each other for territory, with dominant males in particular occupying the choicest stretches of freshwater creeks and streams. Junior crocodiles are thus forced into the more marginal river systems and sometimes into the ocean. This explains the large distribution of the animal (ranging from the east coast of India to northern Australia) as well as its being found in odd places on occasion (such as the Sea of Japan). The saltwater crocodile's underwater speed can be 15 to 18 miles (24 to 29 km) per hour in short bursts, but when cruising it can travel 2 to 3 miles (3.2 to 4.8 km).

FIGURE 12.1
Estuarine or saltwater crocodile.

12.3.2 Morphological Characteristics

The crocodile's body is divided into three major parts, the head, trunk and tail. The largest skull available measures 1 m in length and is believed to have belonged to a specimen of about 7 m in length. Specimens over 5 m (16.5 ft) in length have been obtained in the Sunderbans and Odisha's river estuaries but are now exceedingly rare. A 4.5 m (just under 15 ft) long captive specimen weighed 408 kg. Single individuals can be found some distance from their usual range as they can travel long distances (over 1000 km) by sea. Barnacles have been found on the scales of a few stray individuals. This seafaring ability probably helps to explain their wide distribution.

A healthy adult male saltwater crocodile is typically 4.8 to 5 m (15.75 to 16.6 ft) long and weighs roughly 770 kg, with larger specimens being very rare exceptions. The largest confirmed saltwater crocodile on record was 6.3 m (20.6 ft) long and probably weighed well over 1900 kg. Many larger sizes have been reported, but these have generally been discredited as exaggerations. Examinations of incomplete remains have never suggested a length greater than 7 m (23 ft). Females are much smaller than males, with typical female body lengths in the range of 2.5–3 m.

The saltwater crocodile has a four-chambered heart like birds and mammals. It has salt secreting glands known as lingual glands present at the back of its throat for maintaining osmolarity. Its heart rate is controlled independently of body temperature regardless of outside ambient temperature. Crocodiles can conserve water by excreting excess sodium ions by these glands, but not by their kidney. It has been observed that the epithelium of the lingual salt glands is highly vascularized (with a rich network of blood vessels) and that neurotransmitters act on the salt gland epithelium directly to trigger secretory activity (Franklin et al., 2005).

12.3.3 Appearance

Adult males of the largest living crocodilian species (in fact the largest living reptile in the world) can reach sizes of up to 6 or 7 m (20 to 23 ft). There is always a lot of interest over the largest ever recorded saltwater crocodile. In general, males over 5 m (17 ft) in length are extremely rare. Females are smaller and do not normally exceed 3 m (10 ft), with 2.5 m being considered very large. This is a large-headed species with a heavy set of jaws. A pair of ridges runs from the eye orbits along the centre of the snout. Scales are more oval in shape than in other species, and scoots are relatively small. An average of 31.2 rows of scales present on the bellies of the saltwater crocodile have been reported (Isberg et al., 2005). Juveniles are pale yellow in colour with black stripes and spots on the body and tail. A small percentage of animals in some regions tend to be much lighter in colour depending on the habitat. The juvenile colouration persists for several years, growing progressively pale and less colourful with more indistinct bands, which eventually disappear. Mature adults are generally dark, with lighter tan or grey areas. The ventral surface (belly) is creamy yellow to white in colour, except the tail, which tends to be grey on the underside near the tip. Dark bands and stripes are present on the lower flanks but do not extend onto the belly region.

As its name implies, this species has a high tolerance for salinity, being found in brackish water around coastal areas and in rivers. However, it is rarely present in freshwater rivers and swamps. Movement between different habitats occurs between the dry and wet seasons and as a result of social status – juveniles are raised in freshwater areas, but eventually sub-adult crocodiles are usually forced out of these areas (used for breeding by dominant, territorial adults) into more marginal and saline areas. Subordinate animals unable to establish a territory in a tidal river system are either killed or forced out into the sea, where they move around the coast in search of another river system.

12.3.4 Movement

Crocodylus porosus generally moves at a faster rate in water. It is capable of swimming 19 km to 24 km per hour during an emergency and 2 to 3 miles during a more relaxed time. During swimming, the crocodile folds its limbs against its body and laterally undulates, using its tail and body to move through the water (Pough et al., 2004). Juvenile saltwater crocodiles can swim fast given their body length, but when their body length increases, the swimming capacity decreases (Elsworth et al., 2003). Apart from being an aquatic good swimmer, *C. porosus* can also move quickly on land. While walking in the terrestrial environment, the crocodile keeps its limbs underneath its body and holds its belly above the ground, a type of movement commonly known as a high walk (Pough et al., 2004).

12.3.5 Food and Feeding Habits

Saltwater crocodiles are carnivorous and scavengers. The species lives mainly on fish (predatory fish). It eats a wide variety of prey, although juveniles are restricted to smaller items such as insects, amphibians, crustaceans, small reptiles and fish. The larger the animal grows, the greater the variety of items that it includes in its diet, with only the smaller items taken less frequently. Prey items include crustaceans and vertebrates (e.g., turtles, snakes, shore and wading birds, buffalo and domestic livestock, wild boar, monkeys). It often feeds on carcasses flowing into the area from nearby human habitations and occasionally feeds on, for example, cattle, deer, sambars and wild pigs. Female saltwater crocodiles nest in mangrove forests, preparing a mound nest, unlike other species of crocodiles, which usually dig nests on sandy river banks.

12.3.6 Communication and Behaviour

The saltwater crocodile is one of the most intelligent reptiles. They communicate through barks. The barking sounds are categorized into four different calls: distress call, threat call, hatching call and courtship bellow. A distress call is barked by juveniles in danger. It consists of high-pitched calls but for a short duration. The second is a threat call, which is performed when the crocodiles are defending their territory. The third one is a hatching call, a short bark, performed by newborns. The fourth is a courtship bellow, which is heard as a long, low growl (Pough et al., 2004).

As an ambush predator, it usually waits for its prey to get close to the water's edge before striking without warning and using its great strength to drag the animal back into the water. Most prey animals are killed by the huge jaw pressure of the crocodile, although some animals may be incidentally drowned. The saltwater crocodile is an immensely powerful animal, having the strength to drag a fully grown water buffalo into a river or crush a full-grown bovid's skull in its jaws. Crocodiles cannot see clearly when submerged, but they have a fantastic assortment of adaptations that help them to watch a hunt in dim conditions (Nagloo et al., 2016). Furthermore, the saltwater crocodile (*C. porosus*) and the freshwater crocodile (*C. johnstony*) have the capacity to determine the colour, vision and visual clarity.

12.3.7 Reproduction and Parental Care

Breeding territories are established in freshwater areas. Females reach sexual maturity at 10 to 12 years of age, males mature later, at around 16 years of age, and 40 to 60 eggs are usually laid in mound nests made from plant matter and mud. These are constructed between the months of November and March during the wet season. This serves to raise the eggs above ground to prevent losses due to flooding. Alternatively, if the nest is in danger of getting too dry, the female has been known to splash water onto it from a purpose-dug pool. Juveniles hatch after around 90 days, although this varies with nest temperature. The female digs the young out of the nest when they start their characteristic chirping sounds, assisting them to the water by carrying them in her mouth. Restocking programmes in India (Bhitarkanika National Park in Odisha) have met with success.

12.3.7.1 Mating

Mating takes place during February to April. Nests are made in May. An average of 45 eggs are laid. Hatchlings emerge from the eggs after 80–90 days. The mother crocodile actively guards the nest by remaining in a wallow near the nest. The nests are usually prepared using, for example, mangrove twigs, leaves and mud. Nests are usually made in areas on high ground that will not be inundated during the highest high tide of flood waters during the rainy season and where it can get direct sunlight.

12.4 Why Should We Conserve the Saltwater Crocodile?

12.4.1 Status

The saltwater crocodile is included under Schedule I of the Wildlife (Protection) Act of 1972 and endangered as per Red Data Book categories of the IUCN, and it is also included in Appendix I of the Convention on International Trade in Endangered Species of Wild Fauna and Flora. The river systems of the BKWS and its fringe areas are the last stronghold of the species in Odisha. The estuarine or saltwater crocodile (*C. porosus*) is known to be the largest (7 m) among all species of living crocodiles in the world. It inhabits the deltaic regions of Brahmani, the Baitarani, Dhamra and Mahanadi River systems of the state and in the estuaries of these rivers where there is regular flow of tidal waters from the sea. These rivers and deltaic areas are the most preferred habitats of this species (Nayak and Padhi, 2011).

12.4.2 Reasons for Conservation

Human beings are completely dependent on other living beings. Every breath we take, every bite of food we eat and every drop of water we drink generally comes from the diversity of life. Living things are our resources and life-support systems that maintain conditions in which we can survive and prosper. Living things are the basis of species (Nayak, 2005). Crocodilians will be threatened in India owing to indiscriminate killing for commercial purposes and severe habitat loss until the 1972 Wild Life Protection Act is strictly enforced. Several countries import saltwater crocodile skins, including France, Japan, Singapore and Italy (MacNamara et al., 2003). All three species of crocodile (gharial, mugger and saltwater crocodile) in the river systems of Odisha were on the verge of extinction in the 1970s. Thus, the conservation of crocodiles was a very urgently felt need (Nayak and Padhi, 2011).

12.4.3 Reasons to Conserve Crocodiles

Crocodile populations dropped because of ever-increasing human activity in rivers and their other traditional habitats. Crocodiles are used for food, medicines and other purposes. According to Bustard (FAO 1974), among the world's 22 species of crocodile, the skin of the saltwater crocodile is used to make leather that is unsurpassed. Crocodiles feed on a wide variety of foods like crabs, fish and animal carcasses (Nayak, 2001). They bring a balance to ecosystems and keep the environment clean. They are also an attraction for ecotourists, which means greater earnings for local businesses. Therefore, we should conserve saltwater crocodiles for our own benefit as well as for nature. Apart from this, the Bhitarkanika Mangroves forest area has ecotourism potential. Bhitarkanika has been identified as an ecotourism destination in Odisha, which is a paradise for nature lovers, conservationists and biologists. The ecotourism potential has not yet been fully exploited. Few infrastructures are available in the Bhitarkanika area, such as Chanbali, Dangamal, Dhamra, Habalikhati, Gupti and Ekakulanasi, to meet the needs of tourists (Mohanty et al., 2004).

12.5 Conservation and Management of Saltwater Crocodile in Bhitarkanika Conservation Area

Despite their immense role in protecting human resources as well as biodiversity, the unique mangrove habitats of India have been facing tremendous threats owing to indiscriminate exploitation of mangrove resources for multiple uses like fodder, fuel wood and timber for building material, alcohol, paper, charcoal and medicine (Upadhyay et al., 2002). Apart from those uses, the conversion of forest area to aquaculture and agriculture, construction of ports and harbours, extensions of human habitation, overgrazing, urbanization, industrialization and chemical pollution are major common occurrences that eat into mangrove areas (Blasco and Aizpuru, 1997; Naskar, 2004; Upadhyay et al., 2002). Owing to these threats more than 33% of India's mangrove areas have been lost over the last 15 years, with the east coast area losing about 28%, the west coast area about 44% and Andaman and Nicobar Islands about 32% (Jagtap et al., 1993; Naskar, 2004).

The Brahmani, Baitarni and Mahanadi deltaic region of the Kendrapara district in the north-eastern part of the coastal state of Odisha makes up the Bhitarkanika Conservation Area (BCA). A mangrove wetland ecosystem possessing high genetic and ecological diversity, Bhitarkanika covers a total area of 2154.26 km^2, of which BKWS and the national park cover 672 km^2, the Gahirmatha (Marine) Wildlife Sanctuary covers 1435 km^2 while the buffer zone in the Mahanadi delta covers 47.26 km^2. The natural boundaries of the sanctuary are defined by the Dhamara River to the north, Maipura River to the south, Brahmani River to the west and the Bay of Bengal in the east. The coastline from Maipura to Barunei forming the eastern boundary of the sanctuary is an ecologically crucial habitat.

Bhitarkanika is a wetland of international importance, and 2672 km^2 area of wetland habitat was declared a Ramsar Site in 2002. The site contains 300 plant species belonging to 80 families of both mangroves and non-mangroves (Banerjee, 1984). Bhitarkanika supports one of the areas with the largest mangrove plant diversity in India and has more than 82 species of mangroves and its relatives. Fifty-five of the 58 Indian mangrove species (Banerjee and Rao, 1990) and 3 varieties of Sundari trees (*Heritiera* spp.), including *H. kanikensis*, an endemic species, are found here. The characteristic mangrove species are *Avicennia alba, A. officianalis, Rhizophora mucronata, Excoecaria agallocha, Acanthus illicifolius, Sonneratia apetala* and *Heritiera minor*. About 62 species of invertebrates, 19 species of fish, and 5 amphibian species are recorded in this area. Twenty-nine reptilian species, of which 4 are species of turtle, including the olive ridley sea turtle (*Lepidochelys olivacea*), 1 species of crocodile (*saltwater crocodile, Crocodylus porosus*), 9 species of lizards and 16 species of snakes including the king cobra (*Ophiophagus hannah*) are found here. BCA offers two types of bird habitats: riverine islands or heronry and coastal wetlands along the eastern boundaries. Species diversity of birds at this site is 263 belonging to 63 families, including 147 residents, 99 winter migrants, 15 vagrants and 16 local migrants (Gopi and Pandav, 2007). Included in this list are more than 79 species of migratory waterfowl. In the site community nesting of aquatic birds (heronry) occurs and 12 wetland bird species nest and roost together annually (June–October). This habitat also supports eight varieties of Kingfishers including a sizeable population of endangered brown-winged kingfisher. Thirty-one species of mammals are reported from this area; this includes 5 species of marine dolphins: the humpbacked dolphin (*Sousa*), Irrawady dolphin (*Orcaella brevirostris*), chinensis pantropical spotted dolphin, common dolphin (*Delphinus delphis*) and finless black porpoise (*Neophocaena phocaenoides*).

12.6 Present Status of Saltwater Crocodile in Bhitarkanika

12.6.1 Saltwater Crocodile Project at Dangmal-Bhitarkanika National Park

In Odisha, the estuarine or saltwater crocodile (*C. porosus*) is restricted to the mangrove swamps of the Brahmani–Baitarani Delta of the north-eastern portion of the state, which falls within the Bhitarkanika National Park. Around the mid-1970s, the population of these saltwater crocodiles dropped to a critical level, leaving only a small viable population in the main Bhitarkanika River and a few adjoining creeks. The decline of the population was mostly due to overexploitation, poaching and indiscriminate hunting. To save these greatly endangered species from extinction, a conservation programme was launched by the state government through the Forest Department.

Apart from rearing and releasing 1717 crocodiles in nature, 26 captive reared crocodiles have been supplied to other state projects. The programme for the conservation of the estuarine crocodile and its habitat was introduced in 1975 by Dr. H.R. Bustard, a consultant for FAO/UNDP. The entire mangrove habitat was declared as part of the BKWS on 22 April 1975 to protect the saltwater crocodile. The project started at Dangmal. Illegal trapping and killing of crocodiles was stopped. Every measure was taken to protect the adults, sub-adults and juveniles. The practice of egg collection from the wild and their subsequent incubation technique was launched to build up the depleted population. The reared crocodiles, which measured 1.2 m long, were released into creeks and creeklets.

An annual census of crocodiles is conducted in mid-winter. The population estimation is done by direct sighting in various creeks and rivers both during the day and at night. Night counting gives a better picture of hatchling and yearling populations since identification is easy at night. The crocodiles are classified into different categories as per age gradation: up to 0.6 to 1.2 meter are yearlings, up to 1.2 to 1.8 meter are juveniles are, up

TABLE 12.1

Census of Saltwater Crocodile (*Crocodylus porosus*) Population of Bhitarkanika Mangroves from 1995–1996 to 2014–2015

Year	Hatchling	Yearling	Juvenile	Sub-adult	Adult	Total
1995–96	304	71	34	21	10	511
1996–97	136	232	161	63	68	660
1997–98	252	106	121	113	76	668
1998–99	149	146	160	144	72	672
1999–00	319	181	123	145	146	914
2000–01	341	277	237	136	107	1098
2001–02	431	328	182	138	206	1285
2002–03	484	370	180	82	192	1308
2003–04	525	303	210	100	220	1358
2004–05	681	290	168	106	204	1449
2005–06	657	283	196	121	197	1454
2006–07	503	466	257	132	224	1482
2007–08	538	342	227	139	252	1498
2008–09	538	374	256	144	260	1572
2009–10	519	373	298	150	270	1610
2010–11	531	377	304	166	276	1854
2011–12	489	320	423	154	280	1648
2012–13	486	356	395	128	284	1649
2013–14	504	387	307	142	304	1644
2014–15	522	370	331	167	281	1671
2015–16	608	334	266	172	302	1682

Source: Department of Forest and Environment, Odisha.

to 1.8 to 2.4 meter are sub-adults and beyond 2.4 meter are adults. The presence of all age classes of crocodiles is a healthy sign of a viable population. The population of crocodiles in Bhitarkanika from 1995 to 2016 is given in Table 12.1. The largest crocodile population, at 1854 specimens, was observed during 2010–2011, while the smallest population, 511 specimens, was observed during 1995–1996 (Table 12.1).

12.6.2 Attacks on Humans by Wild Crocodiles

The saltwater crocodile is said to be the most dangerous species of crocodilians from a human standpoint. In the BKWS all manner of human attack exists. In particular, human beings are vulnerable to attack when they intrude illegally into the crocodile habitat, where the male crocodiles predominate. But it has been observed that adult crocodiles under normal circumstances never leave their territory to chase human beings on land, unlike terrestrial predatory animals like tigers. Most incidents occur when the victims enter into the crocodile habitat for illegal fishing, poaching, or collection of wood, honey and Nalia grass from the river or creek banks or to set traps for deer, wild boars and other animals.

12.6.3 Attack on Domestic Livestock by Wild Crocodiles

High tide is usually essential to bring crocodiles waiting calmly in the water within striking distance of animals grazing on the river bank, and indeed 90% of attacks occur during

TABLE 12.2

Human and Cattle Deaths by Crocodile in Bhitarkanika
from 1996–1997 to 2009–2010

Year	No. of Human Beings Killed	No. of Cattle Killed
1996–97	3	12
1997–98	2	4
1998–99	1	3
1999–00	2	4
2000–01	3	2
2001–02	2	1
2002–03	1	5
2003–04	1	3
2004–05	1	5
2005–06	1	3
2006–07	–	1
2007–08	–	3
2008–09	1	5
2009–10	2	2

Source: Department of Forest and Environment, Odisha.

the rainy season when the river banks are flooded by high tide or flood water. Cattle attacks increase the unpopularity of the saltwater crocodile and increase the pressures working against its conservation. The solution lies in maintaining a strip of undisturbed mangrove forests at least 100 m wide along all rivers/creeks adjacent to cultivated land and human habitations inside the sanctuary. Human and cattle deaths by crocodile attack in Bhitarakanika during the 1996–2010 period are given in Table 12.2. The highest number of human deaths by crocodile occurred during the 1996–1997 and 2000–2001 periods, with three occurring in both years. No humans were reported killed during 2006–2007 and 2007–2008 (Table 12.2). The highest number of cattle killed, 12, occurred during 1996–1997, and the lowest number of cattle deaths occurred in 2001–2002 and 2006–2007, with only one death reported (Table 12.2).

12.7 Conclusion

Saltwater crocodiles are considered threatened in India owing to several factors. In the river system of Bhitarkanika in Odisha, *C. porosus* was on the verge of extinction in the 1970s. Now its population is slowly increasing thanks to the programme for conserving estuarine crocodiles at Dangamala, Bhitarkanika, but it remains at unacceptable levels. Therefore, the local people should be educated about the crocodile's important role in eco-tourism and in balancing the ecosystem. If the population of this saltwater crocodile starts to decline, then immediate action with regard to crocodile farming should be undertaken by the government and non-governmental organizations, and the crocodiles should be released to natural water bodies. Currently *C. porosus* is a threatened species, and special programmes and laws should be enacted and strictly enforced to protect and conserve the saltwater crocodile in and around Bhitarkanika National Park.

References

Aubreville, A. 1964. Problemes de la mangrove d'hieret d'aujourd'hui. *Adansonia*, 4: 19–23.

Badola, R. and Hussain, S.A. 2003. Valuation of the Bhitarkanika mangrove ecosystem for ecological security and sustainable resource use. *Wetlands and Biodiversity EERC Working Paper Series: WB-1*; 1–124.

Banerjee, L.K. 1984. Vegetation of Bhitarkanika Sanctuary, Odisha state. *Journal of Economic & Taxonomic Botany*, 5: 1065–1079.

Banerjee, L.K. and Rao, T.A. 1990. *Mangroves of Odisha Coast and their Ecology*. Bishen Singh Mahendra Pal Singh, Dehra Dun, India, pp. 1–118.

Blasco, F. 1977. Outline of ecology, botany and forestry of the mangals of the Indian subcontinent. pp. 241–260. In: *Ecosystems of the World 1: Wet Coastal Ecosystems*. V.J. Chapman (ed.), Elsevier, Amsterdam.

Blasco, F. and Aizpuru, M. 1997. Classification and evolution of the mangroves of India. *Tropical Ecology*, 38: 357–374.

Blasco, F., Chanda, S. and Thanikaimoni, G. 1975. Main characteristics of Indian Mangroves. pp. 71–83. In: *Proceedings of International Symposium on Biology and Management of Mangroves*. G. Walsh, S.C. Snedaker and H.J. Teas (eds.) Institute of Food and Agricultural Science, University of Florida, Florida.

Chadha, S. and Kar, C.S. 1999. *Bhitarkanika, Myth and Reality* Natraj Publisher, Dehradun, 388.

Chapman, V.J. 1976. *Mangrove Vegetation. J Cramer, FL-9490*. Vaduz, Germany.

Clough, B.F. 1982. *Mangrove Ecosystems in Australia: Structure, Function and Management*. Australian National University Press, Canberra.

Dani, C.S. and Kar, S.K. 1999. Bhitarkanika – A unique mangrove ecosystem. In *Bhitarkanika – The wonderland of Orissa Nature and Wildlife Conservation Society of Orissa Bhubaneswar*. Behura (ed.). pp. 30–40.

Elsworth, P.G., Seebacher, F. and Franklin, C.E. 2003. Sustained swimming performance in crocodiles (*Crocodylus porosus*): Effects of body size and temperature. *Journal of Herpetology*, 37(2): 363–368.

Franklin, C.E., Taylor, G. and Cramp, R.L. 2005. Cholinergic and adrenergic innervation of lingual salt glands of the estuarine crocodile, *Crocodylus porosus*. *Australian Journal of Zoology*, 53(6): 345–351.

Gopi, G.V. and Pandav, B. 2007. Observations on breeding biology of three stork species in Bhitarkanika mangroves, *India*. *Indian Birds*, 3(2): 45–50.

Grzimek, B., Illies J. and Klausewitz, W. 1976. *Grzimek's Encyclopedia of Ecology*. Van Nostrand Reinhold Company, New York.

Isberg, S.R., Thomson, P.C., Nicholas, F.W., Barker, S.G. and Moran, C. 2005. Quantitative analysis of production traits in saltwater crocodiles (*Crocodylus porosus*): I. reproduction traits. *Journal of Animal Breeding and Genetics*, 122: 361–369.

Jagtap, T.G., Chavan, V.S. and Untawale, A.G. 1993. Mangrove ecosystems of India: A need for protection (synopsis). *AMBIO*, 22: 252–254.

MacNae, W. 1968. A general account of the fauna and flora of mangrove swamps and forests in the Indowest Pacific region. *Advances in Marine Biology*, 6: 73–270.

MacNamara, K., Nicholas, P., Murphy, D., Riedel, E., Goulding, B., Horsburgh, C., Whiting, T. and Warfield, B. 2003. RIRDC – Publ. No 02/142, Canberra.

Mohanty, S.C., Kar, C.S., Kar, S.K. and Singh, L.A.K. Wild Odisha, 2004. *Wild life Organisation*. Forest Department, Govt. of Odisha, Bhubaneswar, 68–71.

Nagloo, N., Colllin, S.P., Hemmi, J.M. and Hart, N.S. 2016. Spatial resolving power and spectral sensitivity of the salt water crocodile, *Crocodylus johnstoni*. *Journal of Experimental Biology*, 219: 1394–1404.

Naskar, K.R. 2004. *Manual of Indian Mangroves*. Daya Publishing House, New Delhi, India.

Naskar, K.R. and Guha Bakshi, D.N. 1987. *Mangrove Swamps of the Sundarbans – An Ecological Perspective*. Naya Prakash, Calcutta, India.

Nayak, L. and Padhi, P. 2011. Conservation and management of salt water Crocodile (*Crocodylus porosus*) in relation to some physico-chemical parameters from Bhitakanika sanctuary, Odisha. *Nature Environment and Pollution Technology*, 10(3): 389–394.

Nayak, L. 2001. *Recent Trends in Aquaculture. Crocodile culture, 159–162*. Berhampur University Publisher.

Nayak, L. 2005. Loss of marine biodiversity – Conservation of sea turtles along the Odisha coast. *Journal of Indian Ocean Studies*, 13(1): 141–146.

Pough, F.H., Andrews, R.M., Cadle, J.E., Crump, M.L., Savitzky, A.H. and Wells, K.D. 2004. Archosauria: crocodilians (crocodylia). In *Herpetology* (pp. 166–169): Prentice Hall, Upper Saddle River, NJ.

Saenger, P., Hegerl E.J. and Davie, J.D.S. 1983. *Global Status of Mangrove Ecosystems*. Commission on Ecology paper no. 3. IUCN, Switzerland.

Thatoi, H.N. 2004. Study on vegetation in mangrove Forests of Bhitarkanika Wildlife Sanctuary. Project. Report submitted to Forest Department (Mangrove Forest Division, Rajnagar) Government of Orissa, 23–26.

Tomlinson, P.B. 1986. *The Botany of Mangroves*. Cambridge University Press, New York.

UNDP/UNESCO. 1986. *Mangrove of Asia and the Pacific: Status and Management*. Quezon City.

Upadhyay, V.P., Ranjan R. and Singh, J.S. 2002. Human mangrove conflicts: The way out. *Current Science*, 83: 1328–1336.

13

Management Challenges of Sinking of Oil Tanker at Shela Coastal River in Sundarbans Mangroves in Bangladesh

Md. Nazrul Islam, Md. Al Amin and Md. Noman

CONTENTS

13.1 Introduction 323
    13.1.1 Importance of Sundarbans Mangroves 324
        13.1.1.1 Ecosystem Service Values 324
        13.1.1.2 Tourism Sites and Their Importance 325
        13.1.1.3 Ecosystem Health and Its Indicators 326
13.2 Sinking of Oil Tanker in Shela River in the Sundarbans 326
    13.2.1 Significance of the Study 327
    13.2.2 Aims and Objectives 328
    13.2.3 Methods and Data Sources 328
13.3 Results and Discussion 329
    13.3.1 Environmental Impacts 329
    13.3.2 Impact on Water Quality 330
    13.3.3 Impacts on the Biodiversity of the Sundarbans Mangroves 330
        13.3.3.1 Impact on Aquatic Environment 330
        13.3.3.2 Status of Phytoplankton and Zooplankton 331
13.4 Management Challenges of Sundarbans Forest Regeneration 332
    13.4.1 Natural Regeneration Process and Challenges 335
    13.4.2 Challenges Associated with Avoiding Oil Spills in the Future 335
    13.4.3 Lack of Scientific Management Approaches 336
    13.4.4 Other Challenges of Sundarbans 337
13.5 Recommendations for Ecosystem Development 337
13.6 Conclusion 338
References 338

13.1 Introduction

Bangladesh is a low-lying country of South Asia. It is situated between the Himalayas in the north and the Bay of Bengal in the south (Siddique, 2014). The snow melt from the Himalayas runs towards the Bay of Bengal through India and Bangladesh and creates many rivers in Bangladesh. Thus, Bangladesh is known as a riverine country (Uddin, 2011; Hoff, 2012; Chowdhury, 2014; Rahman et al., 2015; Razzaque, 2017). The Ganges (Padma), the Brahmaputra and the Meghna are the main rivers of Bangladesh. These rivers meet at the Bay of Bengal and create many deltas, the deltas of the Ganges, Brahmaputra and Meghna (Karim, 1994). The Sundarbans, one of the largest mangrove forests in the world, lies there. The Sundarbans mangrove forest occupies an area of around 140,000 ha and lies on the delta of the Ganges, Brahmaputra and Meghna Rivers on the Bay of Bengal. It is adjacent to the border of India (Hoff, 2012; Chowdhury, 2014; Rahman et al., 2015; Razzaque, 2017). The Sundarbans was declared a World Heritage Site in 1987. The site is intersected by a complex network of tidal waterways, mudflats and small islands of salt-tolerant mangrove forests and presents an excellent example of ongoing ecological processes (Champion and Seth, 1968). The area is known for its wide range of flora and fauna, including 260 bird species, the Bengal tiger and other threatened species such as the estuarine crocodile and the Indian python (Kajanus et al., 2012).

The Sundarbans is a comprehensive ecosystem comprising one of the three largest single tracts of mangrove forests of the world. But we are losing the value of our natural beauty 'Sundarbans' by doing unwise things (Ellison, 2002). The Padma Oil Company tanker *Southern Star VII*, which sank in the Shela River, is a recent example. It was almost a disaster for the forest and created a grave threat to the habitat and ecosystem (Hoff, 2012; Chowdhury, 2014). It will have adverse effects on the health of the Sundarbans ecosystem. Sweet water dolphins were the first in line of victims because the Shela River is known as a sanctuary for them. A thick layer of oil on river water will reduce the level of dissolved oxygen, causing breathing difficulties for these dolphins (Rahman et al., 2010; Uddin, 2011; Marzan et al., 2017). Because of the high tides in the forest, the spilled oil remains on the vegetation and the forest topsoil, damaging the vegetation, which is the main food for various kinds of deer (Kajanus et al., 2012). A diminishing number of deer in the long run will affect the population of tigers. The reproduction of saltwater trees, such as *sudari, kewra, goran, poshur* and *gol*, will be threatened since they are grown from the seeds that fall on the ground and the oil spill will hinder seed germination. This will further affect the food supply of other animals (Rahman et al., 2015; Razzaque, 2017). The pungas fish and crocodiles will also be under threat because oil has already started settling on the vegetation and soil on the banks of the Shela River and on waterborne moss. A solution to this problem must be devised.

However, the reserve is used by a large number of rural communities located within a 20-km-wide zone outside the forest boundary. The total population living in the identified area around the reserve is estimated to be as high as 855,000 (Rahman et al., 2010; Uddin, 2011; Marzan et al., 2017). It is estimated that approximately 30% of the nearby population, or 300,000 people, are dependent on the solid recovered fuel Sundarbans Reserved Forest (SRF) for their livelihoods, with around 200,000 regularly collecting resources from the Sundarbans. More than a million people depend on the Sundarbans for their livelihoods, with a large part involved in various resource collections, including working seasonally as nipa palm and other non-timber resource collectors, fishermen and honey hunters (Kajanus et al., 2012). The SRF is an important source of revenue for the government. A significant part of its value derives from the extensive shrimp breeding and nursery grounds supporting this important export industry. Approximately 8300 people reside close to the accident site.

13.1.1 Importance of Sundarbans Mangroves

13.1.1.1 Ecosystem Service Values

The importance of the Sundarbans for Bangladesh and its economy is beyond description. The Sundarbans mangrove ecosystem has diverse, valuable goods and services (Rahman et al., 2010; Uddin, 2011; Marzan et al., 2017). The major ecosystem services of the Sundarbans include services related to timber, fish, thatching materials, fuel wood, crab, honey and wax,

cultural services tourism, worship, regulatory services coastal protection, storm regulation, flood regulation, supporting service biodiversity and habitat for wildlife and plants
(Rahman et al., 2015; Razzaque, 2017). This study reveals the economic value of provisioning, cultural and regulatory services of the Sundarbans. The value of regulatory services in
the Sundarbans demonstrates the value of coastal protection by estimating it based on their
value calculated in studies from other countries that are located in the same geographical areas (Chaffey et al., 1985). However, provisioning for example timber, fish and cultural
tourism services has been estimated only for the products extracted and tourist activities
annually, not for the total stock of products, for example timber, fish or the potential value of
tourism in all of the Sundarbans (National Academy of Sciences, 1985). The economic value
of the whole stock of services, tourism potential, various regulatory services and potential
value of the supporting services of the Sundarbans could be much higher than present estimates (Eaton, 1991). Since the forest is located in the southern part of the Tropic of Cancer
and bounded by the northern limits of the Bay of Bengal, it is classified as a tropical moist
forest (IUCN, 1997). The temperatures in the Sundarbans are more equable than those of the
adjacent land areas (Rahman et al., 2010; Uddin, 2011; Marzan et al., 2017).

The average annual maximum and minimum temperatures vary between 30 and 21°C.
High temperatures occur from mid-March to mid-June and low temperatures in December
and January. The mean maximum recorded temperature for the hottest months was 32.4°C
at Patuakhali, in the eastern part of the Sundarbans (Rahman et al., 2015; Razzaque, 2017).
The mean annual relative humidity varies from 70% at Satkhira to 80% at Patuakhali.
Humidity is highest in June–October and lowest in February (Akhter, 2002). Annual rainfall in the Sundarbans is in the range of 1640–2000 mm, and rainfall increases from west to
east. Most rainfall occurs during the monsoon season from May to October. Frequent and
heavy showers occur from mid-June to mid-September. Often storms accompanied by tidal
waves result in widespread inundation and cause damage to vegetation and animal life.

13.1.1.2 Tourism Sites and Their Importance

The enchanting stories about the Sundarbans, its emerald mangrove forests that seem like
an unspoiled showcase for the diversity of life, its wildlife, the rivers, canals and creeks
flowing deep inside the forest have always attracted human interest (Chaudhuri and
Choudhury, 1994). In earlier days, before the advent of power boats, the only means of
transport within the forest was forbiddingly difficult. It was a hazardous journey made
without sufficient drinking water, food and safety measures (Blower, 1985). Starting in
2000 guided tour operators with their custom-built vessels equipped to go inside the forest
allowed tourists to stay in the forest for several days with ample supplies of food, drinking
water and accommodations (Rahman et al., 2010; Uddin, 2011; Marzan et al., 2017). This has
attracted domestic and foreign tourists to behold the beauty of silence. The scenic beauty
with the sandy beach is found and presents an opportunity to see both sunrise and sunset
(Rahman et al., 2015; Razzaque, 2017). Facilities include a rest house, or the facilities of the
nearby town may be used. The important tourist spots in the Sundarbans are as follows:

Katka: Katka is one of the heritage sites in the Sundarbans. Katka of the Bagerhat range
is a special sanctuary administered by the Forest Department. Here, tiger and deer hunting is forbidden. This site attracts tourists for its beauty and wildlife (Ahmed et al., 2004).
At Katka there is a wooden tower 12.19 meters (40 feet) high from where you can enjoy
the scenic beauty of the Sundarbans. A beautiful beach on the ocean called Badamtali Sea
Beach can be enjoyed at Katka while walking to the beach from the watchtower. Katka is
full of different plant species. Blackberry trees are found in abundance in Katka. Sundari,

gewa, kewra, dhundal and other trees also occur in Katka. Tiger, deer, crocodile, monkey, several varieties of birds, wild hen and other animals are also found there. The scenery of Badamtali Sea Beach is very attractive. Katka is famous for its natural beauty and incredible plant species, which give the place extra beauty.

Hiron Point: Hiron Point is another tourist spot in the Sundarbans. It is called the world heritage state. Hiron Point is a great site for spotting tigers and other spectacular and rare wildlife. Also known as Nilkamal, it is well known for its tigers, deer, monkeys, crocodiles and many precious birds. The site attracts many tourists owing to its natural beauty and primal splendor. Here one can watch rare species of wild animals and birds, as noted in the diaries of bird watchers and wildlife enthusiasts.

Dublar Char: Dublar Char, a small, beautiful island in the southern region of the Sundarbans facing the Bay of Bengal, is famous for its beautiful scenic spots. A dry fish processing factory was recently established there. The island has all the natural beauty of any of the world's famous islands (Akhter, 2006). Herds of spotted deer graze on it. On a casual walk around the island tourists will likely see many wild animals (Barbier and Sathirathai, 2004). With water all around and lots of fish fauna, Dublar Char offers visitors a unique experience.

Herbaria: The Herbaria houses tigers. It is the most attractive and dangerous place in the Sundarbans. It is located beside the Poshur River and in the Chadpai range of the Sundarbans. Here one can see the famous Royal Bengal tiger as well as many other wild animals and birds, though the spotted deer is the main attraction. Sundari trees are ubiquitous, and a watchtower affords a view of the entire forest.

13.1.1.3 Ecosystem Health and Its Indicators

While health in human terms is defined as more than merely the absence of symptoms of disease, in common medical practice doctors rely on screening for symptoms of malfunction to assess a person's health status. Similarly, for nature health is commonly taken to be the absence of detectable symptoms of ecosystem pathology (Rahman et al., 2010; Uddin, 2011; Marzan et al., 2017). The ecosystem is something more than a community of species but less than the biosphere (Rapport, 1992). An ecosystem indicator can be any measure that provides information about the quality or conditions of the ecosystem or the effectiveness of its management. Because not all ecosystem components and processes can be measured and evaluated, ecological indicators are used to determine ecosystem health with a reduced set of measurements that can represent or 'indicate' the overall state of the system (Rapport, 1992). The study findings focus on the following areas: oil spill extent, response operations, environmental impacts on aquatic environments, mangroves and wildlife, human and socioeconomic impacts, health impacts and livelihoods (Hoff, 2012; Chowdhury, 2014; Rahman et al., 2015; Razzaque, 2017). This chapter begins with a general overview of the Sundarbans area. It then moves on to an assessment of the oil spill extent, after which the findings and activities related to the response operations are described. Finally, an assessment is given of the potential impacts on the environment and human beings.

13.2 Sinking of Oil Tanker in Shela River in the Sundarbans

On 9 December 2014, an oil tanker accident in the Sundarbans of Bangladesh led to the release of approximately 358,000 litres of heavy fuel oil into the river and mangrove

FIGURE 13.1
On 9 December 2014, an oil tanker accident occurred on the Shela River in the Sundarbans of Bangladesh.

ecosystem (IPCC, 2014). Shela River is near Mongla Port and is about 100 km (62 mi) from Kolkata Port. A number of factors, including timely tidal variations and the decision to ban tanker traffic in the river, minimized the penetration of oil into the mangrove ecosystem (Yender et al., 2002). Nonetheless, the oil spill accident must be considered serious because it occurred in a wildlife sanctuary, a World Heritage Site and a Ramsar Site treasured for its unique biodiversity (Rahman et al., 2010; Uddin, 2011; Marzan et al., 2017). The lack of a formal oil spill contingency plan, which among other things would have designated an appropriate competent overseeing authority as well as the limited experience and response infrastructure, made response and recovery efforts challenging (Hoff, 2012; Chowdhury, 2014; Rahman et al., 2015; Razzaque, 2017). The lack of training, appropriate equipment and experience resulted in unintended negative impacts on the local community with immediate health impacts such as difficulties in breathing, headaches and vomiting reported among community responders. While water and shoreline clean-up operations are over, the removal of oiled debris, the management of response generated waste and the assessment of a final disposal option need to continue (Habib, 1989). As a coastal mangrove forest, the vegetation in the Sundarbans gets inundated twice a day at high tide. Now that there is oil, as water recedes with low tide, the oil will remain on the vegetation and the forest topsoil (Figure 13.1).

The vegetation is the main food of various kinds of deer that live in the dense forest surrounding the river. The deer, in turn, is one of the main foods of Bengal tigers. Thus, in the long run, the population of deer and tigers, the two best known animals from the Sundarbans, will be affected (Banerjee and Santra, 1999). The mangrove ecosystem of the Sundarbans is primarily made up of four kinds of saltwater trees: sundari, kewra, goran, poshur and gol. These trees reproduce from the windfall seeds that fall on the ground (Hoff, 2012; Chowdhury, 2014; Rahman et al., 2015; Razzaque, 2017). As oil settles on the forest topsoil, these seeds will die, and in the long run, the regeneration of the Sundarbans will be adversely affected. That in turn will put at risk the deer and different types of primates that depend on these trees for their lives (Ali, 1998). These windfall seeds are the staple food of pungash fish that also inhabit the Shela waters. This fish, again, is one of the main foods of crocodiles, a famous reptile from these forests. If the pungash have nothing to eat, they will die eventually, putting the lives of crocodiles at risk as well.

13.2.1 Significance of the Study

As an issue, ecosystem health is very complex owing to its different elements, indicators and interrelated relationships. As a result, narrowing down the focus is very difficult but it is important for the sake of the study. This study regards ecosystem health assessment after the sinking of the oil tanker *OT Southern Star VII* in the Shela River in the Sundarbans. This study looks at the health of nature, flora, fauna, plankton, biological oxygen demand (BOD) and aquatic ecosystem based on a data analysis. The assessment was done at both the national and local levels (Sundarbans) and on the basis of water quality, fishing and forest resource data.

13.2.2 Aims and Objectives

The main aim of this study is to evaluate the health of the Sundarbans ecosystem following the sinking of the oil tanker *OT Southern Star VII*. The objective is to determine the path of the pollution, to identify the species affected by the spreading fuel oil in the Shela River in the Sundarbans and to try to determine the ecosystem health gap by comparing conditions before and after the sinking of the oil tanker. To carry out the study, the following objectives are considered:

  i. Investigate the situation before and after the sinking of the oil tanker in the Shela River;
  ii. Examine the present state of ecosystem health in the Shela River in the Sundarbans;
  iii. Assess local people's perceptions about the challenges related to the management of this ecosystem.

13.2.3 Methods and Data Sources

Different kinds of data sources, primary and secondary, were used. The Department of Forest of the Ministry of Environment and Forests (MoEF) of the Government of Bangladesh (GoB) provided oversight and guidance on a community-based response effort. Fishermen and community members were engaged to collect oil. Additionally, fine-mesh nets were placed across the mouths of tributary/distributary creeks and channels to prevent the oil from entering them during rising tides (Balasubramanian and Ajmal Khan, 2002). Fishing nets were locally deployed along the river banks to recover drifting oil. The response operations were carried out over approximately 12 days (from 12 to 22 December), during which a reported 68,200 litres of oil were collected by communities and purchased by the Bangladesh Petroleum Corporation (BPC). On 15 December, the Economic Relations Division (ERD) of the Ministry of Finance submitted a quest to the United Nations Development Programme (UNDP) (Annex 1) to provide technical assistance to (1) assess oil spill containment and clean-up needs and (2) conduct an assessment and draft an action plan for recommended mitigation measures. On 17 December, a United Nations Disaster Assessment and Coordination (UNDAC) team was deployed to support the UNDP in the assessment and coordination, under the leadership of the United Nations Environment Programme (UNEP)/ UN Office for the Coordination of Humanitarian Affairs (OCHA) Joint Environment Unit (JEU). The team was subsequently reinforced by specialists supported by USAID, the French Ministry of Ecology, Sustainable Development and Energy (France) and the European Commission through the European Union Civil Protection Mechanism, as well as by representatives of government, academia, UNDP, other agencies and nongovernmental organizations (NGOs).

13.3 Results and Discussion

The Sundarbans is the largest delta, backwater and tidal area of the region and thus provides diverse habitats for several hundred aquatic, terrestrial and amphibian species. The property is of sufficient size to adequately represent its considerably high floral and faunal diversity, with all key values included within the boundaries. Unfortunately, an empty cargo vessel hit the oil tanker on account of poor visibility from the dense dawn fog. The oil spill spread about 65 km upstream and downstream of the Shela and at least 20 canals linked to that of the river of the Sundarbans (Rahman et al., 2010; Uddin, 2011; Marzan et al., 2017). Different small plants, dolphins, crocodiles and other aquatic animals were adversely affected by this water pollution. Moreover, the world's largest mangrove forest has also been adversely affected by this serious threat to its aquatic life, both flora and fauna. Several impacts can be identified from the sinking of the oil tanker in the Shela River in the mangrove forest of the Sundarbans. These impacts are discussed in the following subsections.

13.3.1 Environmental Impacts

Environmentalists warned that the event was an ecological catastrophe, as the spill occurred in a protected area where rare dolphins were present. Experts expressed concerns that the oil spill would hamper the well-being of the aquatic organisms in the area. Wildlife near the river is at risk of dying because the smell of oil makes breathing difficult. Some images indicate that the disaster killed some animals (Table 13.1).

The oil spill also poses a major threat to the forest's food cycle. Reports from various sources showed that microorganisms, the primary level of the food cycle, are dying. The UN expressed deep concern over the oil spill, urging the GoB to impose a complete ban on

TABLE 13.1

Oil Tanker Spill Affected Species in the Shela River and Mangrove Sundarbans

High Effect/Higher Value/ Higher Risk Species	Higher Value/Lower Risk/ Medium Effect	Low Effect/Higher Value/ Lower Risk
Crab	Bengal tiger (*Panthera tigris*)	Wild boar
Shrimp	Spotted deer (*Axis axis*)	Spotted deer
Faisha fish	King cobra (*Ophiophagus hannah*)	Monkey
Phytoplankton	Wild boar (*Sus scrofa*)	Passerine birds
River terrapin (*Batagur baska*)	Lesser adjutant (*Leptoptilos javanicus*)	Leopard cat (*Prionailurus bengalensis*)
White-rumped vulture (*Gyps bengalensis*)	Oriental magpie robin (*Copsychus saularis*)	Indian crested porcupine (*Hystrix indica*)
Zooplankton	Rhesus monkey (*Macaca mulatta*)	Osprey (*Pandion haliaetus*)
Finless porpoise (*Neophocaena phocaenoides*)	Barking deer (*Muntiacus muntjak*)	Common vine snake (*Ahaetulla nasuta*)
Small-clawed otter (*Aonyx cinerea*)	Marbled toad (*B. stomaticus*)	Brahminy kite (*Haliastur indus*)
Estuarine crocodile (*Crocodylus porosus*)	Brown fish owl (*K. zeylonensis*)	White-bellied sea eagle (*Haliaeetus leucogaster*)
Masked finfoot (*Heliopais personatus*)		Indian flapshell turtle (*Lissemys punctata*)
Fishing cat (*Prionailurus viverrinus*)		Spot-tailed pit viper (*Trimeresurus erythrurus*)
Pallas's fish eagle (*Haliaeetus leucoryphus*)		Dusky eagle-owl (*Bubo coromandus*)

commercial vessels moving through the forest (Rahman et al., 2010; Uddin, 2011; Marzan et al., 2017). Several species that are at risk as a result of the spill have been identified by *National Geographic Traveler*. These species are the Irrawaddy dolphin, Bengal tiger, leopard, great egret, rhesus macaque, northern river terrapin, black-capped kingfisher, chital, saltwater crocodile and horseshoe crab.

13.3.2 Impact on Water Quality

In the Sundarbans the range between high and low tide is 3–4 m. The tidal range at Mongla is higher than at Hiron Point. The mean tidal level is 2.310 m, 1.829 m and 1.700 m at Mongla, Sundrikota and Hironpoint, respectively. The tidal amplitude at spring tide is 2.5–3 times higher than at neap tide (Baksha and Lapis, 2000). Salinity concentration is higher at spring high tide than at neap tide (Rahman et al., 2010; Uddin, 2011; Marzan et al., 2017). Water flow and the tides in small cross-connected channels depend on the timing and magnitude of the high water in the main channels. In addition, there is a time gap between the water flow and tide from one estuary to another (Table 13.2).

The duration of ebb tide is longer than that of flood tide, i.e., water flow is higher during flood tide than ebb tide (Agrawala et al., 2003). The ebb tide generally does not transport back all the silt/oil that is carried in during flood tide and causes sedimentation (Hoff, 2012; Chowdhury, 2014; Rahman et al., 2015; Razzaque, 2017). The oligohaline zone's sodium chloride content is less than 5 parts per thousand (ppt), which occurs in a small area of the north-eastern part of the forest. The mesohaline zone has a sodium chloride content of 5–10 ppt and covers the north-central to south-eastern part of the forest. The polyamine zone's sodium chloride content exceeds 10 ppt and covers the western part. Sundarbans estuaries/rivers are laden with high concentrations of suspended sediment (Hoq et al., 2006). Suspended sediment varies not only spatially but also in the vertical water column. Visibility is very poor and there is high turbulence in the estuary.

Many experts, both national and international, say that any commercial vessel movement, such as from oil tankers, is harmful for the wildlife in the forest. To ensure an ecological balance in the Sundarbans, the UNDP and GoB agreed to work together. The extent of the damage caused by the oil spill will be assessed in the next 5 to 10 years (Kajanus et al., 2012). It can be calculated by investigating how large an amount of phytoplankton and zooplankton remains in the water of the affected river and channel region and compare that with other restricted regions of the Sundarbans (Rahman et al., 2010; Uddin, 2011; Marzan et al., 2017). In fact, water quality is a complex subject, in part because water is a complex medium intrinsically tied to the ecology of the Earth. The water salinity level of the Shela River slowly increases over time and several rivers are affected by the salinity in the Sundarbans. Mangroves in the Sundarbans are spread across the Ganges Delta. The increased salinity is caused by the scarcity of freshwater in the Ganges. The reduction in the flow rate of the Ganges River has resulted in high saline water in upstream areas. The high-salinity zone is situated in the south-western corner of the Sundarbans.

13.3.3 Impacts on the Biodiversity of the Sundarbans Mangroves

13.3.3.1 Impact on Aquatic Environment

Oil consists of hydrocarbons ranging from volatile organic compounds to complex non-biodegradable ones (e.g., asphaltenes). Spilled oil in coastal waters is acted upon by a number of chemical and physical processes, collectively known as weathering. The way in

TABLE 13.2

Oil Collection Time Duration and Agency Achievements

Date	Event
9 December 2014	Oil tanker accident in Chandpai Wildlife Sanctuary of Bangladesh Sundarbans results in release of approximately 358,000 L of heavy fuel oil. UNDP submits proposal to respond to oil spill to MoEF.
10 December	MoEF, Department of Forest and Department of Environment begin response. MoEF forms an expert committee that includes relevant government stakeholders and academics to assess the environmental damage due to the oil spill and to put forward suggestions for reducing such damage in the future.
11 December	Bangladesh Petroleum Corporation starts buying collected oil from community. Vessel salvaged and towed to Joymoni. Department of Environment of MoEF begins collection and analysis of water samples from oil-affected area, continuing until 26 December.
12–13 December	MoEF 13-person assessment team visits accident site.
15 December	GoB convenes and creates inter-ministerial body headed by Ministry of Shipping to address the spill.
16 December	UNDP proposal to respond to the spill approved by Ministry of Finance, Economic Relations Division.
18–20 December	UN-led team of experts arrives in Dhaka from France, Japan, Switzerland, USA and from across Bangladesh.
20 December	Vessel towed to Mongla.
21 December	Oil collection stops; BPC reported purchasing a total of 68,200 L of oil. MoEF assessment team submits its report.
22–27 December	Assessment team conducts field work in and around spill site.
31 December	Assessment team presents preliminary conclusions and recommendations to GoB.
1–15 January 2015	Assessment team incorporates feedback from key stakeholders and submits final report to GoB.

Source: Modified from Hoff, R. (ed.). 2012. *Oil Spill in Mangroves – Planning & Response Considerations.* NOAA Ocean Service, National Oceanic and Atmospheric Administration, Seattle 72 pp.; Chowdhury, A. 2014. *Impact of Oil Spillage on the Environment of Sundarbans (World's Largest Mangrove Forest) in Bangladesh – Draft Report.* Khulna University; Rahman, Md.M. et al. 2015. *Wetlands Ecological Management,* 23: 269–283; Razzaque, J. 2017. *Transnational Environmental Law.* Available at: http://eprints.uwe.ac.uk/30904.

which spilled oil behaves depends largely on how persistent the oil is, and its persistence in the environment depends on a series of factors, including the amount and type of oil spilled along with local meteorological and oceanic conditions (Table 13.3).

13.3.3.2 Status of Phytoplankton and Zooplankton

The Shela River, where the oil tanker sank, and the Pashur, another affected place, are home to several species, including plants, rare plants, Irrawaddy and Ganges dolphins (Arroyo, 1979; Akhter, 2006; Amin et al., 2016). The catastrophe of the oil tanker is unprecedented in the Sundarbans, and it will have a long-term impact on vast populations of small fishes, dolphins, plankton, algae, fungi, plants, animals and mangrove ecology. Because of the oil spill, the water quality has changed (Rahman et al., 2010; Uddin, 2011; Amin et al., 2016; Marzan et al., 2017). Earlier data revealed that there were approximately 300–400 phytoplankton and 20–30 zooplankton species present in a litre of water in the Shela River, but following the sinking of the oil tanker only about 20–30 phytoplankton and 2 zooplankton species have been found in a litre of Shela River water (Akhter, 2006; Hoq et al., 2006). The oil that spread into the Shela River has had a severe impact on the aquatic lives and mangrove plants growing

TABLE 13.3

Oil Tanker Sinking Threatened Species in Shela River and Mangrove Sundarbans

Gravely Threatened	Threatened
Crab	Bengal tiger (*Panthera tigris*)
Shrimp	Spotted deer (*Axis axis*)
Faisha fish	King cobra (*Ophiophagus hannah*)
Phytoplankton	Wild boar (*Sus scrofa*)
River terrapin (*Batagur baska*)	Lesser adjutant (*Leptoptilos javanicus*)
White-rumped vulture (*Gyps bengalensis*)	Oriental magpie robin (*Copsychus saularis*)
Zooplankton	Rhesus monkey (*Macaca mulatta*)
Finless porpoise (*Neophocaena phocaenoides*)	Barking deer (*Muntiacus muntjak*)
Small-clawed otter (*Aonyx cinerea*)	Marbled toad (*B. stomaticus*)
Estuarine crocodile (*Crocodylus porosus*)	Brown fish owl (*K. zeylonensis*)
Masked finfoot (*Heliopais personatus*)	
Fishing cat (*Prionailurus viverrinus*)	
Pallas's fish eagle (*Haliaeetus leucoryphus*)	

along the shorelines where a thick layer of oil was deposited. Fish and other aquatic animals started dying after a day or two, while the mangroves might have started dying after a month. According to many wildlife experts, the spill has also caused a huge problem on the shorelines of canals and rivers from which the animals of the mangrove forest drink (Rahman et al., 2010; Uddin, 2011; Marzan et al., 2017). According to United Nation's erythrocyte protoporphyrin (EP) level, the rates of concentration of dissolved chemicals should not exceed 10 milligrams. Anything higher than this level will threaten animal and plant life. Testing in the Pasur and Shela Rivers showed the presence of 500–1400 mg of dissolved oil. As a result, small fish, phytoplankton and zooplankton cannot live (Amin et al., 2016). A large number of animals died as a result of the oil spill. Other animals moved elsewhere. Oil enters the body of phytoplankton, zooplankton, crab, snail, etc. and as results the breath of these animals has been decreased (Atmadja and Verchot, 2012). As a result they die from oxygen deprivation. In a mangrove forest, tree roots have holes. Spilled oil will clog up these holes and the trees will die (Rahman et al., 2010; Uddin, 2011; Marzan et al., 2017). If less plankton grow, aquatic species such as fish will have less food to eat, which will affect other species dependent on fish. A 15-member team led by Dr. Rouf found in its study that two types of plankton – phytoplankton and zooplankton – have decreased in the river by around 40 and 80 pc, respectively (Rouf et al., 2015). Seven days following the tanker accident, the team found 600 phytoplankton and 100 zooplankton per liter of river water (Rahman et al., 2010; Uddin, 2011; Marzan et al., 2017). In previous years, the number of phytoplankton was 1000 to 2000 and that of zooplankton was 500 to 600 per liter of water during the winter season. Aquatic species that failed to migrate must have consumed oil, which will have a negative effect on their life cycle and their population (Kajanus et al., 2012).

13.4 Management Challenges of Sundarbans Forest Regeneration

DPSIR Framework

In a recommendation to the European Environment Agency (EEA) on how it should proceed with the development of a strategy for an integrated environmental assessment,

RIVM proposed the use of a framework that distinguished between driving forces, pressures, states, impacts and responses. This became known as the DPSIR framework and has since been more widely adopted by the EEA, acting as an integrated approach for reporting, e.g., in the EEA's State of the Environment reports. The framework is seen as giving a structure within which to present the indicators needed to enable feedback to policymakers on environmental quality and the resulting impact of political choices made or to be made in the future (EEA, 1998).

According to the DPSIR framework, there is a chain of causal links starting with *'driving forces'* (economic sectors, human activities) through *'pressures'* (emissions, waste) to *'states'* (physical, chemical and biological) and *'impacts'* on ecosystems, human health and functions, eventually leading to political *'responses'* (prioritization, target setting, indicators). Describing the causal chain from driving forces to impacts and responses is a complex task and tends to be broken down into subtasks, e.g., by considering the pressure-state relationship (EEA, 1998). The components of the DPSIR framework are defined in what follows. Classes of data referring to past and present situations are listed after each definition.

Driving Forces

A driving force is a need. Examples of primary driving forces for an individual are the need for shelter, food and water, while examples of secondary driving forces are the need for mobility, entertainment and culture. For an industrial sector a driving force could be the need to be profitable and to produce at low costs, while for a nation a driving force could be the need to keep unemployment levels low. In a macroeconomic context, production or consumption processes are structured according to economic sectors (e.g., agriculture, energy, industry, transport, households).

- Population (number, age structure, education level, political stability)
- Transport (people, goods, road, water, air, off-road)
- Energy use (energy factors based on type of activity, fuel types, technology)
- Power plants (types of plant, age structure, fuel types)
- Industry (types of plant, age structure, resource types)
- Refinery/mining (types of plant/mining, age structure)
- Agriculture (number of animals, types of crops, stables, fertilizers)
- Landfills (type, age)
- Sewage systems (types)
- Non-industrial sectors
- Land use

Pressures

Driving forces lead to human activities such as transportation or food production, they meets people's needs. These human activities exert pressures on the environment as a result of production or consumption processes that can be divided into three main types: (1) excessive use of environmental resources, (2) changes in land use and (3) emissions (of chemicals, waste, radiation, noise) into the air, water and soil.

- Use of resources
- Emissions (per driving force for numerous compounds)
- Direct emissions into air, water and soil
- Indirect emissions into air, water and soil
- Production of waste
- Production of noise
- Radiation
- Vibration
- Hazards (risks)

States

As a result of pressures, the state of the environment is affected, that is the quality of the various environmental compartments (e.g., air, water, soil) are affected in relation to the functions that these compartments fulfill. The state of the environment is thus the combination of physical, chemical and biological conditions.

- Air quality (e.g., national, regional, local, urban)
- Water quality (rivers, lakes, seas, coastal zones, groundwater)
- Soil quality (national and local regions, natural areas, agricultural areas)
- Ecosystems (biodiversity, vegetation, soil organisms, water organisms)
- Humans (health)
- Soil use

Impacts

Changes in the physical, chemical or biological state of the environment determine the quality of ecosystem health and the welfare of human beings. In other words, changes in the ecosystem state may have environmental or economic impacts on the functioning of ecosystems, their life-supporting abilities and, ultimately, human health and the economic and social performance of society.

Responses

A response by society or policymakers is the result of an undesired impact and can affect any part of the chain between driving forces and impacts. An example of a response related to driving forces is a policy to change modes of transportation, e.g., from private (cars) to public (trains), while an example of a response related to pressures is a regulation concerning permissible SO_2 levels in flue gases.

Figure 13.2 depicts the complete DPSIR framework. In addition to defining the components of DPSIR, it is useful to describe the various cause-and-effect relationships (because it is often difficult to attribute ecosystem changes unambiguously to human pressures). NERI (National Environmental Research Institute, Denmark) has proposed a methodology in which environmental problems are defined and structured in such a way that a clear relationship between them and pressures emerges. This often involves the use of physical

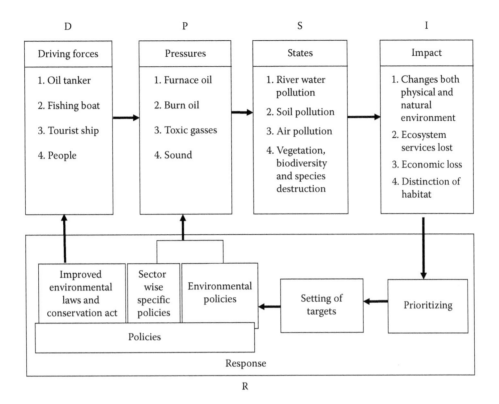

FIGURE 13.2
DPSIR framework for prioritizing possible mitigation measures to protect Shela River and Sundarbans ecosystems.

or chemical state indicators as the target variable, while the associated changes in biological state variables are treated as derived effects. A similar argument can be made for the causal links between driving forces (i.e., basic socio-economic development of different sectors of society) and environmental pressures in terms of emissions, resource use and land use.

13.4.1 Natural Regeneration Process and Challenges

Natural regeneration refers to the renewal of a tree crop by natural means, as opposed to artificial regeneration by means of planting or sowing, as done in mangrove plantations. The mangrove forest of the Sundarbans is dependent on natural regeneration for its existence. Over the greater part of the forest, seedling recruitment was sufficient for replacement of the harvested trees. The average number of seedlings appearing per year was about 27,750/ha, although recruitment densities varied considerably among different parts of the forest. Increasing salinity is likely to adversely affect reproductive cycles and capacities, the size of suitable spawning grounds, and the feeding/breeding/longitudinal migration of freshwater fish species (Dasgupta 2017). This will lead to a noticeable loss of habitat of freshwater fish in the western part of the Sunderbans (Samanta et al., 2013).

13.4.2 Challenges Associated with Avoiding Oil Spills in the Future

People dependent on fishing and crab hunting are directly affected by the oil spill and will remain so for a fairly long time, even after the spill has been dispersed due to the expected large-scale mortality of fish and crabs in the spill-affected area. This spill is a result of sheer negligence (Ellison et al., 1999). During a certain time of year, the Sundarbans forests are enveloped in a thick fog in the morning, with visibility down to zero. Visibility is restored to normal usually no earlier than 8 a.m., and vessel traffic must be halted until visibility is restored. As in India, vessel traffic is not permitted through the tiger reserve or the national park/sanctuaries. Bangladesh could impose the same policy, but this means finding alternative routes, which is not always easy owing to very low flow rates in some of the creeks as a result of siltation.

United Nations Educational, Scientific and Cultural Organization (UNESCO) and RAMSAR Convention add that

> [I]n 2011 United Nations Wetland Organization Authority RAMSAR and UNESCO ordered a cessation of ship and tanker movement inside the Sundarbans waterways. Violating this proposal, BIWTA was permitted to use 10 to 20 ships on this route. But 150 to 200 ships use this route daily. According to the Forest Department and the American organization Wildlife Conservation Society, 8 types of dolphin are found within a 120 km radius of the Bay of Bengal. These types of dolphin are hardly found in other parts of the world (Prothom Alo, 2014).

On 17 December 2014, *Bangladesh Protidin* reported that:

> [N]o tigers are found around 50 km (from the Shela River to the main Sundarbans area). Only around 100 tigers remain in Bangladesh's famed Sundarbans Forest, far fewer of the endangered animals than previously thought, according to one census. Some 440 tigers were recorded during the previous census in 2004 in the Sundarbans. Forest and fishery resources have been degraded, and because of the fish and animal scarcity that resulted from the sinking of the oil tanker, fishermen and foresters have seen their incomes plummet. They thought that because of the spilled oil, the animals and fish might move elsewhere. Owing to a scarcity of drinkable water and movements of large numbers of people, animals have moved into other areas of the forest. Local workers have collected 56,200 litres of fuel oil. Watercress was driven into 18 canals to sweep away the fuel oil yesterday.
>
> [T]here is no policy to stop the oil spreading. A fisherman caught kakra and kept them on his field. After collecting a large amount of kakra, he washed them with water. The staff of Prothom Alo asked the man why he had washed the kakra. He replied that he wanted to remove the oil from the bodies of the kakra so that the kakra could live. He also said that if the kakra were destroyed, he could not earn money by catching and selling them in the market. At Kumirer Char a man said every day they see at least six to seven crocodiles in the Char. But they found no crocodiles in the Char after the oil spread out. In the villages of Tambulbunia, Andharmanik and Joymoni Ghol only two white herons were found. But this time different types of birds are available there. Because of pollution and scarcity of food, the birds do not come to this area (Prothom Alo, 13 December 2014).

13.4.3 Lack of Scientific Management Approaches

The Sundarbans has been under scientific management for over 100 years. So far, Sundarbans management mainly deals with the harvesting of plant resources that could not be regulated properly owing to the demand of people and illicit removal of forest products (Sarker et al., 1985). Other aspects of ecosystems such as wildlife, fishery, tourism, biodiversity conservation and socio-economic aspects of the surrounding population have not been taken into due consideration. Following declaration as a World Heritage

Site, all sorts of commercial operations are banned in heritage areas, but no remarkable development programme has yet been undertaken by authorities. However, as a result of practical management, conservation and development of a World Heritage Site, many more problems await solutions (Saha et al., 2005). These include low community awareness of heritage protection, inadequate manpower and funds to protect illicit felling of timber and poaching in the heritage site, and synchronous legal documents for heritage management and conservation. We must always strengthen and apply the necessary measures to improve the management capacity of the Sundarbans World Heritage Site.

13.4.4 Other Challenges of Sundarbans

Commercial vessels are not the only threat facing the fragile ecology of the area. There are about 20 cement factories in Mongla at the edge of the forest. Green campaigners have also been protesting the 1320 MW Rampal coal-fired power plants, a joint initiative of India and Bangladesh, just 14 km from the forest. Coal is supposed to be carried through the Sundarbans. The government has permitted another 600 MW coal-fired power plant to be built by a private company, Orion Limited. Environmentalists and the civil society of Bangladesh have formed a national committee to save the Sundarbans and have decided to launch a global campaign asking international funding agencies and donors not to finance the power plant (Sarkar et al., 1999). 'Two coal fired power plants and other development activities taken up by the government show how the government cares less about the sensitive Sundarbans forest. Despite repeated requests from environmentalists, the government is going ahead with the power plant,' said Abul Matin, secretary-general of the committee. The unique mangrove forest of the Sundarbans, which stretches for 10,000 ha, with roughly 60% in Bangladesh and the rest in India, is important for both countries. It forms a natural wall against cyclones. About 30% of the area comprises water bodies, including rivers large and small, canals and creeks. However, there is no exact estimate of how many rivers flow through the area.

13.5 Recommendations for Ecosystem Development

Here are some recommendations for tourism development: (1) grant entrance permission letters for tourists; (2) provide for the security of all tourists; (3) hire expert forest wardens; (4) develop a transportation system for tourists. In transport vehicles named was written only in Bengali language. The name of each tourist spot and others facilities have been written in local language with Bengali alphabet so that the foreign tourists cannot read the information and not understand the situation. Thus, it must be written all the information like tourist spot location and directions in English or both Bengali and English. The engine boats are the most popular mode of transportations for medium to short and long distance traveling. Such boats are wooden made which are often known as superficial engines. Due to lack of roads and network a very few others transportation facilities have been found in this area. In every tourist spot, there are needed some tourist community police or security police because it will be better for tourist safety in developing country like Bangladesh (Le Floch et al., 2002). The Department of Environment has now formulated a draft contingency plan that will be discussed with different agencies before going to the cabinet for approval. The draft plan suggests the government should establish a separate entity that will lead the response to future oil spills. At the same time, the draft suggests that the government make the

Bangladesh Coast Guard the operational entity as the only organization that has the access and the capacity to work across coastal areas (Karim, 2009). The Department of Environment would remain the lead agency under the contingency plan as it has sole authority to check environmental pollution in Bangladesh. Since Bangladesh does not have expertise on the issue, the other signatories, who have already formulated their own national contingency plans, have offered to help. For example, Sri Lanka invited Bangladeshi government officials to join in a week-long workshop on oil spill management. Another form of regional cooperation, though, has raised questions about Bangladesh's commitment to keeping its waterways clean. On 15 November 2015 India and Bangladesh signed a standard operating procedure (SOP) to launch the Agreement on Coastal Shipping between the two countries. This agreement was part of a series of initiatives between the two countries when the Indian prime minister visited Bangladesh in June 2015. While the agreement may boost trade, commentators have noted that the route – so far unspecified – is likely to be through the Sundarbans using the Shela River, endangering the mangrove forests again.

13.6 Conclusion

The research presented here is on challenges associated with managing an oil spill following the sinking of an oil tanker on the Shela River in the Sundarbans Mangroves in Bangladesh. The affected area was in the southern part of Bangladesh and the heart of the Sundarbans (Shela River). The oil spread out over a vast area of the Sundarbans and throughout the Shela River. The families living near the oil spill area, in addition to the vegetation, fish, animals and many species that inhabit the Sundarbans, were affected by fuel oil. To perform this study, both primary and secondary data were collected and analysed. Primary data were collected through questionnaire surveys and focus group discussions. Secondary data were from UNDP and GoB surveys. These data were analysed using a DPSIR model. The study looked at the impacts of the oil spill in the Sundarbans Mangrove Forest. Has the government compensated the affected families and fishermen properly? This study mainly identifies the path of the pollution and the affected families and species. Many people are involved in river- and forest-related work such as fishing, rowing, honey and comb collection, wood and crab selling, and others. When the problem began, 2500 people lost their jobs. This research report will help to decrease this type of accident by creating awareness and by helping to develop policies and systems to protect the Mangrove Forest and its waterways. But it is also necessary to compensate the affected people. This research will also serve as an example to guide policymakers, those who implement the policies, environmental experts and geographers, and above all, it will support researchers' efforts to assess integrated phenomena like oil spills in a meaningful way.

References

Agrawala, S., T. Ota, A.U. Ahmed, J. Smith and M. van Aalst. 2003. *Development and Climate Change in Bangladesh: Focus on Coastal Flooding and the Sundarbans.* Report COM/ENV/EPOC/DCD/DAC(2003)3/FINAL. Organisation for Economic Cooperation and Development, Paris.

Ahmed, M.F., Q.K. Ahmad and Md. Khalequzzaman. 2004. Regional Cooperation on Transboundary Rivers: Impact of the Indian River Linking Project. *Proceedings of an International Conference.* Bangladesh Poribesh Andolan and Bangladesh Environment Network, Dhaka, Bangladesh.

Akhter, D.M. 2002. *Observations of a Mangrove Forest, Bangladesh from Landsat Imagery.* Bangladesh Forest Department, Dhaka.

Akhter, M. 2006. *Remote sensing for developing an operational monitoring scheme for the Sundarbans Reserve Forest, Bangladesh.* Technische Universität Dresden, Germany.

Ali, S.S. 1998. *Sundarbans: its resources and ecosystem. Integrated Management of the Ganges flood plains and Sundarbans ecosystem.* Khulna University, Khulna, Bangladesh, 1: 38–49.

Amin, M.Z., A. Habib, K. Nafiz and A.M. Swaraz. 2016. Assessment of the water quality, some trace elements content of *Heritierafomes* fruits and *Pneumatophores* of Sundari after oil spill in Shela River at Sundarban, Bangladesh. *International Journal of Environmental Monitoring and Analysis,* 4(6): 167–173. doi: 10.11648/j.ijema.20160406.15

Arroyo, C.A. 1979. Flora of the Philippines mangrove. *BIOTROP Special Publication* 10: 33–44.

Arroyo, C.A. 1997. Vegetation Structure of Mangrove Swamp. *Proceedings of the National Symposium and Workshop on Mangrove Research and Development.* Philippine Council for Agricultural and Resources Research, Los Banos, Laguna, Philippines.

Atmadja, A. and L. Verchot. 2012. A review of the state of research, policies and strategies in addressing leakage from reducing emissions from deforestation and forest degradation (REDD+). *Mitigation and Adaptation Strategies for Global Change,* 17: 311–317.

Baksha, M.W. and E.B. Lapis. 2000. Insects Pests of the Sundarbans Mangroves of Bangladesh. Baksha, M. W. (eds), Bangladesh Forest Research Institute, Chittagong, *Bangladesh Journal of Forest Science,* 2: 25–34.

Balasubramanian, T. and S. Ajmal Khan. 2002. *Mangroves of India. State of the Art Report.* Environmental Information System Center, Parangipettai, India, 140 p.

Banerjee, A. and S.C. Santra. 1999. Plankton composition and population density of the Sundarbans mangrove estuary of West Bengal (India). In: D.N. Guha Bakshi, P. Sanyal and K.R. Naskar (eds.), *Sundarbans Mangal.* Naya Prokash, Calcutta, pp. 340–349.

Barbier, E. and S. Sathirathai (eds.). 2004. *Shrimp Farming and Mangrove Loss in Thailand.* Edward Elgar Publishing, Cheltenham, UK.

Blower, J.H. 1985. *Sundarbans Forest Inventory Project. Bangladesh Wildlife Conservation in the Sundarbans.* Report 151. Overseas Development Administration, Land Resources Development Center, Surbiton, UK.

Chaffey, D.R., F.R. Miller and J.H. Sandom. 1985. *A Forest Inventory of the Sundarbans, Bangladesh.* Overseas Development Administration, London.

Champion, H.G. and S.K. Seth. 1968. *Revised Forest Types of India.* Manager of Publications, Govt of India, New Delhi.

Chaudhuri, A.B. and A. Choudhury. 1994. *Mangroves of the Sundarbans. Volume 1: India.* World Conservation Union, Gland.

Chowdhury, A. 2014. *Impact of Oil Spillage on the Environment of Sundarbans (World Largest Mangrove Forest) in Bangladesh – Draft Report.* Khulna University.

Dasgupta, S. 2017. Increasing salinity in a changing climate likely to alter Sundarban's ecosystem, The feature today 22 January 2017, The World Bank, IBRD-IDA. http://www.worldbank.org/en/news/feature/2017/01/22/increasing-salinity-in-a-changing-climate-likely-to-alter-sundarbans-ecosystem

Eaton, R. 1991. Human settlement and colonization in the Sundarbans, 1200–1750. In: J. Seidensticker, R. Kurin, and A.K. Townsend (Eds.). *The Commons in South Asia: Societal Pressure and Environmental Integrity in the Sundarbans.* The International Center, Smithsonian Institution, Washington, DC, USA.

EEA (European Environment Agency). 1998. Guidelines for Data Collection and Processing -EU State of the Environment Report. Annex 3. National Institute of Public Health and Environment, Bilthoven, Netherlands.

Ellison, A.M. 2002. Macroecology of mangroves: Large scale patterns and processes in tropical coastal forests. *Trees*, 16: 181–194.

Ellison, A.M., E.J. Farnsworth and R.E. Merkt. 1999. Origins of mangrove ecosystems and the mangrove biodiversity anomaly. *Global Ecology and Biogeography*, 8: 95–115. doi:10.1046/j.1466-822X.1999.00126.x

Habib, M.G. 1989. Wildlife management of the Sundarban – a case study. In: G.M.M.E. Karim, A.W. Akonda and Sewitz, P. (eds.), *Conservation of Wildlife in Bangladesh*. German Cultural Institute, Forest Department, Dhaka University, Wildlife Society of Bangladesh, UNESCO, Dhaka, pp. 161–168.

Hoff, R. (ed.). 2012. *Oil Spill in Mangroves – Planning & Response Considerations*. NOAAOcean Service, National Oceanic and Atmospheric Administration, Seattle 72 pp.

Hoq, M.E., M.A. Wahab and M.N. Islam. 2006. Hydrographic status of Sundarbans mangrove, Bangladesh with special reference to post-larvae and juvenile fish and shrimp abundance. *Wetlands Ecology and Management*, 14: 49–93.

IPCC. 2014. Climate Change 2014: Synthesis Report. *Contribution of Working Groups I, II and III to the Fifth Assessment Report of the Intergovernmental Panel on Climate Change* [Core Writing Team, R.K. Pachauri and L.A. Meyer (eds.)]. IPCC, Geneva, Switzerland, 151 p.

IUCN. 1997. *World Heritage Nomination – IUCN Technical Evaluation*, IUCN, Sundarban Wildlife Sanctuaries, Bangladesh.

Kajanus, M., P. Leskinen, M. Kurttila and J. Kangas. 2012. Making use of MCDS methods in SWOT analysis—Lessons learnt in strategic natural resources management. *Forest Policy and Economics*, 20, 1–9.

Karim, A. 1994. The physical environment. In: Z. Hussain and G. Acharya (eds.), *Mangroves of the Sundarbans*, Volume II: Bangladesh. IUCN, Bangkok, pp. 11–42.

Karim, S. 2009 Implementation of the MARPOL Convention in Bangladesh. *Macquarie Journal of International and Comparative Environmental Law*, 51, 51–82.

Le Floch, S., J. Guyomarch, F. Merlin, P. Stoffyn-Egli, J. Dixon and K. Lee. 2002. The influence of salinity on oil-mineral aggregate formation. *Spill Science & Technology Bulletin*, 8, 65–71.

Marzan, L.W. et al. 2017. *Journal of Genetic Engineering and Biotechnology* http://dx.doi.org/10.1016/j.jgeb.2017.02.002.

National Academy of Sciences. 1985. *Oil in the sea: inputs, fates and effects*. National Academy Press, Washington, DC, USA.

Rahman, M.M., M.M. Rahman and K.S. Islam. 2010. The causes of Sundarban mangrove forest ecosystem of Bangladesh: conservation and sustainable management issue. *AACl Bioflux* 3(2): 77.

Rahman, Md.M., Md.N.I. Khan, A.K.F. Hoque and I. Ahmed. 2015. Carbon stock in the Sundarbans mangrove forest: spatial variations in vegetation types and salinity zones. *Wetlands Ecological Management*, 23: 269–283.

Rapport, D.J. 1992. Evolution of indicators to assess the state of health of regional ecosystems under stress and recovery. *Proc. Internat. Symp. Ecological Indicators*, Fort Lauderdale, Florida, USA.

Razzaque, J., 2017. Payments for ecosystem services in sustainable mangrove forest management in Bangladesh. *Transnational Environmental Law*. Available from: http://eprints.uwe.ac.uk/30904.

Rouf, M.A., A. Bouazza, R.M. Singh, W.P. Gates and R.K. Rowe. 2015. Gas diffusion coefficient and gas permeability of an unsaturated geosynthetic clay liner, *Geosynthetics Conference 2015*, February 15–18, Portland, Oregon, USA.

Saha, S.K., K. Roy, P. Banerjee, A. Al Mamun, Md. Arifur Rahman and G.C. Ghosh. 2005. Technological and environmental impact assessment on possible oil and gas exploration at the Sundarbans coastal region. *International Journal of Ecology and Environmental Sciences*, 31: 255.

Samanta, A., K. Chakraborti, M. Bandyopadhyay and R. Sengupt. 2013. *Moule, Honey collectors of Sundarbans and their ITKs*, 1: No-2.

Sarker, S.U. 1985. Density, productivity and biomass of raptoral birds of the Sundarbans, Bangladesh. *Proceedings of SAARC Seminar on Biomass Production*, 15 April 1985, Dhaka, pp. 84–92.

Sarkar, D., G.N. Chattopadhyay and K.R. Naskar. 1999. Nature and properties of coastal saline soils of Sundarbans with relation to mangrove vegetation. In: D.N. Guha Bakshi, P. Sanyal and K.R. Naskar (eds.), *Sundarbans Mangal*. Naya Prokash, Calcutta, pp. 199–204.

Siddique, A.B. 2014. The Sundarbans in big trouble. Retrieved 10 December 2014, from Dhaka Tribune: http://www.dhakatribune.com/bangladesh/2014/dec/10/sundarbans-big trouble#sthash. JXpSUQgh.dpuf

The Independent, Dublar Char prepares for 'Rash Mela' amid security, POST TIME: 23 November, 2015 11:32:53 PM.

Townsend, A.K. 1992. Human Use and Conservation of the Indian Sundarbans. Master in Marine Affairs thesis, School of Ocean and Fishery Sciences, University of Washington, Seattle, USA.

Uddin, M.S. 2011. Economic valuation of Sundarbans mangrove ecosystem services - a case study in Bangladesh. MSc thesis, UNESCO-IHE, the Netherlands.

Yender, R., J. Michel and C. Lord. 2002. *Managing Seafood Safety after an Oil Spill*. Seattle: Hazardous Materials Response Division, Office of Response and Restoration, National Oceanic and Atmospheric Administration.

Index

Note: Page numbers followed by *"fn"* indicate footnotes.

A

Abiotic components, 159
Accounting for externalities in ecosystem-based fishery management, 185–186
Accounting system for fish stock, 169
Acropora cervicornis, see Staghorn coral
Acropora spp., 217
Acute effects, 127–129
Adaptation(s)
    of mangrove wetlands, 274–275
    of rock patches, 279
    of sandy shores, 276
    solutions for adaptation to climate change, 226–232
Ad hoc fashion, 14
Advection-diffusion equations, 149
Aerobic bacteria, 150
Agriculture, 246–247
Aleutian Low Pressure system (ALP system), 101
Algae, 220
Algal blooming, 147–148
Aliphatic hydrocarbons, 122
Aliphatic hydrocarbon source indicators, 125
Alkylated phenanthrenes, 124, 128
ALP-Pacific Decadal Oscillation pattern, 102
ALP system, *see* Aleutian Low Pressure system
AMBI, *see* AZTI Marine Biotic Index
American Petroleum Institute gravity (API gravity), 131*fn*
Amoco Cadiz spill, 139
Anaerobic bacteria, 150
ANN, *see* Artificial neural network
Annual Report Card, 24
Anthropogenic chemical toxins, 4
Anthropogenic interference, 282
AO, *see* Arctic Oscillation
AOGCM, *see* Atmosphere–ocean general circulation models
API gravity, *see* American Petroleum Institute gravity
Applied Simulations and Integrated Modelling for Understanding of Toxic and Harmful Algal Blooms (ASIMUTH), 71–72

Aquatic ecosystem, 328
Aquatic environment, tidal impact on, 330–331
Aquatic species, 332
ARCHER Cray XC30 system, 71
Arctic Oscillation (AO), 102
Arctic Oscillation–ALP-SHP-EAWM coupled system, 102
Aromatics, 124
Artificial neural network (ANN), 294
Artificial upwelling technology, Triple I application to, 53–56
ASIMUTH, *see* Applied Simulations and Integrated Modelling for Understanding of Toxic and Harmful Algal Blooms
Atmosphere–ocean general circulation models (AOGCM), 106
Atmospheric forcing, 101, 102
Atoll reef, 211
    in Spratly Islands, 212
At-risk populations, 311
Australian Ecosystem Health Monitoring Program (EHMP), 15
Australian Integrated Marine Observing System, 10
Autonomous underwater vehicles (AUVs), 10
AZTI Marine Biotic Index (AMBI), 296

B

Back reef, 211–212
Bacteria, 147–148
Badamtali Sea Beach, 325
Bagerhat range, 325
Balanus balanoides (*B. balanoides*), 278
Bang-bang, 199, 200
Bangladesh, 323
    management challenges of sinking of oil tanker at Shela Coastal River, 323–338
    management strategies of St. Martin's Coral Island at Bay of Bengal in, 238–257
Bangladesh Petroleum Corporation (BPC), 328
BaU, *see* Business-as-usual
Baula kumbhira, 310
Bay Health Index, 17

BC, see Biocapacity
BCA, *see* Bhitarkanika Conservation Area
BD, *see* Biodiversity
Beaches, 137
    beach profile of sandy shores, 276
Benefit accounting (B accounting), 39
Benthic
    animals, 18
    communities, 294
    monitoring, 16
    organisms, 127–128
    submodel, 149–150
Benthic index of biotic integrity (B-IBI), 23
Benthos, 45
Benzanthracene, 122
Benzene, 122
Bhitarkanika Conservation Area (BCA), 316
    conservation and management of saltwater
        crocodile in, 316–317
Bhitarkanika National Park, 310
    attack on domestic livestock by wild
        crocodiles, 318–319
    attacks on humans by wild crocodiles, 318
    distribution of saltwater crocodiles in
        Odisha, 311–312
    saltwater crocodile project, 317–318
    saltwater crocodile status in, 317
Bhitarkanika Wildlife Sanctuary (BKWS), 309
B-IBI, *see* Benthic index of biotic integrity
Biocapacity (*BC*), 33, 39
    in Inclusive Impact Index, 36–40
Biodiversity (BD), 43, 159
    impact on aquatic environment, 330–331
    impacts on BD of sundarbans mangroves, 330
    oil collection time duration and agency
        achievements, 331
    solutions for protection, 229
    status of Phytoplankton and Zooplankton,
        331–332
Bioeconomic models, 180–181
    and management, 180–181
Biogeochemical processes, 120
Bioindicators, 290, 291
Biological communities, 276
Biological diversity, 266–267
Biological factors, 277–279
Biological models, 74
    general larval dispersal model, 75–78
    harmful algal bloom model, 74–75
    integration results, 83
Biological oxygen demand (BOD), 328
Biology of saltwater crocodile, 312
    appearance, 313–314

communication and behaviour, 314
    food and feeding habits, 314
    habitat, 312
    morphological characteristics, 313
    movement, 314
    reproduction and parental care, 315
Biomarkers, 124–125
Biomonitoring
    to assess coastal environmental status in
        Malaysia, 290
    concepts in use of indicator species, 290–291
    eutrophication, 291–292
    heavy metal, 292–293
    modelling technology in, 294
Biomonitoring ecosystem health
    ecological modelling as tool to characterize
        coastal ecosystem health, 294–299
    environmental issues of Malaysian Coastal
        Waters, 289–290
Biophysical model framework, 68, 69
    biological models, 74–78
    hydrodynamic model, 70–74
    meteorological model, 69
Bioremediation, 134–135
Biotic communities, 271
    of sandy shores, 277
Biotic components, 159
Biotic structure of estuaries, 272
Birds in St. Martin's Island, 253
BKWS, *see* Bhitarkanika Wildlife Sanctuary
Blooming of jellyfish, 289–290
BOD, *see* Biological oxygen demand
Booms, 132–133, 136
Boundary conditions, 11, 151
Boundary forcing, 71–72
BPC, *see* Bangladesh Petroleum Corporation
Braer oil spill, 130–131
Breeding territories, 315
Bruguiera, 274
Burning, 134
Business-as-usual (BaU), 41
Buttress zone, 211–212

C

Calcification, 267–269
Carbazole, 122
Carbon, 121
    cycle, 152–154
    emissions, 33
Carbon dioxide (CO_2)
    concentration, increases impacts in, 222
    emissions, 38, 53, 54

Carbon dioxide Capture and Storage process
        (CCS process), 42
Castillo de Bellver spill, 135
Catchment Source, 23
Cause-effect relationships, 44–45, 334
CBA, *see* Cost-benefit analysis
CBMP, *see* Chesapeake Bay Monitoring
        Program
CBOS, *see* Chesapeake Bay Observing System
CBP, *see* U.S. Chesapeake Bay Program
CBRSP, *see* Chesapeake Bay Remote Sensing
        Program
CCS process, *see* Carbon dioxide Capture and
        Storage process
CEA, *see* Cost-effectiveness analysis
Central coastal lagoons, 223
CEs, *see* Choice experiments
CGE, *see* Computable general equilibrium
        model
Challenges associated with avoiding oil spills
        in future, 335–336
Chemical oxygen demand (COD), 38,
        151, 152
Chemistry of crude and refined oils, 121–126
Cheradia, 239, 241
Chesapeake Bay Journal, 24
Chesapeake Bay Local Government Advisory
        Committee (LGAC), 19*fn*
Chesapeake Bay Monitoring Program (CBMP),
        16, 23–24; *see also* Ecosystem health
        monitoring program (EHMP)
    problems and challenges, 24–26
    programmatic strengths, 23–24
Chesapeake Bay Observing System (CBOS),
        16–17
Chesapeake Bay Remote Sensing Program
        (CBRSP), 17
Chocolate mousse, 132
Choice experiments (CEs), 179
Cholestane, 122
Chthalamus stellatus (C. stellatus), 278
CICES, *see* Common International
        Classification of Ecosystem Services
Citizens Advisory Committee, 18*fn*
Clean Water Act, 17
Climate change, 41, 234
    on coral reef ecosystem, 211–217
    on estuary ecosystem, 222–224
    impacts of climate change on marine
        ecosystems, 211
    on mangrove ecosystems, 217–220
    relationship between climate change and
        marine ecosystems, 211

scenario and marine ecosystems in 21st
        century Vietnam, 224–226, 227–228
    on seaweed and seagrass ecosystem,
        220–222
    solutions for adaptation and resilience to
        climate change, 226–232
    in Vietnam, 210
Climate change, education, training and
        raising awareness of, 232
Climate regime shifts (CRSs), 100
Climate regulation, 4
Climate variabilities, 101
    correlation statistics, 102, 103
    decadal-scale ecosystem changes in KOE
        region, 105
    teleconnected pattern, 102
    time series, 104
Closed coral reefs, 212
Closed-form management solutions, 176
Closing stocks records, 167
Coastal and Wetland Biodiversity
        Management Project (CWBMP), 254
Coastal areas, 289
Coastal dunes, 241
Coastal ecosystem health
    and dynamics, 287–288
    ecological modelling as tool to
        characterizing, 294
    introduction of modelling technology in
        biomonitoring, 294
    modelling of macrobenthos communities
        along Penang coastal waters, 294–299
Coastal ecosystems, 23, 223, 264, 266
    coral reefs, 266–270
    estuaries, 270–272
    mangrove wetlands, 272–275
    pressures on, 6–7
    protection against rise of sea level and
        erosion, 230
    rock patches, 277–281
    sandy shores, 275–277
    services, 4
    states and indicators, 4–6
Coastal environmental status
    biomonitoring to assess coastal
        environmental status in Malaysia,
        290–293
    environmental issues of malaysian coastal
        waters, 289–290
Coastal reef lagoon, 211–212
Coastal resources management, 282
    approach, 282–284
    need for management, 282

Coastal waters management
   biophysical model framework, 68, 69–78
   challenges, 66–67
   general particle-tracking model, 83–91
   HAB model, 82–83
   hydrodynamic model output and
      validation, 78–82
   meteorological model validation, 78
   modelling approach, 67
   results, 78
Coastal zone, 287–288
COD, *see* Chemical oxygen demand
Collaborative governance, 253
Co-management strategy, integration of, 253
   co-management framework, 253–254
   managed resource zone, 255–257
   zoning, 254–255
Common International Classification of
      Ecosystem Services (CICES), 162–163
Competition in species, 278
Component method, 33
Computable general equilibrium model (CGE),
      183
Computational condition, 150
   boundary conditions, 151
   grid system of Tokyo Bay, 150
   numerical methods, 150–151
   simulation cases, 151
Con Co Island, research results on, 216
Con Dao Island, research results on, 214–215
Conductivity, Temperature, Depth profiler
      (CTD), 72
Connectivity matrices, 77
Containment booms, 133
Contingency planning, 135–136
Contingent valuation method (CVM), 179
Conventional bioeconomic models, 180–181
Conventional fisheries economics modelling,
      186–189
Conventional fisheries management, 189
   devastating competition, 196
   development of biotic stocks exposed to
      fishery, 193
   development of profit and effort of two
      fleets, 194
   fishery simulation model, 191–192
   maximization of resource rent, MSY and
      MEY concepts, 194
   phase diagram showing relationship
      between fishing effort, 195
   stock size with natural species interaction,
      190
Conventional microeconomics modeling, 186

Conventional static fisheries economics
      analyses, 196–197
Conversion factor (γ), 39
CO_2 ocean sequestration (COS), 32, 41
Coquina, 240
Coral, 252
Coral reef ecosystem, 232–233; *see also* Estuary
      ecosystem; Mangrove ecosystems;
      Seaweed and seagrass ecosystem
   impact of El Niño and La Niña, 217
   impacts of floods, 216–217
   impacts of increasing water temperature,
      212–213
   impacts of sea-level rise, 214
   impacts of storms, 214–216
   types of coral reef related to climate change,
      211–212
Coral reefs, 140, 217, 242, 266, 288
   calcification, 267–269
   in central Vietnam, 212
   ecosystem function, 270
   nutrition, 269–270
   primary productivity, 267
   at Western Tonkin Gulf, 212
COS, *see* CO_2 ocean sequestration
Cost, 34, 35, 39
Cost-benefit analysis (CBA), 44, 163, 182–183
Cost-effectiveness analysis (CEA), 163
Critical tide level, 277
Cross-impact method, occurrence probability
      by, 45–48
CRSs, *see* Climate regime shifts
Crude oils, 119–120, 121–126, 128, 134
Crustaceans (Lobster, Crabs, Shrimp), 251
CTD, *see* Conductivity, Temperature, Depth
      profiler
Cultural services, 162, 165, 170, 179
Cultural supporting and regulating services,
      175
Currents, 72, 79–80
CVM, *see* Contingent valuation method
CWBMP, *see* Coastal and Wetland Biodiversity
      Management Project
Cycloalkanes, 122
Cyclohexane, 122
Cytochrome P450, 129

D

Daisy chain simulations, 82
Dakhin Para, 241
Dangmal-Bhitarkanika National Park,
      saltwater crocodile project at, 317–318

Data assimilation techniques, 13
Data management and communications
(DMAC), 8, 10, 18
Decision support tools, 23
Deep fore reef zone, 211–212
Deep ocean water utilization, 32
Deepwater Horizon spill (DWH spill), 126
Defensive expenditure method (DE method),
179
Delta estuary, 223
Demand and supply of ecosystem services, 186
DE method, *see* Defensive expenditure method
Department of Environment (DoE), 244
Desalinization in mangrove species, 219
Devastating competition, 196
Diagonal matrix, 54
Dibenzothiophenes, 124, 128
Diffusion computation, 87–88
2,5-Dimethyl naphthalene, 122
Dinophysis, 66–67
Direct climate forcing, 106; *see also* Lag oceanic
forcing
correlation statistics, 107
decadal scale of time series, 107
Discretization method, 148
Dispersal kernels, 77
Dispersal measures, 77
Dispersants, 133–134, 136
Dissolved organic matter, 152
Dissolved oxygen (DO), 243
DMAC, *see* Data management and
communications
DO, *see* Dissolved oxygen
DoE, *see* Department of Environment
Dolphins (*Tursiops truncatus*), 128
DPSIR model, *see* Driver-pressure-state-
impact-response model
Dredging, 185
Drifter tracking simulations, 82–83
Drifter tracks, 72
Driver-pressure-state-impact-response model
(DPSIR model), 3, 163–164, 332–333,
335
Drivers, 3
Driving force of system, 171
Driving forces, 333, 334
Drought impacts, 221–222
Dry fish processing factory, 326
Dublar Char, 326
Dunes of St. Martin's Island, 241
DWH blowout, 137
DWH spill, *see* Deepwater Horizon spill
Dynamic ecosystem modeling, 196–201

E

EAFM, *see* Ecosystem approach to fishery
management
Earth's coastal marine ecosystems, 4
East Asian winter monsoon pattern (EAWM
pattern), 101, 103
East China Sea (ECS), 99, 110, 112
East Korea Warm Current (EKWC), 101
East Sea/Sea of Japan, 100–101, 106, 107, 110
EAWM pattern, *see* East Asian winter monsoon
pattern
Ebb tide, 330
EBFM, *see* Ecosystem-based fishery
management
EBM, *see* Ecosystem-based management
EC, *see* Executive Council
ECC, *see* Environmental Clearance Certificate
Ecological economics, 161
Ecological footprint (EF), 33–34, 38
calculation, 43
LCA of transport and injection of CO_2,
42–43
in life of SPAR-type wind turbines, 57–59
Ecological Footprint Network (EFN), 42, 43
Ecological modelling, 294
of macrobenthos communities along
Penang coastal waters, 294–299
technology in biomonitoring, 294
Ecological risk (ER), 34, 35, 39, 44
calculation, 49
cause-effect relationships, 44–45
COS, 50
cost and benefit, 44
ecological footprint, 42–43
evaluation in Inclusive Impact Index, 41
occurrence probability by cross-impact
method, 45–48
quantifying endpoint with environmentally
changed area, 48
Triple I, 41, 49–50
Ecological values, scaling of, 50–53
Ecology, 159
ECOM, *see* Estuaries and Coastal Ocean
Model
Economic accounting systems, 172
Economic analyses
methods for, 171–172
tools for, 163–164
Economic growth, 159
Economic impacts, 334
Economic relations division (ERD), 328
Economic sectors, 333

Economics of ecosystem-based fisheries
     management
  accounting for externalities in ecosystem-
      based fishery management, 185–186
  bioeconomic modelling and management,
      180–181
  challenges and role of economics, 159–161
  classification of ecosystem services, 161–163
  conventional fisheries economics
      modelling, 186–189
  conventional fisheries management, 189–196
  dynamic ecosystem modeling, 196–201
  ecosystem-based management, 181–183
  green accounting, 164–168, 168–171
  inclusion of ecosystem services in green
      accounting, 172–174
  methods for economic analyses, 171–172
  microeconomic modeling, 174–176
  predator–prey relationships and species
      interactions, 183–185
  prices and valuation, 177–180
  tale, 157–158
  tools for economic analyses, 163–164
Economic value scaling, 50–53
Economy, 159, 191–192
Ecosystem-based fishery management (EBFM),
     181–182
  accounting for externalities in, 185–186
Ecosystem-based management (EBM), 2,
     181–183
Ecosystem(s), 4, 5, 11, 158, 159, 264, 336
  changes, 12
  crowds of tourists, 248
  drivers and water quality, 246
  dynamics response to climate forcing,
      110–114
  ecosystem-based approaches, 3
  externalities, 185–186
  function, 270
  health and indicators, 326
  indicator, 326
  management plans for, 135–140
  overbuilding, 247
  population growth, 247–248
  processes, 159
  protecting and sustaining ecosystem
      services, 15
  recommendations for development, 337–338
  recovery targets, 17–18, 21
  states, 3
  submodel, 148–150
Ecosystem approach to fishery management
     (EAFM), 182

Ecosystem health monitoring program
     (EHMP), 19
  ecosystem recovery targets, 21
  governance for integration, 22–23
  integrating programs, 21–22
  M&E program, 20–21
  problems and challenges, 26
  programmatic strengths, 24
  SEQ, 19–20
Ecosystem services, 181
  classification, 161–163
  and goods, 162
  inclusion in green accounting, 172–174
  values, 324–325
ECS, *see* East China Sea
EEA, *see* European environment agency
EEC, *see* Exclusive Economic Zone
EF, *see* Ecological footprint
EFN, *see* Ecological Footprint Network
Egretta sacra, *see* Pacific reef heron
EHMP, *see* Australian Ecosystem Health
     Monitoring Program; Ecosystem
     health monitoring program
EIA, *see* Environmental impact assessment
EKWC, *see* East Korea Warm Current
El Niño and La Niña impact, 217
El Niño Southern Oscillation (ENSO), 106
EMSS, *see* Environmental Management
     Support System
End-to-end system, 9–10
Endogenous simulations, 176
ENSO, *see* El Niño Southern Oscillation
Environmental Clearance Certificate (ECC),
     254, 255
Environmental effects and management of oil
     spills
  average annual releases, 120
  basic methodology of mitigation, 131–135
  management plans for different ecosystems,
      135–140
  physics and chemistry of crude and refined
      oils, 121–126
  toxicity of oil to marine life, 126–131
Environmental impact(s), 329–330, 334
  simulation, 36–38
Environmental impact assessment (EIA), 163
Environmental issues of Malaysian Coastal
     Waters, 289–290
Environmental Management Support System
     (EMSS), 20
Environmental mismanagement in St. Martin's
     Island, 246
Environmental pollution, 211

causes on St. Martin's Island, 245
Environmental problems, 334
Environmental processes, 124–125
Environmental Protection and Biodiversity
    Conservation Act (1999), 310
Environmental risk assessment method, 34
Environmental Sensitivity Index (ESI), 136–137,
    138
Environmental sensitivity maps, 136–137
Environments, 137
    beaches, 137
    coral reefs, 140
    exposed tidal flats, 138
    mangroves, 139–140
    marshes, 139
    rocky shores, 138–139
EP level, *see* United Nation's erythrocyte
    protoporphyrin level
ER, *see* Ecological risk
ERD, *see* Economic relations division
Erosion impacts, 219, 224
ESI, *see* Environmental Sensitivity Index
Estuaries, 270
    biotic structure, 272
    salinity adaptations, 271
    types, 271
Estuaries and Coastal Ocean Model (ECOM),
    148
Estuarine, *see* Saltwater crocodile (*Crocodylus
    porosus*)
Estuary ecosystem, 222, 234; *see also* Coral reef
    ecosystem; Mangrove ecosystems;
    Seaweed and seagrass ecosystem
    estuary types relating to geo-climate, 222–223
    impacts of erosion, 224
    impacts of sea-level rise, 223
    impacts of temperature, 223
    impacts of typhoons and floods, 223–224
    vulnerability of estuarine ecosystems to
        climate change, 223
Ethane, 122
Ethylthiophene, 122
Euclidian distances, 294
European environment agency (EEA), 332
Eutrophication phenomenon, 221, 291–292
Exclusive Economic Zone (EEC), 288
Executive Council (EC), 16
Exogenous simulations, 176
Exposed tidal flats, 138
Externalities, 158
    in ecosystem-based fishery management,
        185–186
Exxon Valdez spill, 128, 129, 137

F

Facultative halophytes, 274
FASTNEt project, *see* Fluxes Across Sloping
    Topography in North East Atlantic
    project
Fauna, 251
    birds, 253
    coral, 252
    Crustaceans, 251
    fish, 252–253
    mollusks, 251–252
    reptile, 253
    rocks, 252
    seaweeds, 251
Final benefit (B), 34
Findlater jet, 265
Finite-difference scheme, 150–151
Finite Volume Community Ocean Model
    (FVCOM), 70, 148
Fish, 45, 252–253
Fishery, 168–169
    management, 166
    monitoring economic importance, 165–166
    policies, 166
    productivity, 110–114
    simulation model, 191–192
Fishery Survey of India (FSI), 308
Fishing
    effects on marine ecosystems, 185
    effort, 197
    restrictions, 66
Fish stock, 157
    production factor, 169
Flatform reefs, 212
Fleets, fishery simulation model, 191–192
Floating offshore wind turbine, spatial
    analysis of III, 60–61
Flooding events, 4
Floods impacts, 216–217, 221–222, 223–224
Flood tide, 330
Flora, 250–251
Fluxes Across Sloping Topography in North
    East Atlantic project (FASTNEt
    project), 72
Food production, 333
Food web, 220
Forest protection and development of forests,
    229
Fossil fuels depletion, 56
Freshwater, 243–244
Fringing reefs in Vietnam, 211–212
FSI, *see* Fishery Survey of India

Fuel-burning electricity generation systems, 59
Full residence time simulations, 82
Fundulus grandis (*F. grandis*), 130
Fundulus heteroclitus (*F. heteroclitus*), 130
Funnel-shaped estuary, 223, 224
FVCOM, *see* Finite Volume Community Ocean
          Model

G

Galaxea genus, 217
Gastropods, 129, 130, 251–252, 293
GCMS, *see* Global climate model system
GDP, *see* Gross domestic product
GEEM, *see* General equilibrium ecosystem
          model
Gelling agents, 132, 134
General equilibrium ecosystem model (GEEM),
          183
General larval dispersal model, 75; *see also*
          General particle-tracking model;
          Harmful algal bloom model (HAB
          model); Hydrodynamic model
     development, 76
     dispersal measures, 77
     habitat sites and wind scenarios, 77–78
     horizontal movement, 76
     larval duration, 77
     mortality, 76
     vertical movement, 76
General Ocean Turbulence Model (GOTM), 71
General particle-tracking model, 83; *see also*
          General particle-tracking model;
          Harmful algal bloom model (HAB
          model); Hydrodynamic model
     diffusion and velocity computation, 87–88
     habitat structure and larval duration, 85–86
     intertidal organisms and novel offshore
          habitat, 90–91
     salinity driven mortality, 88–90
     vertical migration, 88
     wind forcing, 87
Geo-climate, estuary types relating to, 222–223
Geographical location of Saint Martin's Island,
          239
     area, 239
     geomorphology, 241–242
     physiographic characteristics, 239
     soil characteristics, 239–241
Geographic information system database (GIS
          database), 36
Geomorphology of Saint Martin's Island,
          241–242

GHGs, *see* Greenhouse gases
GIS database, *see* Geographic information
          system database
Global climate model system (GCMS), 223
Global Footprint Network, 33
Global hectares (gha), 41
Global Positioning System (GPS), 14
Global warming, 56
Global warning, 43
GNP, *see* Gross national product
Goan estuaries, 272
GoB, *see* Government of Bangladesh
Golachipa, 239, 241, 255
Gordon-Schaefer model, 180–181
GOTM, *see* General Ocean Turbulence Model
Government of Bangladesh (GoB), 328
GPS, *see* Global Positioning System
Grain size of sandy shores, 276
Grazing, 278
Green accounting
     ecosystem services inclusion, 172–174
     in practice, 168–171
     systems, 174
     in theory, 164–168
Greenhouse gases (GHGs), 41
Green-lipped mussel, 293
Gridding method, 148
Grid system of Tokyo Bay, 150
Gross domestic product (GDP), 35
Gross national product (GNP), 159
Growth, 191–192

H

Habitat complexity
     coastal ecosystems, 264, 266–281
     ecosystem concept, 264
     management of coastal resources, 282–284
     oceanography off central west coast of
          India, 264–265
     physiographic and climatological setting of
          Goa, India, 265–266
Habitat sites, 77–78
Habitat structure, 85–86
HAB model, *see* Harmful algal bloom model
Harmful algal bloom model (HAB model), 66,
          74, 82; *see also* General larval dispersal
          model; General particle-tracking
          model; Hydrodynamic model
     biological model integration results, 83
     drifter tracking simulations, 82–83
     K. mikimotoi simulations, 83
     population growth and mortality, 74–75

scenarios and simulations, 75
transport, 75
Harmful algal species, 4
Health-e-Waterways information management
  system, 21–22
Health impacts, 327
Health risk to humans (HR), 34
Healthy Waterways Ltd. (HWL), 22
Healthy Waterways Network, 22
Healthy Waterways Strategy, 21
Heavy metal, 292–293
Hedonic price method (HPM), 179
Herbaria, 326
Heterotrophic production of mangrove
  wetlands, 273
Hexane, 122
High dunes, 241
High walk, 314
Hindcasting, 13
Hiron Point, 326
Hopane, 122
Horizontal movement, 76
HPM, *see* Hedonic price method
HR, *see* Health risk to humans; Human risk
Human activities, 333
Human health, 23
Human responses, 3
Human risk (HR), 35, 39, 43
  COS, 50
  cost and benefit, 44
  evaluation in Inclusive Impact Index, 41
  Triple I, 41, 49–50
HWL, *see* Healthy Waterways Ltd.
Hybrid Coordinate Ocean Model (HYCOM), 148
Hydrocarbon
  distribution, 121
  fractions, 121
  mixture, 121
  molecular markers, 123
Hydrodynamic model, 70; *see also* General
  larval dispersal model; General
  particle-tracking model; Harmful
  algal bloom model (HAB model)
  boundary forcing, 71–72
  evaluation tools, 73–74
  output and validation, 78–82
  parameterisations, 71
  validation data, 72–73
  west coast of Scotland FVCOM model, 70
Hydrodynamic submodel, 148–149
Hydrogen, 121
Hypoxia, 4
Hypoxic problems, 147–148

I

IBMs, *see* Individual biological models
IEAs, *see* Integrated ecosystem assessments
Immunotoxic effects, 130
IMPACT, *see* Inclusive Marine Pressure
  Assessment and Classification
  Technology
Inclusion of ecosystem services in green
  accounting, 172–174
Inclusive Impact Index, 32, 34
  application to artificial upwelling
    technology, 53–56
  EF, 33–34
  environmental risk, 34
  evaluation of ER and HR, 41–50
  scaling of ecological and economic values,
    50–53
  simulation-based evaluation of EF and
    biocapacity, 36–40
  spatial analysis, 56–61
  sustainable technologies, 36
  Triple I, 32–33, 34–35
  Triple I light, 35
  Triple I star, 35
Inclusive Marine Pressure Assessment
  and Classification Technology
  (IMPACT), 32
Index of sustainable economic welfare
  (ISEW), 165
India
  conservation and management of saltwater
    crocodile, 307–319
  mangroves, 272
  oceanography off central west coast,
    264–265
  physiographic and climatological setting of
    Goa, 265–266
Indicators, 17
Indicator species, basic concepts in use of,
  290–291
Individual biological models (IBMs), 74
Ingestion of oil-contaminated prey, 127–128
Input-output analysis (I/O analysis), 53
In situ sensors, 11
Integrated coastal governance, 14–15
Integrated coastal zone monitoring
  CBP, 16–19
  EHMP, 19–23
  framework for designing and assessing
    performance of SoS, 3–7
  informing IEAs, 7–9
  integrated SoS, 9–15

Integrated coastal zone monitoring (*Continued*)
    problems and challenges, 24–26
    programmatic strengths, 23
    sustained and integrated coastal zone
        observing SoS, 15
Integrated ecosystem assessments (IEAs), 2, 3,
    13, 21
    informing IEAs, 7–9
Integrated SoS, 9
    boundary conditions and time scales, 11
    end-to-end system, 10
    feedbacks, 9–10
    integrated coastal governance, 14–15
    linking observations and models, 12–13
    performance assessments, 13–14
    sampling ecosystems and problem of
        undersampling, 11–12
Integrated Urban Water, 23
Interactions, fishery simulation model,
    191–192
Intergovernmental Panel on Climate Change
    (IPCC), 41
Intergovernmental Platform on Biodiversity
    and Ecosystem Services (IPBES), 161
International cooperation, 229
International Union for Conservation of
    Nature (IUCN), 256, 283, 309–310
Intertidal boulder reef, 249
Intertidal estuarine marshy ecosystem, 265
Intertidal organisms and novel offshore
    habitat, 90–91
Intertidal sand flats, 275–276
Intrinsic growth rate, 184
Inventory analysis, 33
I/O analysis, *see* Input-output analysis
IPBES, *see* Intergovernmental Platform on
    Biodiversity and Ecosystem Services
IPCC, *see* Intergovernmental Panel on Climate
    Change
Irish Marine Institute's NE Atlantic Model
    (NEA-ROMS), 72
ISEW, *see* Index of sustainable economic
    welfare
Isobutane, 122
Isoosmotic body fluids, 271
Isoprenoids, 125
IUCN, *see* International Union for
    Conservation of Nature

J

Japanese yen (JPY), 41
Joint environment unit (JEU), 328

K

Karenia brevis (*K. brevis*), 67
Karenia mikimotoi (*K. mikimotoi*), 67, 74
    simulations, 83
Katka, 325–326
KBC, *see* Kuroshio Branch Current
KBCNT, *see* KBC to north of Taiwan
KBC to north of Taiwan (KBCNT), 108
KBC to west of Kyushu (KBCWK), 108, 110
KBCWK, *see* KBC to west of Kyushu
KC, *see* Kuroshio Current
Keystone species, 278
Killifish (*Fundulus heteroclitus*), 128–129
Korean marine environment, teleconnected
        pattern of climate forcing in, 101–110
Korean marine waters, 99
Korean waters, 105; *see also* Coastal waters
        management
    direct climate forcing, 106–107
    ecosystem structure change pattern, 113
    lag oceanic forcing, 108–110
Korea Strait (KS), 103
    seasonal volume transport through, 108
Korea/Tsushima Strait (KTS), 101
KS, *see* Korea Strait
KTS, *see* Korea/Tsushima Strait
Kuji Khumbhiora, 310
Kuroshio Branch Current (KBC), 99
Kuroshio Current (KC), 100
Kuroshio geostrophic transport, 103

L

Lag oceanic forcing, 108–110; *see also* Direct
        climate forcing
Land, 249
Large marine ecosystem (LME), 99
Larval
    duration, 85–86
    settlement, 279
LCA technique, *see* Life-cycle assessment
        technique
LCC, *see* Liman Cold Current
Legal regulations and implementation,
        283–284
Leontief inverse matrix, 54
LGAC, *see* Chesapeake Bay Local Government
        Advisory Committee
Life-cycle assessment technique (LCA
        technique), 33
    comparison of III_{light} by LCA with power
        generating systems, 59–60

Life-cycle impact assessment method (LIME), 43
Life cycle of mangrove wetlands, 273–274
Liman Cold Current (LCC), 101
LIME, *see* Life-cycle impact assessment method
Lingual glands, 313
Living marine resources (LMRs), 4, 17
LME, *see* Large marine ecosystem
LMRs, *see* Living marine resources
Lotka-Volterra model, 183
Lotka-Volterra predator–prey equations, 184
Low dunes, 241

M

Machine learning, 294
Macrobenthic communities
    on rocky habitats, 294–296
    on soft-bottom habitats, 296–299
Macrobenthos
    communities modelling along Penang coastal waters, 294
    macrobenthic communities on rocky habitats, 294–296
    macrobenthic communities on soft-bottom habitats, 296–299
    types, 150
Macroeconomic(s), 159
    accounting problems, 158
    accounting systems, 171
Madhya Para, 241
Malaysia, biomonitoring to assess coastal environmental status in, 290
    basic concepts in use of indicator species, 290–291
    eutrophication, 291–292
    heavy metal, 292–293
Malaysian Coastal Waters, environmental issues of, 289–290
Managed resource zone, 255
    restricted access zone, 256–257
    sustainable-use zone, 256
Management plans for different ecosystems
    contingency planning, 135–136
    ESI, 136–137, 138
    specific environments, 137–140
Management simulations, 176
Management Strategy Evaluation Action Plan, 21
M&E Program, *see* Monitoring and Evaluation Program

Mangrove ecosystems; *see also* Coral reef ecosystem; Estuary ecosystem; Seaweed and seagrass ecosystem
    impacts of erosion, 219
    impacts of salinity rise, 219
    impacts of sea-level rise, 218–219
    impacts of storms, 219
    increase in estuarine current, 220
    sensitivity of mangrove ecosystems to climate change, 217–218
Mangrove(s), 139–140, 288, 308–309
    adaptations, 274–275
    crab *Scylla olivacea*, 251
    distribution, 308–311
    ecosystems, 233
    forests, 309
    heterotrophic production, 273
    life cycle, 273–274
    primary production, 273
    vegetation, 273
    wetlands, 272
Marginal seas, 99
Marine ecosystems, 211; *see also* Estuary ecosystem
    climate change impacts on, 211
    management, 181–182
    in twenty-first century Vietnam, 224–226, 227–228
Marine Environmental Committee (MEC), 148
Marine herbivores, 220
Marine mollusks, 251–252
Marine oil spills, 126
Marine organisms, 129
Marine pollution, 147–148
Marine protected areas (MPAs), 66
Marine sedimentary rocks, sequence of, 239
Marine spatial planning, 2
Marshes, 139
Mating, 315
MAUT, *see* Multi-attribute utility theory
Maximum economic yield (MEY), 169
Maximum sustainable yield (MSY), 169
MEA, *see* Millennium Ecosystem Assessment
Mean sea level (MSL), 239
Mean trophic level (MTL), 112
Measure of economic welfare (MEW), 165
MEC, *see* Marine Environmental Committee
Mekong Delta, 218
Meteorological model, 69
    validation, 78
Methane, 122
Methylfluorenone, 122
MEW, *see* Measure of economic welfare

MEY, *see* Maximum economic yield
Microbial oxidation, 124
Microeconomic modeling, 174–176
Millennium Ecosystem Assessment (MEA), 159, 160, 161
Ministry of Environment and Forests (MoEF), 328
Mitigation methodology, 131
    bioremediation, 134–135
    booms and skimmers, 132–133
    burning, 134
    Castillo de Bellver spill, 135
    dispersants, 133–134
    gelling agents and sorbents, 134
Mixed estuaries, 271
Mixed layer depths (MLD), 264
MLD, *see* Mixed layer depths
Modular Ocean Model (MOM), 148
MoEF, *see* Ministry of Environment and Forests
Molluscs, 138, 251–252, 273, 277, 296
Mollusks, *see* Molluscs
MOM, *see* Modular Ocean Model
Monitoring and Evaluation Program (M&E Program), 20
Moreton Bay, 21
Mortality, 68, 74–75, 76
MPAs, *see* Marine protected areas
MSL, *see* Mean sea level
MSY, *see* Maximum sustainable yield
MTL, *see* Mean trophic level
Multi-attribute utility theory (MAUT), 163
Multi-scale Ultra-high Resolution Sea Surface Temperature (MUR-SST), 72

N

Naphthalenes, 122
Narikel Jinjira, *see* St. Martin's Island
National accounting system for ecosystem, 173
National Centers for Environmental Prediction (NCEP), 69
National Environmental Research Institute, Denmark (NERI), 334
National Geographic Traveller, 330
National Institute for Environmental Studies (NIES), 55
Natural factors, 282
Natural gas, 119–120
Natural regeneration, 335
Navy Coastal Ocean Model (NCOM), 148

NCEP, *see* National Centers for Environmental Prediction
NCOM, *see* Navy Coastal Ocean Model
NERI, *see* National Environmental Research Institute, Denmark
NES, *see* Northern East Sea/Sea of Japan
Net growth of breeding stock, 167
Net present value (NPV), 172
Net primary production (Net PP), 4
Neural networks, 294
NGO, *see* Nongovernmental organization
NIES, *see* National Institute for Environmental Studies
Nilkamal, 326
Nitrogen, 121
NKCC, *see* North Korea Cold Current
Nongovernmental organization (NGO), 22, 33, 283, 328
Nonpetroleum, 123
Non-supervised ANN algorithm, 297
Non-supervised artificial neural network, 294
Northern East Sea/Sea of Japan (NES), 106
North Korea Cold Current (NKCC), 101
North Pacific gyre oscillation, 102
NPV, *see* Net present value
Numerical methods, 150–151
Numerical models
    ecosystem submodel, 149–150
    hydrodynamic submodel, 148–149
    structure, 148
Numerical simulation, 148
Nutrients, 16
    cycling, 159
    debris, 220
Nutrition, 269–270

O

Observing system experiments (OSEs), 13
Observing system simulation experiment (OSSEs), 13
Occurrence probability by cross-impact method, 45–48
Ocean, 32
    fertilization, 32
    ocean-ecosystem coupled model, 36
    utilization technologies, 32, 35
Oceanic and fisheries response in Northwest Pacific Marginal Seas, 99
    current systems of three different marginal seas, 100
    East Sea/Sea of Japan, 100–101

fishery productivity and ecosystem
dynamics response to climate forcing,
110–114
teleconnected pattern of climate forcing in
Korean marine environment, 101–110
Oceanic forcing, 102
Oceanography off central west coast of India,
264–265
Ocean surface acidification (OSA), 41, 42
OCHA, *see* UN office for coordination of
humanitarian affairs
Odisha, distribution of saltwater crocodiles,
311–312
OECD, *see* Organization for Economic
Co-operation and Development
Offshore wind turbine types, 56
Oil, 330
acute effects, 127–129
collection time duration and agency
achievements, 331
concentration of oil causing deleterious
effects, 127
oiling effects of mangroves, 139–140
sublethal effects, 129–131
tankers, 330
toxicity to marine life, 126
Oil spill, 131–132, 329
contingency plan, 327
sensitivity maps, 135–136
Open coral reefs, 212
Opening stocks records, 167
Operational maps, 137
Optimal fleet, 199, 201
Optimal solutions, 176
Optimal stock biomass, 199, 201
Optimal sustainable development, 190
Oregon State University Tidal Prediction
Software (OTIS), 71
Organization for Economic Co-operation and
Development (OECD), 163–164
OSA, *see* Ocean surface acidification
OSEs, *see* Observing system experiments
Osmoconformation, 271
Osmoregulation, 271
Osmosis, 271
OSSEs, *see* Observing system simulation
experiment
OTIS, *see* Oregon State University Tidal
Prediction Software
Overbuilding, 247
Oxidase enzymes, 129
Oxygen, 121
Oyashio region, 112

P

Pacific Decadal Oscillation pattern (PDO
pattern), 102, 104
Pacific reef heron (*Egretta sacra*), 253
Padma Oil Company tanker Southern Star VII,
324
PAHs, *see* Polynuclear aromatic hydrocarbons
PAH source indicators, 126
Parameterisations, 71
Particle tracking model, 83
Particulate organic matter, 152
Parts per thousand (ppt), 243
PDO pattern, *see* Pacific Decadal Oscillation
pattern
Pedogenesis, 159
Pelagic submodel, 149–150
Penang coastal waters, macrobenthos
communities modelling along, 294
macrobenthic communities on rocky
habitats, 294–296
macrobenthic communities on soft-bottom
habitats, 296–299
Performance assessments, 13–14
Perna viridis (*P. viridis*), 293
Petroleum, 119–120, 121
average annual releases, 120
hydrocarbons, 123
PFM, *see* Production function method
Phase-diagrams, 197
Photochemical oxidation, 124
Photosynthesis process, 140
Phu Quoc coral reef ecosystem, 217
PHY, *see* Phytoplankton
Physical factors, 277–279
Physics of crude and refined oils, 121–126
Physiographic characteristics of Saint Martin's
Island, 239
Phytoplankton (PHY), 45, 152, 154
biomass, 273
status, 331–332
Pisaster ochraceus (*P. ochraceus*), 278
Plankton, 18
Planning and policy, solutions for, 226–229
Platform reef, 211
Policymakers, 21, 22, 232, 334
Political willingness to pay (PWTP), 179
Polyalkylated benzenes, 122
Polycyclic aliphatics, 124
Polynuclear aromatic hydrocarbons (PAHs),
124
POM, *see* Princeton Ocean Model
Population growth, 74–75, 247–248

Porphyrins, 125
Power generating systems
   comparison of III_{light} by LCA with, 59–60
ppt, *see* Parts per thousand
Predation, 74, 278, 294
   ratio-dependent, 184
   risk, 86
Predators, 279
   predator–prey models, 181
   predator–prey relationships and species
      interactions, 183–185
Predicted equilibrium, 198–199
Preservation and promotion of marine
      biodiversity, 229
Pressure(s), 3, 171, 333–334
   on coastal ecosystems, 6–7
   pressure-state relationship, 333
Pressure-State-Response method (PSR
      method), 163–164
Prices and valuation, 177–180
Primary production of mangrove wetlands, 273
Primary productivity, 267
Princeton Ocean Model (POM), 148
Pristane, 122
Production function method (PFM), 179
Property value method, *see* Hedonic price
      method (HPM)
Protected sand flats, 275–276
Protection and development of biodiversity, 229
Provisioning, ecosystem services, 162
Pseudo-nitzschia, 67
PSR method, *see* Pressure-State-Response
      method
Public awareness and training, 283
PWTP, *see* Political willingness to pay

R

Random utility models (RUM), 163, 164
Rayleigh distribution, 59
Receiving Water Quality Model (RWQM), 20
Reclaiming shallow waters effects, 150
Recruitment minus mortality, 187
Red River Delta, 218
Reef-building corals, 267
Reef crest, 211–212
Refined oils, 121–126
Regional Ocean Modeling System (ROMS), 148
Regulating, ecosystem services, 162, 165, 179
Remote sensing, 11
Renewable natural resource, 159
Renewable resources, 161
Reptiles

saltwater crocodile, 314
   in St. Martin's Island, 253
Research Institute for Marine Fisheries (RIMF),
      211
Research Institute of Innovative Technology
      for Earth (RITE), 44, 49
Residual current, 151–152
Resources and management options in St.
      Martin's Islands, 249
   development, 248–249
   fauna, 251–253
   flora, 250–251
   land, 249
   water, 249–250
Restricted access zone, 256–257
Restrictive environmental factor, 220–221
Revealed preferences (RPs), 179
Rhizophora mangrove species, 274, 275
RIMF, *see* Research Institute for Marine
      Fisheries
Risk, 44
   ecological, 44
   environmental, 34
   human, 43–44
   ratio, 34
RITE, *see* Research Institute of Innovative
      Technology for Earth
Riverine country, 323
RMS, *see* Root-mean-square
Rock(s), 252
   adaptation, 279
   patches, 277
   physical and biological factors, 277–279
   species diversity, 279–281
   zonation pattern, 279
Rocky habitats, macrobenthic communities on,
      294–296
Rocky shores, 138–139, 279
ROMS, *see* Regional Ocean Modeling System
Root-mean-square (RMS), 73
RPs, *see* Revealed preferences
RUM, *see* Random utility models
RWQM, *see* Receiving Water Quality Model

S

Salinity, 72, 81, 274–275
   adaptation of estuaries, 271
   driven mortality, 88–90
   rise impacts, 219
Saltwater crocodile (*Crocodylus porosus*),
      310–312, 317
   biology, 312–315

conservation, 315, 316–317
distribution, 311–312
distribution of mangroves, 308–311
fauna, 272
increase in Estuarine current, 220
reasons for conservation, 315–316
status, 315, 317–319
Saltwater trees, 324, 327
Sampling ecosystems, 11–12
Sandy shores, 275
adaptations, 276
biotic communities, 277
grain size and beach profile, 276
SAR, *see* Species-area relationship
Science and engineering, solutions for, 229–230
Science and Technical Advisory Committee
(STAC), 19*fn*
Scientific Expert Panels (SEPs), 22
Scientific management approaches, lack of, 336
Scottish Environment Protection Agency
(SEPA), 72
Seagrass beds, 147–148, 288
Sea level (SL), 4, 210, 214, 218, 220, 233
rise impacts, 214, 218–219, 220, 223
Sea surface height (SSH), 103
Sea surface temperature (SST), 69, 100, 264
Sea surface temperature anomaly (SSTA), 106
Seawater, 242–243, 270
Seaweed and seagrass ecosystem, 234; *see
also* Coral reef ecosystem; Estuary
ecosystem; Mangrove ecosystems
impacts of floods and drought, 221–222
impacts of increases in CO_2 concentration,
222
impacts of sea-level rise, 220
impacts of storm, 221
impacts of temperature rise, 220–221
impacts of turbidity and sedimentation
increase, 222
roles relating to climate change, 220
Seaweeds, 147–148, 251
Sector-specific approaches, 14
Sedimentation
increase, 222
phenomena, 224
Sediment pollution on Con Co Island, 216
SEEA, *see* System for Integrated Environmental
and Economic Accounting
SEEAF, *see* System for Integrated
Environmental and Economic
Accounting for Fisheries
Self-organizing map (SOM), 294
Semiclosed bay, 148

carbon cycle, 152–154
computational condition, 150–151
example of results, 151
numerical models, 148–150
residual current, 151–152
water quality, 152
Semiclosed coral reefs, 212
Sensitivity of mangrove ecosystems to climate
change, 217–218
Sentinel site, 12*fn*
SEPA, *see* Scottish Environment Protection
Agency
SEPs, *see* Scientific Expert Panels
SEQ, *see* South East Queensland
SEQ Healthy Waterways Strategy, 20, 24
SES, *see* Southern East Sea/Sea of Japan
Sessile organisms, 277
Shallow waters, 147–148
effects of reclaiming, 150
shallow-water benthic species, 128
Shannon's Diversity Index, 296
Shela River, 324
oil tanker sinking threatened species in, 332
oil tanker spill affected species in Shela
River and Mangrove Sundarbans, 329
Shelly limestone, *see* Coquina
SHP system, *see* Siberian High Pressure system
Siberian High Pressure system (SHP system),
101
in wintertime, 102
Simpler inclusive impact index, *see* Triple I
light (III_{light})
Simpson's biodiversity index, 114
Simulation-based evaluation of EF, 36
action plans and impact simulation, 36
environmental impact simulation, 36–38
self-cleaning technologies, 40
water purification effects and Triple I, 38–40
water purification technology application
plan, 36
Sinking of oil tanker in Shela River in
Sundarbans, 326
aims and objectives, 328
methods and data sources, 328
significance, 327–328
Skimmers, 132–133, 136
SL, *see* Sea level
SMBCP, *see* St. Martin's Biodiversity
Conservation Project
SMEW, *see* Sustainable measure of economic
welfare
SNA, *see* Standard National Accounts
Socioeconomics, 23

Soft-bottom habitats, macrobenthic
    communities on, 296–299
Soil characteristics of Saint Martin's Island,
    239–241
Solar radiation, 151, 273, 277
Solutions for adaptation and resilience to
    climate change, 226
    education, training and raising awareness
        of climate change, 232
    for planning and policy, 226–229
    for protection of biodiversity, 229
    for protection of coastal ecosystems against
        rise of sea level and erosion, 230
    for science and engineering, 229–230
    for strengthening international cooperation,
        232
    to strengthen resilience of mangroves to
        climate change, 230–232, 233
SOM, see Self-organizing map
SOP, see Standard operating procedure
Sorbents, 134
SoS, see System of systems
South East Queensland (SEQ), 19–20
Southern East Sea/Sea of Japan (SES), 106
SPAR-type offshore wind turbine, 56–57
    EF in life of, 57–59
Spatial analysis of Inclusive Impact Index, 56
    comparison of III_{light} by LCA with power
        generating systems, 59–60
    EF in life of SPAR-type wind turbines, 57–59
    object, 56–57
    spatial analysis of III of floating offshore
        wind turbine, 60–61
Spatial resolution, 11
Species-area relationship (SAR), 48
Species diversity, 4, 6, 279–281
Species richness, 5fn
Spilled oil in coastal waters, 330
Spring bloom, 112, 114
SPs, see Stated preferences
SRF, see Sundarbans reserved forest
SSH, see Sea surface height
SST, see Sea surface temperature
SSTA, see Sea surface temperature anomaly
STAC, see Science and Technical Advisory
    Committee
Staghorn coral (Acropora cervicornis), 268
Standard National Accounts (SNA), 164–165
Standard operating procedure (SOP), 338
Stated preferences (SPs), 163, 179
Steranes, 125
St. Martin's Biodiversity Conservation Project
    (SMBCP), 254

St. Martin's Island, 238
    careless littering by tourists creates waste
        on, 244
    causes of environmental pollution on St, 245
    drivers of ecosystem and water quality,
        246–248
    dumping place of wastes, 245
    environmental mismanagement, 246
    geo-environmental settings, 238
    geographical location, 239–242
    integration of co-management strategy,
        253–257
    resources and management options, 249–253
    resources promoting development, 248–249
    water characteristics, 242–244
Stochastic simulations, 176
Stock biomass, 169
Storms impacts, 214, 219, 221
    research results on Con Co Island, 216
    research results on Con Dao Island, 214–215
Strait of Malacca, 288
Strategic sensitivity maps, 136–137
Stratified estuaries, 271
Strengthening international cooperation, 232
Strengthen resilience of mangroves to climate
    change, 230–232, 233
Strong sustainability, 161
Sublethal effects, 129–131
Sulfur, 121
Sundarbans, 324, 329
    challenges, 335–336, 337
    ecosystem, 328
    lack of scientific management approaches,
        336
    management challenges, 332
    natural regeneration process and
        challenges, 335
Sundarbans Mangroves in Bangladesh, 323
    ecosystem health and indicators, 326
    ecosystem service values, 324–325
    environmental impacts, 329–330
    impact on water quality, 330
    impacts on biodiversity, 330–332
    importance, 324
    recommendations for ecosystem
        development, 337–338
    results and discussion, 329
    sinking of oil tanker in Shela River, 326–328
    tourism sites and importance, 325–326
Sundarbans reserved forest (SRF), 324
Superficial engines, 337
Supervised analysis, 294
Supporting services, 162, 165, 179

Surface Velocity Program (SVP), 72
Sustainability, 160, 183, 193
  of ecosystems, 162
  environmental, 32, 36
  of human activities, 33
  measure, 160
  weak, 160
Sustainable development, 2, 190
Sustainable measure of economic welfare
     (SMEW), 165
Sustainable-use zone, 256
Sustained coastal zone observing SoS, 15–23
SVP, *see* Surface Velocity Program
System for Integrated Environmental and
     Economic Accounting (SEEA),
     162–163, 165
System for Integrated Environmental and
     Economic Accounting for Fisheries
     (SEEAF), 161, 165
System of systems (SoS), 2, 15
  coastal ecosystem services, 4
  coastal ecosystem states and indicators, 4–6
  framework for designing and assessing
     performance of, 3
  pressures on coastal ecosystems, 6–7
  sustained and integrated coastal zone
     observing, 15–23
System performance, 13–14

T

TACs, *see* Total allowable catches
Tactical maps, 137
Taiwan Current (TC), 100
Taiwan–Tsushima Warm Current (TTWC), 108
Tale, 157–158
Taylor skill score (S), 73
TC, *see* Taiwan Current
TCM, *see* Travel Cost Method
Technical Guidelines for Responsible Fisheries,
     168
TEEB, *see* The Economics of Ecosystems And
     Biodiversity
Teleconnected pattern of climate forcing in
     Korean marine environment, 101
  climate variabilities, 101–106
  direct climate forcing, 106–107
  lag oceanic forcing, 108–110
  lag phase between PDO and SWT, ALP and
     SWT, 109
Temperature, 72, 81
  impacts, 223
  rise impacts, 220–221

The Economics of Ecosystems And
     Biodiversity (TEEB), 161
Thermal power, 59
Thermal regulation, 127–128
Threatened species, 324
Tidal exposure/amplitude, 277
Tides, 72, 78–79
Time-lapse photography, 269
Time scales, 7, 10, 11
TKE, *see* Turbulent kinetic energy
TMDLs, *see* Total Maximum Daily Loads
T-N, *see* Total nitrogen
Tokyo Bay, 36, 149–150
  carbon cycle in, 153
  grid system, 150
  marine environment in, 148
Torrey Canyon spill, 133
Total allowable catches (TACs), 176
Total Maximum Daily Loads (TMDLs), 18
Total nitrogen (T-N), 151
Total phosphorus (T-P), 151
Tourism, 248
  sites and importance, 325–326
  strengths of tourism development, 248
Tourists, crowds of, 248
Toxicity of oil to marine life, 126–131
T-P, *see* Total phosphorus
Transportation, 14, 38, 58, 333, 337
Travel Cost Method (TCM), 39, 179
Trawling, 185
Triple I light (III_{light}), 32, 34–35, 38–41, 49–50, 56
  comparison of III_{light} by LCA with power
     generating systems, 59–60
  spatial analysis of floating offshore wind
     turbine, 60–61
Triple I light star ($III_{light}{}^*$), 35
Triple I star (III^*), 35
Tropical coastal areas, ecosystems in, 288
Tsushima Warm Current (TWC), 100, 101, 108
TTWC, *see* Taiwan–Tsushima Warm Current
Turbidity, 222
Turbulent diffusion, 87
Turbulent kinetic energy (TKE), 74
TWC, *see* Tsushima Warm Current
Twenty-first century Vietnam, marine
     ecosystems and climate change
     scenario in, 224–226, 227–228
Typhoons, 223–224

U

UK National Ecosystem Assessment (UK
     NEA), 161

UK NEA, *see* UK National Ecosystem
       Assessment
U-matrix, *see* Unified distance matrix
Uncertainty radius (UR), 75
UNDAC, *see* United Nations Disaster
       Assessment and Coordination
Undersampling, problem of, 11–12
UNDP, *see* United Nations Development
       Programme
UNEP, *see* United Nations Environment
       Programme
UNESCO, *see* United Nations Educational,
       Scientific and Cultural Organization
UNFCCC, *see* United Nations Framework
       Convention on Climate Change
Unified distance matrix (U-matrix), 294
United Nations Development Programme
       (UNDP), 328
United Nations Disaster Assessment and
       Coordination (UNDAC), 328
United Nations Educational, Scientific and
       Cultural Organization (UNESCO),
       336
United Nations Environment Programme
       (UNEP), 328
United Nation's erythrocyte protoporphyrin
       level (EP level), 332
United Nations Framework Convention on
       Climate Change (UNFCCC), 211, 228
UN office for coordination of humanitarian
       affairs (OCHA), 328
Unsupervised analysis, 294
Upwelling stream, 213–214
UR, *see* Uncertainty radius
U.S. Chesapeake Bay Program (CBP), 15, 16,
       23–24
    data management and communications, 18
    ecosystem recovery targets, 17–18
    governance for integration, 18–19
    Real-time *in situ* observations, 16–17
User satisfaction, 13–14
U.S. Integrated Ocean Observing System, 10
Uttar Para, 239, 241

V

Validation data, 72–73
Vegetation, 324, 327
    aquatic, 251
    coastal, 216
    mangrove, 265
    natural, 272
    Pandanus beach, 249

Velocity computation, 87–88
Vertical migration, 88
Vertical movement, 76
Vietnam, climate change impacts on marine
       ecosystems in, 210–234
Virgin ecosystem, 187
Virtual particles (VPs), 75
Volatile organic compounds (VOCs), 44
VPs, *see* Virtual particles
Vulnerability of estuarine ecosystems to
       climate change, 223

W

Walleye Pollock, 110, 112
Waste
    careless littering by tourists creates waste
       on St. Martin's Island, 244
    water characteristics, 242–244
Water, 249–250
    characteristics, 242
    freshwater, 243–244
    impacts of increasing water temperature,
       212–213
    seawater, 242–243
    water-logged conditions, 275
    water-quality parameters, 16
Water purification
    effects, 38
    technology application plan, 36
Water quality (WQ), 4, 5*fn*, 152, 290, 330
    crowds of tourists, 248
    ecosystem drivers and, 246
    overbuilding, 247
    population growth, 247–248
    tidal impact on, 330
WAVES, *see* Wealth Accounting and Valuation
       of Ecosystem Services
Weak sustainability, 160
Wealth Accounting and Valuation of
       Ecosystem Services (WAVES), 161
Weathering, 330
    characteristic effects, 123–124
    effects, 132
Weather Research and Forecasting model
       (WRF model), 69
Weight belt, 276
West Scottish Coastal Ocean Modelling System
       'WeStCOMS-FVCOM', 71
Wild crocodiles
    attack on domestic livestock by, 318–319
    attacks on humans by, 318
Wild Life Protection Act (1972), 315

Wind
   forcing, 87
   scenarios, 77–78
World heritage state, 326
Worldwide distribution of saltwater
      crocodiles, 311
World Wide Fund for Nature (WWF), 33
WQ, *see* Water quality
WRF model, *see* Weather Research and
      Forecasting model
WWF, *see* World Wide Fund for Nature

Y

Yellow Sea (YS), 99, 101, 110
Yield factor, 33

Z

Zonation pattern, 279
Zoning, 254–255
Zooplankton (ZOO), 45, 112, 152, 332
   communities in coral reefs, 269
   status, 331–332
Zooxanthellae
   extracellular products, 269
   relationship of calcification to, 268

9 780367 571948